Handbook of Environmental Control

HANDBOOK

of

ENVIRONMENTAL CONTROL

VOLUME V: HOSPITAL and HEALTH
CARE FACILITIES

EDITOR

Conrad P. Straub, Ph.D.

Professor and Director
Environmental Health
School of Public Health
University of Minnesota

Published by

CRC PRESS, Inc.
18901 Cranwood Parkway · Cleveland, Ohio 44128

Library of Congress Cataloging in Publication Data

Main entry under title:

Hospital and health care facilities.

(CRC handbook of environmental control; v. 5)
Bibliography: p.
Includes index.
1. Health facilities – Sanitation – Handbook, manuals,
etc. 2. Health facilities – Safety measures – Handbooks,
manuals, etc. I. Straub, Conrad P. II. Series.
TD145.C2 vol. 5 [RA969] 620.8'08s [648] 72-92118
ISBN 0-87819-275-1

© 1975 by CRC Press, Inc.

International Standard Book Number 0-87819-275-1

Library of Congress Card Number 72-92118
Printed in the United States

ADVISORY BOARD

PREFACE TO THE ENVIRONMENTAL CONTROL SERIES

Increasing population, developing industry, changing agricultural patterns, and growing community concern, along with a desire for a higher standard of living — physically, materialistically, emotionally, and sociologically — have focused attention on the multiple sources of contamination, pollution, and stress and the need for extensive control of these sources. Furthermore, the recognition that these problems involve many disciplines and that their resolution calls for evaluation of the benefits vs. the risks of the specific actions proposed has prompted development of a series of handbooks on environmental control. The aim of the series is to bring together pertinent information in tabular form that will be useful in evaluating the environment, not only from the standpoint of effects on the ecosystem, aquatic and terrestrial, but also on man's relationship to the environment and the environment's relationship to man.

The division of subject matter is along three basic lines: the macroenvironment, which deals primarily with the air, soil, and water milieus; the microenvironment, which is concerned primarily with the community — home, school, work, recreation — and with the factors responsible for stress effects such as noise, temperature, vibration, and radiation; and the personal environment, which relates to the products used and consumed by man. The first segment, dealing with the macroenvironment, consists of four volumes, one each on the air environment and the soil environment, and two on the water environment — water and wastewater.

The series is designed in a logical sequence to answer the questions *what, why, when,* and *where,* and *how* to treat them. In raising and answering these questions perhaps a broader, more comprehensive evaluation of the solutions offered and the consequences of particular actions can be reached.

It is my hope that this series on environmental control will prove useful to a wide range of disciplines, both technical and scientific, and at the same time be of interest and use to social and political scientists, economists, and attorneys. Hopefully, it will be of value to all decision-makers regardless of their backgrounds, concerns, and environmental affiliations. And, in the final analysis, the series should be of use to all students in the environmental and ecological disciplines.

Information in the series of handbooks will be continuously updated, and comments, criticisms, and suggestions regarding scope, format, and selection of subject matter are invited. Any errors or omissions should be brought to the attention of the editor and publisher.

I acknowledge the assistance of the Advisory Board and various contributors and wish especially to recognize permission of the many authors, journals, and books for use of published material. Special acknowledgments are made to the members of the editorial staff of CRC Press, Inc. for their intelligent and invaluable assistance.

C.P. Straub
Minneapolis, Minnesota
September, 1975

PREFACE TO VOLUME V: HOSPITAL AND HEALTH CARE FACILITIES

In contrast to the first four volumes of the *CRC Handbook of Environmental Control*, which were concerned primarily with the macroenvironment — air, water, and land (solid wastes), this fifth volume is concerned with the microenvironment — hospital and health care facilities. As in the previous volumes, selected data having relevance to this field have been collected and tabulated. The data presented provide information on the hospital environment, indicating microbiologic aspects, disinfection and cleaning techniques, ventilation, toxic agents, radiological health, noise, safety, food hygiene, solid wastes, liquid wastes, and water supply. Some data on nursing homes are also included.

The information in this volume should be of use to administrators of hospital and health care facilities, to environmental practitioners concerned with this particular environment, medical staff, nurses, engineers, laboratory personnel, and others concerned with the basic principles of environmental health and safety in hospital and health care facilities.

In collecting information for this volume, I wish to acknowledge the assistance of Messrs. Paul Mills and Yung Ahn, graduate students, University of Minnesota, Minneapolis, Minnesota.

C.P. Straub
Minneapolis, Minnesota
September 1975

THE EDITOR

Conrad P. Straub, Ph.D., is Professor and Director, Environmental Health, School of Public Health, University of Minnesota, Minneapolis.

Dr. Straub graduated from the Newark College of Engineering with a B.S. degree in Civil Engineering in 1936 and a C.E. degree in 1939. He received his M.C.E. and Ph.D. degrees in sanitary engineering in 1940 and 1943, respectively, from Cornell University.

Dr. Straub served for 2 years with the Corps of Engineers before joining the U.S. Public Health Service, with which he was associated for 25 years. While in the Public Health Service he served in China and Poland, and was assigned to Oak Ridge National Laboratory (1948–1956) and the Robert A. Taft Sanitary Engineering Center (1956–1966), where he held the positions of Chief, Radiological Health Research Activities, Deputy Director, and Director. He serves as a consultant to various groups including the World Health Organization.

Dr. Straub is a member of the American Society of Civil Engineers, American Water Works Association (life member), Federation of Sewage Works Association, American Public Health Association, the Health Physics Society, Sigma Xi-RESA, Phi Kappa Phi, Chi Epsilson, and other societies. He has published approximately 160 papers and reports, written chapters in several books, has authored the book *Low Level Radioactive Wastes*, has served as coeditor of the initial four volumes of the *CRC Handbook of Environmental Control*, and is editor of *CRC Critical Reviews in Environmental Control*.

TABLE OF CONTENTS

Section 1

Kinds and Numbers of Institutions

1. KINDS AND NUMBERS OF INSTITUTIONS

1-1   BASIC PRINCIPLES OF HEALTHFUL HOUSING

A. Fundamental Physiological Needs

1. Maintenance of a thermal environment that will avoid undue heat loss from the human body
2. Maintenance of a thermal environment that will permit adequate heat loss from the human body
3. Provision of an atmosphere of reasonable chemical purity
4. Provision of adequate daylight illumination and avoidance of undue daylight glare
5. Provision for admission of direct sunlight
6. Provision of adequate artificial illumination and avoidance of glare
7. Protection against excessive noise
8. Provision of adequate space for exercise and for the play of children

B. Fundamental Psychological Needs

9. Provision of adequate privacy for the individual
10. Provision of opportunities for normal family life
11. Provision of opportunities for normal community life
12. Provision of facilities that make possible the performance of the tasks of the household without undue physical and mental fatigue
13. Provision of facilities for maintenance of cleanliness of the dwelling and of the person
14. Provision of possibilities for esthetic satisfaction in the home and its surroundings
15. Concordance with prevailing social standards of the local community

C. Protection Against Contagion

16. Provision of a water supply of safe sanitary quality, available to the dwelling
17. Protection of the water supply system against pollution within the dwelling
18. Provision of toilet facilities of such a character as to minimize the danger of transmitting disease
19. Protection against sewage contamination of the interior surfaces of the dwelling
20. Avoidance of insanitary conditions in the vicinity of the dwelling
21. Exclusion from the dwelling of vermin which may play a part in the transmission of disease
22. Provision of facilities for keeping milk and food undecomposed
23. Provision of sufficient space in sleeping rooms to minimize the danger of contact infection

D. Protection Against Accidents

24. Erection of the dwelling with such materials and methods of construction as to minimize danger of accidents due to collapse of any part of the structure
25. Control of conditions likely to cause fires or to promote their spread
26. Provision of adequate facilities for escape in case of fire
27. Protection against danger of electrical shocks and burns
28. Protection against gas poisonings
29. Protection against falls and other mechanical injuries in the home
30. Protection of the neighborhood against the hazards of automobile traffic

Source: Committee on the Hygiene of Housing, *Basic Principles of Healthful Housing,* American Public Health Association, New York, 1939. With permission.

1–2   REQUIREMENTS FOR ACCEPTING GENERAL HOSPITALS FOR REGISTRATION[a]

Function: The primary function of the institution is to provide patient services, diagnostic and therapeutic, for a variety of medical conditions, both surgical and nonsurgical.

1. The institution shall maintain at least six inpatient beds, which shall be continuously available for the care of patients who are nonrelated, and who stay on the average in excess of 24 hr per admission.
2. The institution shall be constructed, equipped, and maintained to ensure the health and safety of patients, and provide uncrowded, sanitary facilities for the treatment of patients.
3. There shall be an identifiable governing authority, legally and morally responsible for the conduct of the hospital.
4. There shall be a chief executive to whom the governing authority delegates the full-time responsibility for the operation of the hospital in accordance with established policy.
5. There shall be an organized medical staff that includes doctors of medicine, and that may include, but shall not be limited to, doctors of osteopathy and dentists. The medical staff shall be accountable to the governing authority for maintaining proper standards of medical care, and it shall be governed by bylaws adopted by said staff and approved by the governing authority.
6. Each patient shall be admitted on the authority of a member of the medical staff who shall be directly responsible for the patient's diagnosis and treatment.
 Any graduate of a foreign medical school who is permitted to assume responsibilities for patient care shall possess a valid license to practice medicine or shall be certified by the Educational Council for Foreign Medical Graduates.
7. Registered nurse supervision and other nursing services are continuous.
8. A current and complete[b] medical record shall be maintained by the institution for each patient, and shall be available for reference.
9. Pharmacy services shall be maintained in the institution and supervised by a registered pharmacist.
10. The institution shall provide patients with food service that meets their nutritional and therapeutic requirements; special diets shall also be available.
11. The institution shall maintain diagnostic X-ray service, with facilities and staff for a variety of procedures.
12. The institution shall maintain clinical laboratory service, with facilities and staff for a variety of procedures. Anatomical pathology services shall be regularly and conveniently available.
13. The institution shall maintain operating room services with facilities and staff.

 The American Hospital Association may, at the sole discretion of its Board of Trustees, grant, deny, or withdraw the registration of an institution.

[a] Requirements approved by the Board of Trustees, May 1955, and adopted by the House of Delegates, September 1955. Amended June 1956; February 1957; August 1959; May 1963; May 1964; February 1965; and February 1970.
[b] The completed records in general shall contain at least the following: the patient's identifying data and consent forms, medical history, record of physical examination, physician's progress notes, operative note, nurse's notes, routine X-ray and laboratory reports, doctor's orders, and final diagnosis.

Source: *Hospitals,* Guide issue, Part 2, August 1, 1971. With permission.

1−3 TYPES OF INSTITUTIONS AND POPULATIONS SERVED IN THE UNITED STATES

Educational Institutions

Level of instruction	Number of schools	Enrollment, 1970−1971	Number of teachers
Kindergarten	−	2,821,213	−
Elementary and secondary	90,821	32,579,586 13,329,785	1,131,774 928,341
Higher education	−	7,136,075	−

Health-care Institutions

Kind of institution	Number of institutions	Admissions, 1970	Staff
Hospitals	7,123	31,759,000	2,537,000
Nursing homes	12,182	674,667	387,986

Miscellaneous Institutions

Kind of institution	Number of institutions	Population, 1970	Staff
State and federal prisons	−	200,000	−
Hotels	23,625	−	−
Motels, etc.	41,954	−	−
Recreation camps	8,990	−	−

Adapted from D. Golenpaul, Ed., *Information Please Almanac, Atlas and Yearbook, 1972,* 26th ed., Simon & Schuster, New York, 1971.

1–4   GROWTH OF SKILLED AND
INTERMEDIATE CARE FACILITIES,[a] 1963–1973

	Number of facilities	Number of beds	Average size	Number of residents	Utilization rate
1963	13,086	507,530	39		
1967	14,489	765,148	52		
1969	14,998	879,091	59	793,074	90.2%
1971	16,849	1,141,652	68	1,024,254	89.7%
1973	15,737	1,175,865	74	1,074,480	91.4%

Note: This table provides details on the growth of skilled care and intermediate care facilities. It shows a 20% increase in the number of facilities between 1963 and 1973 and over a 100% increase in the number of beds for the same time period. It also indicates an increase of 89% in the average size of facility, from 39 beds to 74 beds, between 1963 and 1973.

[a] These figures reflect totals for HEW counts of facilities providing nursing care and personal care with nursing, which are "roughly" comparable to skilled and intermediate care.

Source: American Health Care Association, *Long Term Care Facts,* AHCA, Washington, D.C., 1975, 8. With permission.

1–5   DISTRIBUTION OF OTHER INPATIENT HEALTH FACILITIES BY BED SIZE, 1969

	Number							
Bed size	Total other inpatient facilities	Deaf or blind	Unwed mothers	Physically handicapped	Mentally retarded	Emotionally disturbed	Orphanage or dependent children	Other
Total	4300	120	170	40	924	500	946	1600
Less than 25 beds	2107	18	93	12	538	345	345	756
25–49	789	13	57	15	110	71	270	253
50–74	405	7	16	6	46	37	137	156
75–99	216	8	1	–	29	17	66	95
100–199	365	37	3	2	60	21	89	153
200–499	233	30	–	4	32	4	35	128
500 beds or more	185	7	–	1	109	5	4	59

Source: Department of Health, Education, and Welfare, *Health Resources Statistics, 1971 ed.,* DHEW Pub. No (HSM) 72–1509, National Center for Health Statistics, Health Service and Mental Health Administration, Public Health Service, U.S. Department of Health, Education and Welfare, Rockville, Md., 1972.

1–6   CLASSIFICATION OF HOSPITALS ACCORDING TO SERVICES AND CATEGORIES OF FACILITIES REPORTED IN HOSPITALS

Services	Facilities
General medical and surgical	Postoperative recovery room
Hospital unit of an institution (prison hospital, college infirmary, etc.)	Intensive-care unit
	Intensive cardiac-care unit
Hospital unit with a mental retardation school	Open-heart-surgery facilities
	Pharmacy with FT-registered pharmacist
Psychiatric	Pharmacy with PT-registered pharmacist
Tuberculosis and other respiratory diseases	X-ray therapy
	Cobalt therapy
Narcotic addiction	Radium therapy
Geriatric	Radioisotope facility
Maternity	Histopathology laboratory
Eye, ear, nose, and throat	Organ bank
Rehabilitation	Blood bank
Orthopedic	Electroencephalography
Chronic disease	Inhalation-therapy department
Other specialty	Premature nursery
Children's general	Self-care unit
Children's hospital unit of an institution	Extended-care unit
	Inpatient renal dialysis
Children's psychiatric	Outpatient renal dialysis
Children's tuberculosis and other respiratory diseases	Physical therapy department
	Occupational therapy department
Children's eye, ear, nose, and throat	Rehabilitation inpatient unit
Children's rehabilitation	Rehabilitation outpatient unit
Children's orthopedic	Psychiatric inpatient unit
Children's chronic disease	Psychiatric outpatient unit
Children's other specialty	Psychiatric partial hospitalization program
Institution for mental retardation	Psychiatric emergency services
Epilepsy	Psychiatric foster and/or home care
Alcoholism	Social work department
	Family planning service
	Home-care department
	Hospital auxiliary
	Organized outpatient department
	Major emergency department
	Basic emergency department
	Provisional emergency unit
	Emergency referral service

Source: *Hospitals,* Guide issue, Part 2, August 1, 1971. With permission.

1–7   CLASSIFICATION OF HOSPITALS IN THE UNITED STATES ACCORDING TO CONTROL

I.   Governmental
    A.   Federal
        1.   Air Force
        2.   Army
        3.   Navy
        4.   Public Health Service, Indian Service
        5.   Public Health Service, other than Indian Service
        6.   Veterans Administration
        7.   Justice Department
        8.   Other Federal
    B.   Nonfederal
        1.   State
        2.   City
        3.   County
        4.   City-county
        5.   Hospital district or authority

II. Nongovernmental
    A.   Nonprofit
        1.   Church operated
        2.   Other nonprofit
    B.   For profit
        1.   Individual
        2.   Partnership
        3.   Corporation

III. Osteopathic hospitals
    A.   Nonprofit
        1.   Church operated
        2.   Other nonprofit
    B.   For profit
        1.   Individual
        2.   Partnership
        3.   Corporation

Source: *Hospitals,* 1969 Guide issue. With permission.

Section 2

Microbiological Considerations

---

2. MICROBIOLOGICAL CONSIDERATIONS

2.1 ENVIRONMENTAL MICROBIOLOGY

2.1—1   AIRBORNE CONTAMINATION IN HOSPITALS

| | Microbial numbers | | | Microbial types[a] | | | |
| | | Frequency of counts | | | | | |
Location	Mean count per ft³	<10/ft³, %	>20/ft³, %	Gram + cocci, %	Gram + rods, %	Gram − rods, %	Molds %
Patient rooms	20.0	35.1	36.5	41.4	12.9	13.6	26.7
Corridors	20.6	26.0	39.6	32.5	18.9	26.8	15.4
Central supply	10.2	73.8	14.1	49.6	14.3	5.9	11.8
Operating rooms	10.5	59.6	7.8	66.7	14.3	6.7	6.5
OB delivery rooms	4.5	93.3	0.0	44.4	24.9	13.3	9.1
Laundry handling	27.5	35.5	41.0	24.8	32.5	10.2	17.2
Waste storage and disposal	72.4	9.8	59.3	34.9	30.5	12.6	12.2
Utility rooms	18.1	34.2	29.2	28.0	21.0	11.8	32.3
Substerilizing rooms	19.5	17.0	47.9	−	−	−	−
Induction rooms	13.0	52.8	23.6	−	−	−	−
Incinerator rooms	18.6	24.3	37.8	35.2	37.1	11.1	6.2

[a] Remaining isolates (to total 100%) consist of diptheroids, yeasts, and actinomycetes.

Source: R.G. Bond et al., Eds., *Environmental Health and Safety in Health-Care Facilities,* Macmillan, New York, © 1973, 32. With permission.

REFERENCES

V.W. Greene et al., *Appl. Microbiol.,* 10, 561, 1962.
V. W. Greene et al., *Appl. Microbiol.,* 10, 567, 1962.

2.1–2   CALCULATED REDUCTIONS IN EXPOSURE TO AIRBORNE PARTICLE TRANSFER

Condition	u	v	α	β
A. Balanced supply (no airlocks)	1000	1000	11 (6)	2 (1.7)
	500	500	28 (11)	4 (2)
	200	200	80 (34)	7.5 (4)
	100	100	230 (90)	19 (9)
	50	50	800 (270)	60 (21)
B. Directional flow (no airlocks)	500	10	0.9×10^3	65
	200	10	1.4×10^3	100
	60	10	3.3×10^3	220
	10	500	0.6×10^3	73
	10	200	1.0×10^3	100
	10	60	2.9×10^3	220
C. Airlocks				
Unventilated				
Balanced	[1.5	1.5	6.7×10^5	4.7×10^4]
With directional	{ 50	1.5	2.0×10^4	1.4×10^3
air flow	} 1.5	50	2.0×10^4	1.4×10^3
(All with allowance for sedimentation in the airlock)				
Ventilated				
Input or extract	5	5	6.0×10^4	4.2×10^3
With allowance for sedimentation	2.9	2.9	1.8×10^5	1.3×10^4
D. Airlocks with recirculation				
Input ventilation				
Rapid passage	0.75	0.75	2.7×10^6	1.9×10^5
0.01 hr dwell	0.24	0.24	2.6×10^7	1.8×10^6
Extract ventilation				
Rapid passage	2.9	2.9	1.8×10^5	1.3×10^4
0.01 hr dwell	1.5	1.5	6.7×10^5	4.7×10^4
Directed ("laminar") flow	$\to 0$	$\to 0$	$\to \infty$	$\to \infty$

Notes: The table shows values of the ratio, α, of particle concentration in the source room to that in other patient rooms, and the ratio, β, of the concentration if all the rooms were combined into one space with uniform mixing throughout to the concentration in a nonsource patient room for the particular system, i.e., the larger the figure the better the isolation. The air flows from each room into the communications area and in the reverse direction, u and v, are given in cubic meters per hour.

The following values have been used in the evaluation of α and β: particle settling velocity 0.3 m/min, ten patient rooms each with floor area 15 m², and air input at 300 m³/hr (this corresponds to a ventilation rate of 6/hr for a room height of 3 m). Settling in these rooms is then equivalent to an additional clean air supply of 0.3 × 60 × 15 = 270 m³/hr, hence v_1 = 300 + 270 and has been taken as 600 m³/hr. The communications area covers 80 m² with an air input of 1200 m³/hr (this corresponds to a ventilation rate of 5/hr for a ceiling height of 3 m). Settling in this area is equivalent to an additional clean air supply of 0.3 × 60 × 80 = 1440 m³/hr, hence v_2 = 1200 + 1440 and has been taken as 2500 m³/hr. Entries and exits to and from each room have each been taken as 5/hr (m = 5) and 1 m³ of air has been assumed to be exchanged across the door opening each time a door is opened and shut. The volume of an airlock has been taken as 10 m³ with a floor area of 4 m² and with 100 m³/hr input or extract when appropriate.

2.1–2   CALCULATED REDUCTIONS IN EXPOSURE TO AIRBORNE PARTICLE TRANSFER (continued)

When the airlock is ventilated the air has been considered as delivered to or taken from both sides of the lock in equal amounts. Recirculation has been taken as 2000 m³/hr and assumed to come into operation only on entering the locks. The figures in parentheses in Group A relate to conditions of negligible ventilation supply to the system. The values for the unventilated airlock have been enclosed in square brackets since this condition is practically unrealizable unless sealed doors are used.

Source: O. Lidwell, *J. Hyg.*, 70, 287, 1972. With permission of Cambridge University Press.

2.1–3   THE EXPOSURE TO AIRBORNE PARTICLE TRANSFER OBSERVED IN SOME HOSPITAL SITUATIONS

Ref.	System	Conditions	Tracer	v_1	v_2	u or v	α
			A. Natural and Balanced				
1	Partitioned ward	Natural ventilation	*Staphylococcus aureus*	–	–	(1400)	5
2	Divided ward	Natural ventilation	*S. aureus*	–	–	(600)	10
3	Divided ward	Natural ventilation	*S. aureus*	–	–	(?00)	40
4	Divided ward	Natural ventilation	*S. aureus*	–	–	(400)	18
5	Divided ward	Part mechanical ventilation	Nitrous oxide	50–200	4000	100–170	70–80
5	Divided ward	Air conditioned	Nitrous oxide	700	4000	300	80
			B. Directional Flows				
6	Isolation unit	Positive rooms	*S. aureus*	350	?1500	20	500
7	Divided ward	Positive rooms	*S. aureus*	800	2500	a	14
7	Divided ward	Positive rooms	Freon® gas	800	2500	50	100
7	Divided ward	Positive rooms	Potassium iodide particle	800	2500	50	400
			C. Airlocks				
8	Isolation unit	Extract airlocks	Potassium iodide particle	200	1100	2.9	3.4×10^5
8	Isolation unit	Extract airlocks	Potassium iodide particle	200	1100	17	9.6×10^3

Notes: The table shows the observed values of α, the ratio of the particle concentration in the source room to that in other patient rooms. The values of v_1, the ventilation supply to a patient room, and v_2, that to the communication area, both given in cubic meters per hour, are usually only rough estimates based on limited information. Values of u or v, also in m³/hr, have been deduced from the above using the following equation:

$$c_s/c_1 = \alpha = 1 + (v_1 v_2 + v_2 v + n u v_1)/uv$$

where:

c = equilibrium concentration in source room

2.1–3   THE EXPOSURE TO AIRBORNE PARTICLE TRANSFER OBSERVED IN SOME HOSPITAL SITUATIONS (continued)

c_1 = equilibrium concentration in patient room
α = ratio of c_s/c_1
u = ventilation supply from patient room to communications area
v = ventilation supply from communications area to patient room
v_1 = ventilation supply to patient room
v_2 = ventilation supply to communications area

For the balanced and natural ventilation system u and v are assumed equal. For References 6 and 7 u has been taken as 200 and 400 m³/hr, respectively, and the values given in the seventh column are those calculated for v. Where the tracer is particulate v_1 and v_2 have been doubled to allow for sedimentation (this is true if s = 0.3 m/min, h = 3 m, and ventilation is at six changes/hr) before applying the formula. No estimates of v_1 or v_2 can usefully be made for the first four situations listed. The values of u and v given in parentheses are deduced from those given in Table 2.1–2 for the no ventilation condition.

The two values of u, v, and α for Reference 8 refer to 10 and 60 entries per exits per hour, respectively (m = 5 and m = 30).

[a] The discrepancy between the value of α in this row and that for the same system with a similar-sized tracer particle in the row below suggests that the microorganisms reached the nonsource rooms by other routes than air transfer.

Source: O. Lidwell, *J. Hyg.,* 70, 287, 1972. With permission of Cambridge University Press.

REFERENCES

1. O.M. Lidwell, et al., *J. Hyg.,* 69, 113, 1971.
2. O.M. Lidwell, et al., *J. Hyg.,* 64, 321, 1966.
3. R.E.O. Williams, *J. Hyg.,* 65, 207, 1967.
4. P.N. Edmunds, *J. Hyg.,* 68, 531, 1970.
5. G. Baird, *Science,* 3, 113, 1969.
6. R.E.O. Williams and L. Harding, *J. Hyg.,* 67, 649, 1969.
7. N. Foord and O.M. Lidwell, *J. Hyg.,* 70, 279, 1972.
8. A. Hambraeus and H.F. Sanderson, *J. Hyg.,* 70, 299, 1972.

2.1–4 PARTICLE TRANSFER WITHIN THE BURNS UNIT

7.5a = 7.5 m up the passage
7.5b = 7.5 m down the passage
2 = 2 m across the passage from the room door
15 = 15 m up or down the passage

Numbers	Transfer	Ventilation	Activity	α' at sampling position				α' Mean	α''	$\alpha = (\alpha' \times \alpha'')^a$ To patient room	To bathroom
				7.5a	2	7.5b	15b				
1–4	Patient room/passage	Correct	None	$> 10^5$	$> 10^4$	$> 10^5$	$> 10^5$	$(> 10^5)$	$(> 10^5)$	$> 10^9$	—
		Correct	In/out through airlock	204	45	505	818	245	39	9.6×10^3	—
		Faulty	None	248	58	226	1200	251	$(> 10^5)$	$> 2.5 \times 10^7$	—
		Faulty	In/out through airlock	71	20	46	780	83	66	5.5×10^3	—
5–10	Bathroom/passage	Correct	None	88	22	297	1050	155	$(> 10^5)$	$> 1.5 \times 10^7$	$> 10^{10}$
		Correct	In/out through airlock	44	19	173	562	96	215	3.8×10^3	5.3×10^1
		Correct	In/out through direct door	39	10	292	741	96	63	3.8×10^3	1.5×10^4
		Faulty	None	38	21	151	751	98	$(> 10^5)$	$> 4.1 \times 10^7$	$> 2.5 \times 10^7$
		Faulty	In/out through airlock	39	21	48	480	66	400	4.3×10^3	3.3×10^4
		Faulty	In/out through direct door	31	15	79	202	52	66	3.4×10^3	5.5×10^3

Notes: $\alpha = \dfrac{\text{conc. of particles in source room}}{\text{conc. of particles in receiving room}}$, $\alpha' = \dfrac{\text{conc. of particles in source room}}{\text{conc. of particles in passage}}$, $\alpha'' = \dfrac{\text{conc. of particles in passage}}{\text{conc. of particles in receiving room}}$.

The values of α' and α'' are, in each case, the log mean values of 5 to 7 experiments. α' mean = α' averaged over all sampling positions, log mean of four values.

[a] Left-hand column α values for transfer to a patient room; first four lines, from another patient room; last six lines, from the bathroom. Right-hand column α values for transfer from a patient room to the bathroom. In all cases the same activity and type of ventilation is assumed to hold for both source and receiving room.

Source: A. Hambraeus and H.F. Sanderson, *J. Hyg.*, 70, 299, 1972. With permission of Cambridge University Press.

2.1−5 COMPARISON OF OBSERVED AND PREDICTED VALUES
(FOR DESIGNED VENTILATION)

		α'^a	u^b	α''^c	v^d
Room (airlock)	Observed value	143	19	39	13
	Predicetd value	160	17	30	17
Bathroom (airlock)	Observed value	170	16	215	9
	Predicted value	640	5	450	5
Bathroom (direct door)	Observed value	59	41	63	21
	Predicted value	45	60	23	60

Notes: The α' values have been derived as the arithmetic average of the particle concentrations at the different sampling positions, i.e., by summation and averaging of $1/\alpha'$.

Predicted values of u and v assume a value of 1 m^3 for w, the volume of air transferred through a door on opening it, passing through, and shutting it again.

[a] α' = ratio of concentration of particles in source room to concentration of particles in passage.

[b] u = rate of air movement from room to passage.

[c] α'' = ratio of concentration of particles in passage to concentration of particles in receiving room.

[d] v = rate of air movement from passage to room.

Source: A. Hambraeus and H.F. Sanderson, *J. Hyg.,* 70, 299, 1972. With permission of Cambridge University Press.

2.1–6 AIRBORNE BACTERIAL COUNTS OF CARPETED AND TILED AREAS OF THE HOSPITAL SHOWING MONTHLY AVERAGES CALCULATED FROM WEEKLY SAMPLES – 1967

Month	Average per cubic foot of air[a]		
	Entire hospital (32 sites)	Tile (14 sites)	Carpet (14 sites)
January	11	9	13
February	8	8	10
March	10	10	11
April	10	9	13
May	11	11	11
June	12	11	13
July	14	14	14
August	13	13	13
September	14	15	12
October	13	13	15
November	12	11	13
December	18	17	21
Total	147	141	159
Mean	12	12	13

[a] Each site was sampled weekly, 20 ft^3 of air per sample, and the averages calculated from the total sampling for the total period as indicated.

Source: J.G. Shaffer and I.D. Key, *Health Lab. Sci.*, 6, 215, 1969. With permission.

2.1–7 AIRBORNE BACTERIAL COUNTS OF CARPETED AND TILED AREAS OF THE HOSPITAL SHOWING MONTHLY AVERAGES CALCULATED FROM WEEKLY SAMPLES – 1968

Month	Average per cubic foot of air[a]		
	Entire hospital (32 sites)	Tile (14 sites)	Carpet (14 sites)
January	11	11	11
February	11	12	10
March	12	14	11
April	10	10	11
May	12	13	12
June	13	13	14
Total	69	73	69
Mean	12	12	12

[a] Each site was sampled weekly, 20 ft^3 of air per sample, and the averages calculated from the total sampling for the total period as indicated.

Source: J.G. Shaffer and I.D. Key, *Health Lab. Sci.*, 6, 215, 1969. With permission.

2.1–8 TOTAL NUMBER OF *STAPHYLOCOCCUS AUREUS* ISOLATED FROM CARPETED AND UNCARPETED AREAS BY MONTHS – 1967

Number of *S. aureus* colonies[a]

Month	Entire hospital (32 sites)	Tile (14 sites)	Carpet (14 sites)
January	49	12	37
February	48	23	18
March	48	22	18
April	63	35	26
May	128	33	88
June	44	19	13
July	46	6	13
August	60	29	11
September	57	15	17
October	38	15	15
November	99	50	35
December	53	11	37
Total	733	270	328
Mean	61	23	27

[a] Each site was sampled weekly, 20 ft³ of air per sample, and the total number of *S. aureus* colonies recorded.

Source: J.G. Shaffer and I.D. Key, *Health Lab. Sci.,* 6, 215, 1969. With permission.

2.1–9 TOTAL NUMBER OF *STAPHYLOCOCCUS AUREUS* ISOLATED FROM CARPETED AND UNCARPETED AREAS BY MONTHS – 1968

Number of *S. aureus* colonies[a]

Month	Entire hospital (32 sites)	Tile (14 sites)	Carpet (14 sites)
January	48	22	24
February	30	10	11
March	40	25	9
April	54	42	10
May	116	15	97
June	33	13	19
Total	321	127	170
Mean	54	21	28

[a] Each site was sampled weekly, 20 ft³ of air per sample, and the total number of *S. aureus* colonies recorded.

Source: J.G. Shaffer and I.D. Key, *Health Lab. Sci.,* 6, 215, 1969. With permission.

2.1–10 AVERAGE COLONY COUNT AND PERCENT CHANGE OF AIR COUNTS OBTAINED FROM 30 PATIENT ROOMS WITH TILE FLOORS AND FROM 30 PATIENT ROOMS WITH CARPETS

Colonies per ft³ of air

Activity	Tile	Carpet	Percent difference
Before vacuuming	9.0	11.6	29

Source: B.Y. Litsky, *Health Lab. Sci.,* 10, 28, 1973. With permission.

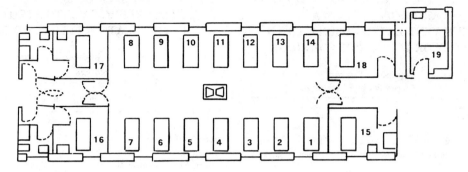

2.1–11 Plan of ward: 1–14, beds in open ward; 15 and 18, recirculation-ventilated cubicles; 16 and 17, plenum-ventilated cubicles; 19, window-ventilated cubicle.[a]

[a] A plenum-ventilation system supplies warmed, humidified, and filtered air to both cubicles at a rate of 20 air changes per hour and to the annexes at a rate of 10 changes per hour. The air is passed through a pre-filter which has an efficiency of 99% for 3-μm particles. There is also a pre-heater, a capillary washer type of humidifier, and a fan which has an output of about 1100 ft^3/min and drives the air along ducts to the two cubicles and the annexes.

Source: G.A.J. Ayliffe et al., *J. Hyg.*, 67, 417, 1969. With permission of Cambridge University Press.

2.1–12 CLEARANCE OF *BACILLUS SUBTILIS* VAR. *GLOBIGII* FROM VENTILATED CUBICLES

Time after release, minutes	Cubicle	Number of *B. subtilis* var. *globigii* in 12 ft^3 of air		
		Ventilation switched on after release of organisms	Ventilation on throughout	Ventilation switched off
1	Plenum	267	70	150
5		101	35	156
10		20	11	146
20		2	2	119
30		1	1	80
1	Recirculation 1	260	130	150
5		168	94	110
10		120	65	91
20		60	31	82
30		30	20	80
1	Recirculation 2	378	317	185
5		245	160	189
10		189	106	160
20		80	50	141
30		48	8	121

Source: G.A.J. Ayliffe et al., *J. Hyg.*, 69, 511, 1971. With permission of Cambridge University Press.

2.1–13   TRANSFER OF *BACILLUS SUBTILIS* VAR. *GLOBIGII* FROM A PLENUM-VENTILATED CUBICLE TO AN AIRLOCK OR TO OUTSIDE CORRIDOR

Time after release of organism, minutes	Number of *B. subtilis* var. *globigii* in 12 ft³ of air			
	Inside cubicle	In corridor outside cubicle	Inside cubicle	In airlock
1	292	8	265	115
2	318	20	230	160
	Doors opened[a]		Inner door opened (30 sec)	
3	220	32	152	144
4	133	18	120	119
5	120	12	68	84

[a] The inner door was opened for 30 sec, then closed; outer door was opened for 30 sec.

Source: G.A.J. Ayliffe et al., *J. Hyg.,* 69, 511, 1971. With permission of Cambridge University Press.

2.1–14   TRANSFER OF *BACILLUS SUBTILIS* VAR. *GLOBIGII* FROM VENTILATED CUBICLE (RECIRCULATION 1) TO CORRIDOR

Time after release of organism, minutes	Number of *B. subtilis* var. *globigii* in 12 ft³ of air			
	Inside cubicle		In corridor outside cubicle	
	Vent off	Vent on	Vent off	Vent on
1	240	208	5	42
2	260	140	2	38
	Door opened (30 sec)			
3	222	120	50	72
4	184	109	23	68
5	160	80	20	62

Source: G.A.J. Ayliffe et al., *J. Hyg.,* 69, 511, 1971. With permission of Cambridge University Press.

2.1–15   TRANSFER OF *BACILLUS SUBTILIS* VAR. *GLOBIGII* FROM CORRIDOR OUTSIDE A PLENUM-VENTILATED CUBICLE WITH AIRLOCK TO INSIDE OF CUBICLE

Time after release of spores, minutes	Number of *B. subtillis* var. *globigii* in 12 ft³ of air			
	Inside cubicle		Outside cubicle	
	Ventilation off	Ventilation on	Ventilation off	Ventilation on
1	1	0	174	145
2	3	4	160	122
	Doors opened for 30 sec			
3	7	8	155	100
4	11	5	149	75
5	16	4	104	53
6	26	1	68	28
7	22	0	72	22
8	21	0	52	21
9	20	3	35	12
10	21	1	37	14

Source: G.A.J. Ayliffe et al., *J. Hyg.,* 69, 511, 1971. With permission of Cambridge University Press.

2.1–16 TRANSFER OF *BACILLUS SUBTILIS* VAR. *GLOBIGII* FROM CORRIDOR OUTSIDE VENTILATED CUBICLE (RECIRCULATION 1) TO INSIDE

	Number of *B. subtilis* var. *globigii* in 12 ft^3 of air			
Time after release of spores, minutes	Inside cubicle		Outside cubicle	
	Ventilation off	Ventilation on	Ventilation off	Ventilation on
1	0	1	400	500
2	1	2	200	300
	Door opened for 30 sec			
3	0	1	180	242
4	12	4	132	202
5	8	3	140	184
6	8	8	159	105
7	5	6	143	132
8	12	5	148	120
9	9	1	112	80
10	14	3	75	75

Source: G.A.J. Ayliffe et al., *J. Hyg.,* 69, 511, 1971. With permission of Cambridge University Press.

2.1–17 *KLEBSIELLA* ISOLATIONS FROM HOSPITAL AIR

Month	Total samples	Total positive	Percent positive	Total positive patient isolates during month	Relative humidity, %		Temperature range, °F
					Mean	Median	
March	9	1	11	5	24.3	22.4	67–75
April	30	2	7	64	37.2	37.8	70–78
May	30	2	7	38	42.4	43.1	69–75
June	26	14	54	65	46.8	46.7	70–75

Source: A.G. Turner and J.G. Craddock, *Hospitals,* 47, 79, 1973. With permission.

2.1–18 THE RELATIONSHIP OF ANTIBIOTIC THERAPY TO UPPER RESPIRATORY COLONIZATION WITH *KLEBSIELLA*

	Patient carrier status		
Therapy	Colonized in intensive care unit	Never colonized	Total patients
Antibiotics	31	93	124
No antibiotics	1	23	24

Source: A.G. Turner and J.G. Craddock, *Hospitals,* 47, 79, 1973. With permission.

2.1–19   MONTHLY AVERAGES OF TOTAL BACTERIAL COUNT AND TOTAL STAPHYLOCOCCUS COUNT PER CUBIC FOOT FOR FOUR MONTHS (NOVEMBER 1961 AND 1962 AND DECEMBER 1961 AND 1962) AT EACH SITE ROUTINELY AIR SAMPLED AT LUTHERAN GENERAL HOSPITAL, PARK RIDGE, III.

| | November | | | | December | | | |
| | 1961 | | 1962 | | 1961 | | 1962 | |
	Total bacteria	*S. aureus*	Total bacteria	*S. aureus*	Total bacteria	*S. aureus*	Total bacteria	*S. aureus*
Surgery major hall	24	0.00	10	0.03	15	0.16	7	0.03
Surgery minor hall	24	0.00	8	0.03	13	0.01	2	0.02
Nursery 1	8	0.01	7	0.03	5	0.00	2	0.00
Nursery 2	8	0.00	8	0.04	4	0.00	3	0.03
Nursery 3	5	0.04	5	0.01	4	0.01	2	0.03
Nursery 4	2	0.01	2	0.00	3	0.00	2	0.00
2nd floor, site 1	10	0.00	6	0.00	6	0.01	4	0.03
2nd floor, site 2	13	0.10	7	0.03	6	0.03	4	0.00
2nd floor, site 3	7	0.01	6	0.01	4	0.01	5	0.00
2nd floor, site 4	9	0.04	6	0.01	6	0.03	2	0.00
3rd floor, site 1	14	0.11	9	0.21	8	0.01	5	0.01
3rd floor, site 2	19	1.50	16	0.05	13	0.01	9	0.10
3rd floor, site 3	19	0.03	9	0.04	9	0.01	7	0.01
3rd floor, site 4	16	0.86	11	0.05	11	0.03	12	0.05
4th floor, site 1	11	0.10	9	0.03	10	0.02	7	0.03
4th floor, site 2	18	0.01	15	0.03	21	0.06	10	0.03
4th floor, site 3	12	0.10	12	0.06	10	0.04	15	0.01
4th floor, site 4	19	0.03	9	0.26	12	0.01	5	0.00
6th floor, site 1	15	0.05	13	0.09	14	0.02	12	0.03
6th floor, site 2	17	0.03	19	0.06	16	0.12	13	0.12
6th floor, site 3	12	0.07	12	0.03	12	0.00	9	0.00
6th floor, site 4	22	0.05	21	0.06	16	0.03	8	0.05
7th floor, site 1	25	0.03	9	0.01	14	0.07	7	0.01
7th floor, site 2	10	0.03	9	0.04	7	0.01	10	0.03
7th floor, site 3	22	0.01	10	0.09	10	0.07	7	0.00
7th floor, site 4	6	0.01	10	0.03	5	0.00	7	0.01
X-ray walting room	10	0.01	6	0.03	4	0.01	4	0.00
X-ray main desk	13	0.00	7	0.00	3	0.01	6	0.01
Bacteriology laboratory	3	0.01	3	0.00	5	0.01	3	0.01
Physiotherapy	13	0.01	13	0.03	8	0.03	4	0.00
Average	13.7	0.12	9.6	0.04	9.8	0.04	6.4	0.02

Source: J.G. Schaffer and J. McDade, *Hospitals,* 38, 40, 1964. With permission.

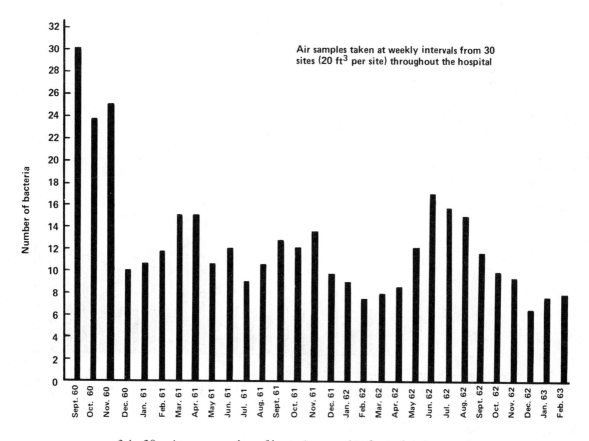

2.1–20   Average number of bacteria per cubic foot of air by months.

Source: J.G. Schaffer and J. McDade, *Hospitals,* 38, 40, 1964. With permission.

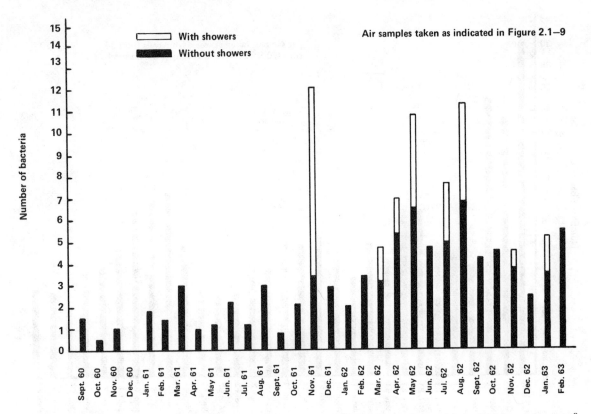

2.1—21   Average number of *Staphylococcus aureus* per 100 ft^3 of air by months with and without showers.[a]

[a] A "shower" has been defined, based on experience, as the isolation of 20 or more *S. aureus* in any given sample at any given site.

Source: J.G. Schaffer and J. McDade, *Hospitals,* 38, 40, 1964. With permission.

2.1—22   RESULTS OF WEEKLY AIR SAMPLING IN AREAS INDICATED FOR FEBRUARY AND MARCH 1962 SHOWING AVERAGE NUMBER OF BACTERIA PER CUBIC FOOT OF AIR SAMPLES AND TOTAL *STAPHYLOCOCCUS AUREUS* ISOLATED

Nursery 2			Area 621-3			Area 621-3		
Date	Bacteria[a]	*S. aureus*	Date	Bacteria[a]	*S. aureus*	Date	Bacteria[a]	*S. aureus*
2-1	2 F	—	2-2	15 FH	—	3-2	7 F	2
2-8	2 VL	—	2-9	15 FH	1	3-9	30 FH	2
2-15	3 F	1	2-16	18 L	4.5	3-16	20 F	9
2-22	8 FH	3	2-23	27.5 EH	2.5	3-23	19 FH	2
						3-30	30 F	20

[a] Activity scale: EH — extremely high; FH — fairly high; F — fair; L — low; VL — very low.

Adapted from J.G. Schaffer and J. McDade, *Hospitals,* 38, 40, 1964.

2.1–24 NUMBERS OF COLONIES OF
STAPHYLOCOCCUS AUREUS FROM WARD AIR

	Number of colonies per 24-hr plate (0.36 m² hr)							
	0	1–	5–	11–	21–	51–	101+	Total
Occupied patient rooms								
Number with 0 type	98	–	–	–	–	–	–	98
Number with 1 type	–	77	11	13	10	18	20	149
Number with 2 types	–	24	15	3	8	1	4	55
Number with 3 types	–	0	3	3	1	0	0	8
Number with 4 types	–	0	0	0	0	0	1	1
Total	98	101	29	19	19	20	25	311
Service areas								
Number with 0 type	84	–	–	–	–	–	–	84
Number with 1 type	–	84	10	5	1	1	0	101
Number with 2 types	–	37	12	4	1	1	0	55
Number with 3 types	–	8	12	3	0	0	0	23
Number with 4 types	–	3	6	2	0	0	0	11
Number with 5 types	–	–	5	1	0	0	0	6
Number with 6 types	–	–	1	0	0	0	0	1
Total	84	132	46	15	2	2	0	281
Empty bedrooms								
Total	73	14	0	0	0	0	0	87
Grand total	255	247	75	34	21	22	25	679

Source: R.E.O. Williams and L. Harding, *J. Hyg.*, 67, 649, 1965. With permission of Cambridge University Press.

2.1–23 Results of continuous air sampling in Operating Room 2 during cholecystectomy Nov. 11, 1960, showing average number of bacteria per cubic foot of air.

Notes: (1) Patient brought in and operating team enters. (2) Circulating nurse leaves and reenters room. (3) Nurse enters with sterile packs. (4) Patient is removed and operating team leaves. (5) Cleanup period begins.

Source: J.G. Schaffer, and J. McDade, *Hospitals*, 38, 40, 1964. With permission.

2.1–25   NUMBERS AND APPARENT SOURCES OF
STAPHYLOCOCCUS AUREUS FROM WARD AIR

| | Number of single strains | | | | |
| | On plates with (colonies/24 hr or 0.36 m² hr)[a] | | | | |
	Total	1–5	6–20	21–100	101 +
Occupied bedrooms, total	291	183	47	37	24
S. aureus of type carried					
By patient in room	127	36	31	37	23
By patient in other room	99	88	10	0	1
By patient in other room or staff	1	1	0	0	0
By staff, not patient	22	20	2	0	0
Source not found	38	38	4	0	0
Service areas, total	362	327	31	4	0
S. aureus of type carried					
By patient in ward	195	163	28	4	0
By patient or staff	8	8	0	0	0
By staff, not patient	35	33	2	0	0
Source not found	124	123	1	0	0
Empty bedrooms, total	14	14	0	0	0
S. aureus of type carried by patient in ward	11	11	0	0	0

[a] About half the plates were exposed for only 12 hr; the counts on these plates have been doubled for entry into this table.

Source: R.E.O. Williams and L. Harding, *J. Hyg.*, 67, 649, 1965. With permission of Cambridge University Press.

2.1–26   SPREAD FROM ROOMS OCCUPIED BY
STAPHYLOCOCCUS AUREUS SHEDDERS[a]

	Plates examined	Percentage yielding *S. aureus* of patient's type
Occupied bedrooms		
At the same end of corridor	106	29.3
At other end of corridor	129	17.1
Corridor or "sluice"		
At the same end of corridor	46	74.0
At other end of corridor	47	25.6

[a] For this table, "shedders" were defined as patients who were nasal carriers of *S. aureus,* and whose staphylococci were found in the air of their own room and at least one other room or ward area.

Source: R.E.O. Williams and L. Harding, *J. Hyg.*, 67, 649, 1965. With permission of Cambridge University Press.

2.1–27 NASAL CARRIER RATES FOR
STAPHYLOCOCCUS AUREUS

Week of swabbing	Number of patients examined	Percentage carrying *S. aureus*
1	197	38.1
2	117	33.3
3	90	32.2
4	60	31.7
5	41	41.5
6	30	36.7
7	24	33.3
8	20	35.0
9, 10	27	37.0

Source: R.E.O. Williams and L. Harding, *J. Hyg.*, 67, 649, 1965. With permission of Cambridge University Press.

2.1–28 NUMBER AND SOURCES OF AIRBORNE STRAINS
OF *STAPHYLOCOCCUS AUREUS*

	Number of possible sources						
	Number settling per 1000 ft² min			Carriers		Number settling per possible source per 1000 ft² min	
Probable source	All strains	T[a] strains	Persons	All strains	T strains	All strains	T strains
Patient(s) in same room or bay							
Single rooms	48.2	3.9	1.0	0.4	0.1	120	39
Four-bed bays	15.6	1.5	3.56	1.2	0.3	12.9	5.0
Patient(s) in other rooms or bays	16.6	9.0	22.3	7.0	1.7	2.4	5.3
Staff carriers	9.0	4.0	–	10.3	2.5	0.9	1.6
No known source	8.7	3.7	–	–	–	–	–
Total	55.1	18.6	–	19.0	4.4	3.1	4.2

Notes: The numbers of possible sources are average over the period of observation.
In forming the totals the values for the four-bed bays and the single rooms have been weighed in proportion to the number of patient-weeks of experience in the different locations. 1000 ft²/min = 93 m²/min.

[a] T strains = strains resistant to tetracycline.

Source: O.M. Lidwell et al., *J. Hyg.*, 69, 113, 1971. With permission of Cambridge University Press.

2.1–29 RATES OF NASAL ACQUISITION OF *STAPHYLOCOCCUS AUREUS*

Probable source	Acquisition rate per 1000 patient-weeks		Number of possible sources (carriers)		Acquisition rate per possible source per 1000 patient-weeks	
	All strains	T[a] strains	All strains	T strains	All strains	T strains
Other patients in same bay	3.4	1.3	0.8	0.2	4.3	6.5
Patients in other rooms or bays	22.0	11.0	7.0	1.7	3.2	6.5
Staff carriers	24.6	10.7	10.3	2.5	2.4	4.3
All known possible sources	50.0	22.9	18.0	4.4	2.8	5.2
No known source	27.3	6.3	–	–	–	–
Probably real	6.9	3.3	–	–	–	–
Probably spurious	20.4	3.0	–	–	–	–
Total	77.4	29.0	–	–	–	–

Notes: The total experience comprised 3327 patient-weeks of which 2750 were in four-bed bays (the interval between the admission swabs and the first regular weekly swab has been counted as a full week). There were 257 apparent acquisitions.

[a] T strains = strains resistant to tetracycline.

Source: O.M. Lidwell et al., *J. Hyg.,* 69, 113, 1971. With permission of Cambridge University Press.

2.1–30 NUMBER OF AIRBORNE CONTAMINANTS PER 100 ft³ OF AIR AT CRITICAL SITES IN A LAMINAR CROSSFLOW OPERATING ROOM

Day	Surgical draping	Wound	Back instrument table	Instrument tray	Filter wall	Manifold
Monday a.m.	Draped	0.0	7.3	1.8	0	98
Monday p.m.	Undraped	0.0	3.6	1.8	0	116
Tuesday a.m.	Undraped	3.6	0.0	11	0	67
Tuesday p.m.	Undraped	5.4	1.8	1.8	0	111
Wednesday a.m.	Draped	3.6	1.8	3.6	0	131
Wednesday p.m.	Undraped	0.0	0.0	1.8	0	87
Thursday a.m.	Draped	3.6	5.4	0.0	0	73
Thursday p.m.	Draped	3.6	5.4	5.4	0	84
Friday a.m.	Draped	1.8	0.0	5.4	0	44
Friday p.m.	Undraped	1.8	5.4	1.8	0	47

Note: Data based on ten neurosurgical cases.

Source: D.G. Fox and M. Baldwin, *Hospitals,* 42, 108, 1968. With permission.

2.1–31   EFFECT OF DRAPING ON NUMBER OF AIRBORNE CONTAMINANTS PER 100 ft³ OF AIR

Condition	Wound	Back instrument table	Instrument tray	Filter wall	Manifold
Draped	1.4	2.2	1.8	0	8.6
Undraped	1.2	1.2	1.8	0	8.6

Source: D.G. Fox and M. Baldwin, *Hospitals,* 42, 108, 1968. With permission.

2.1–32   AVERAGE NUMBER OF SURFACE CONTAMINANTS ON FALLOUT PLATES, PER CASE

Condition	Back instrument table	Instrument tray	Pack table
Draped	0.0	0.4	1.4
Undraped	0.4	0.0	2.4

Source: D.G. Fox and M. Baldwin, *Hospitals,* 42, 108, 1968. With permission.

2.1–33   COMPARISON OF MICROBIAL CHARACTERISTICS OF A LAMINAR CROSSFLOW OPERATING ROOM AND CONVENTIONALLY VENTILATED OPERATING ROOMS

Operating room	Air samples		Surgical instruments		Floor samples taken before surgery
	At wound	In room	Used	Unused	
	Organisms per 100 ft³		Percent	Positive	Organisms per square inch
Laminar flow room	2.4	86	10	0	11.7
New surgical wing	60.0	120	50	38	2.3
Old surgical area	189.0	174	21	10	4.9

Source: D.G. Fox and M. Baldwin, *Hospitals,* 42, 108, 1968. With permission.

2.1–34   PERCENTAGE REDUCTION OF BACTERIAL AND PARTICLE COUNT AT DIFFERENT VELOCITIES OF LAMINAR FLOW COMPARED WITH CONVENTIONAL VENTILATION

	Air velocity, meters per second					
	0.1	0.2	0.3	0.4	0.5	0.6
Bacteria						
Downflow	79.5	90.0	97.1	98.9	98.8	–
Crossflow	39.0	79.4	90.2	90.5	94.6	93.9
Particles $\geqslant 0.5\ \mu$m						
Downflow	88.6	91.7	94.8	96.9	98.0	–
Crossflow	74.8	87.1	90.1	91.1	92.0	94.5

Source: W. Whyte et al., *J. Hyg.,* 71, 559, 1973. With permission of Cambridge University Press.

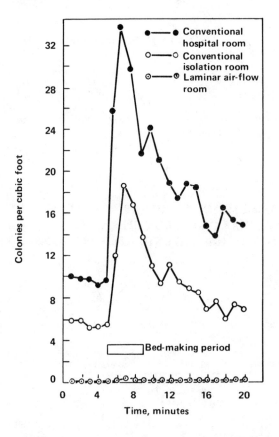

2.1–35   Air contamination in laminar airflow room and conventional isolation and hospital rooms measured by slit-samplers placed 3 ft inside the room entrances. Curve for laminar airflow room is mean of four experiments, others are mean of two experiments.

Source: C.O. Solberg et al., *Appl. Microbiol.,* 21, 209, 1971. With permission.

2.1—36   AIR CONTAMINATION IN LAMINAR AIRFLOW ROOM AND CONVENTIONAL ISOLATION AND HOSPITAL ROOMS

Colonies per cubic foot

Experiment	Laminar airflow room[a] (downstream) at airflow velocity of		Conventional isolation rooms[b]	Conventional hospital rooms[b]
	90 ft/min	60 ft/min		
1	0.047	0.050	8.16	15.90
2	0.039	0.068	9.61	21.56
Mean	0.043	0.059	8.89	18.73

[a] Total of 660 ft^3 sampled.

[b] Mean of three experiments, one in each of three different rooms. Total of 696 ft^3 sampled.

Source: C.O. Solberg et al., *Appl. Microbial.*, 21, 209, 1971. With permission.

2.1—37   IDENTIFICATION OF ENVIRONMENTAL CONTAMINANTS FROM LAMINAR FLOW ROOM AND CONVENTIONAL ISOLATION AND HOSPITAL ROOMS

Number of colonies

Organisms	Laminar airflow room		Conventional isolation rooms		Conventional hospital rooms	
	Blankets and pillows	Floors, walls, furniture, and air	Blankets and pillows	Floors, walls, furniture, and air	Blankets and pillows	Floors, walls, furniture, and air
Bacillus sp.		16	14	81	10	63
Candida sp.			2	4	1	5
Clostridium sp.			1			
Diphtheroids	2	14	3	28	4	12
Enterobacter			2	2		1
Enterococcus	3	3	2	2	3	3
Escherichia coli	4	2	1	2	4	6
Hereliea sp.			1	1		1
Klebsiella sp.	4	1	2		2	9
Lactobacillus sp.			1	2	1	
Neisseria sp.		9	2	16	2	21
Proteus mirabilis	1	1	2	4	1	3
Pseudomonas aeruginosa	3	2	1	8	3	2
Saccharomyces				1		2
Staphylococcus aureus			2	9	4	11
S. epidermidis	33	68	20	61	22	79
Viridans group streptococcus				3		2
Unidentified molds		2	4	16	3	20
Total	50	118	60	240	60	240

Source: C.O. Solberg et al., *Appl. Microbiol.*, 21, 209, 1971. With permission.

2.1–38 PERCENTAGE OF PATIENT IDENTICAL STRAINS IN SAMPLES FROM ENVIRONMENT OF LAMINAR AIRFLOW ROOM AND CONVENTIONAL ISOLATION AND HOSPITAL ROOMS[a]

Room	Blankets and pillows	Floors, walls, tables, lamp, chair	Air
Laminar airflow room	96(50)	52(61)	46 (57)
Conventional isolation rooms	78(60)	36(90)	33(150)
Conventional hospital rooms	67(60)	20(90)	29(150)

[a] Values in parentheses express numbers of colonies examined.

Source: C.O. Solberg et al., *Appl. Microbiol.*, 21, 209, 1971. With permission.

2.1–39   MICROORGANISMS ISOLATED FROM THE PATIENT DURING STAY IN LAMINAR AIRFLOW UNIT

Samples from	Organisms	\multicolumn{13}{c}{Number of organisms isolated on day[a]}
		1[b]	3	8	11	15	22	26	29	32	36	40	43	45
Nose	Diphtheroids	+										+		++
	Escherichia coli										++++	+	++	++++
	Klebsiella sp.								+++		++++	++	++++	+++
	Pseudomonas aeruginosa	+	+	++	+++	++	++	++	++	++	++++	++	++	+++
	Staphylococcus epidermidis							++	++			+++	++	+++
	Viridans group streptococcus							+						
Throat	Diphtheroids	+												
	E. coli	++	+++				++	++	++++	+	++++	++++	++++	++++
	Klebsiella sp.	++	+				+++			+	+++	++	+++	++
	Neisseria sp.		+++	+++	++++	+++								
	P. aeruginosa	++	+	++	++	++	++++	++	+++	++	++	++++	++	+++
	S. epidermidis								++++	+		+	+	++
	Viridans group streptococcus	++	+++	+	+++	+++	+++	++	++++	+++	+++	++++	++	+++
Perineum	*E. coli*	++	++++	++	+	++++	++++	++	+++	++	+++	++++	+++	++++
	Klebsiella sp.	++	+++	++	++++		++++		++	+	+++	+++	++	++
	Proteus mirabilis	++++	+++	+++	++++		++++	+	++++	++	+++	+++	+	+++
	P. aeruginosa									++	+++	++	++	++
	S. epidermidis			++		++		++		+	+++	+++	+++	+++
	Enterococcus	++++	+++	+++	++++	++	++++	++	+++	+++	++++	++++	++++	++++
Stool[c]	*E. coli*	++++	++++	++++	++++	++++	++++	++++	++++	++++	++++	++++	++++	++++
	Clostridium perfringens	+	+		+									
	Klebsiella sp.	++	+++	+++	+++	+++	+++	+	++++	++++	++	++++	+++	++++
	P. mirabilis	++++	++++	+++	+++	++	++	++++	++	++	++	+++	++	++++
	P. aeruginosa					++								
	Enterococcus	++++	++++	++++	++++	++++	++++	++++	++++	++++	++++	++++	++++	++++

a   Symbols: +, fewer than 10 colonies; ++, 10 to 30 colonies; +++, more than 30 colonies but no growth in the third segment of streaking; ++++, more than 30 colonies and growth in all streaking segments.

b   Immediately before entering the laminar airflow unit.

c   Anaerobic cultures also demonstrated *Bacteroides* sp., peptostreptococci, *Lactobacillus* sp., and diphtheroids.

Source: C.O. Solberg et al., *Appl. Microbiol.*, 21, 209, 1971. With permission.

2.1–40   *CLOSTRIDIUM WELCHII* IN THE AIR (SLIT SAMPLING) AND ON THE FLOOR OF OPERATING THEATRES

Hospital	Plenum ventilation	Clean zone	Transfer area	*Cl. welchii* per 100 ft^3 of air	Mean *Cl. welchii* per 100 cm^2 (ten plates)
1	Yes	Yes	Yes	0.7	2.67
2A	Yes	Yes	No	0.4	2.33
2C	No	No	No	3.6	2.33
3	Yes	Yes	No	0	0.67
4	Yes	No	No	0.3	1.67
5	No	No	No	1.3	—[a]
6	No	No	No	3.4	—
7	Yes	Yes	No	0.5	—
8	Yes	Yes	No	0.8	—
9	No	No	No	1.4	—
10	No	Yes	No	0	—

[a] Dash = not tested.

Source: G.A.J. Ayliffe et al., *J. Hyg.,* 67, 417, 1969. With permission of Cambridge University Press.

2.1–41   STUDIES ON THE REDISPERSAL OF BACTERIA FROM THE FLOOR INTO THE AIR

Total counts and counts of *Staphylococcus aureus* in 50 ft^3 of air (slit sampler)

Time of sampling	Experiment 1 (vinyl, blowing)		Experiment 2 (vinyl, blowing)		Experiment 3 (terrazzo, blowing)		Experiment 4 (vinyl, sweeping)		Experiment 5 (vinyl, exercise)		Experiment 6 (terrazzo, exercise)	
	Total	*S. aureus*	Total	*S. aureus*	Total	*S. aureus*	Total	*S. aureus*	Total	*S. aureus*	Total	*S. aureus*
Preliminary sample	358	0	370	0	137	0	720	0	101	0	235	0
During contamination[a]	+ + +[b]	8160	7900	2000	8720	2000	5800	2000	2400	460	1136	288
1 hr after contamination	652	15	364	1	286	8	432	20	92	7	48	0
During floor disturbance	704	24	332	1	458	27	3004	158	169	7	64	1
1 hr after floor disturbance	464	0	180	0	97	2	32	0	56	0	2	0

a   Contamination by shaken blanket in Experiments 2–6; exercise by staphylococcal disperser in Experiment 1.
b   Too numerous to count.

Source: G.A.J. Ayliffe et al., *J. Hyg.*, 65, 515, 1967. With permission of Cambridge University Press.

2.1–42   SURFACE CONTAMINATION IN HOSPITALS[a]

Surface	Mean count per Rodac®[a] plate[b]	Range
Floors (prior to cleaning)	235	54–682
Floors (after wet mopping)	99	17–443
Floors (after wet vacuum cleaning)	39	6–162
Overbed tables (prior to cleaning)	75	22–230
Overbed tables (after cleaning)	32	10–90
Surgical suite floors (prior to cleaning)		
Operating rooms	49	5–106
Interior corridors	95	22–183
Scrub and substerilizing rooms	83	16–187
Exterior corridors	131	39–297
Work rooms	114	43–622
Dressing rooms	326	137–586

[a] Approximately 4 in.^{2}.

[b] Falcon Plastics, Los Angeles.

Source: R.G. Bond et al., Eds., *Environmental Health and Safety in Health-Care Facilities,* Macmillan, New York, © 1973, 32. With permission.

REFERENCES

R.G. Bond et al., *Survey of Microbial Contamination in the Surgical Suites of 23 Hospitals,* Final Report, Research Contract PH–86–63–96, Division of Hospital and Medical Facilities, Bureau of State Services, Public Health Service, U.S. Department of Health, Education, and Welfare, Washington, D.C., March 1964.

A.K. Pryor et al., *Health Lab. Sci.,* 4, 153, 1967.

D. Vesley, *Health Lab. Sci.,* 7, 256, 1970.

2.1−43   RODAC[®] PLATE BACTERIAL COLONY COUNTS OBTAINED FROM HOSPITAL PATIENT ROOM FLOORS IN 1964 AND 1965

| | 1964 | | | 1965 | | | | |
| | Before cleaning | | | | Mean colony counts | | Pooled standard deviation | |
Hospital	Number of samples	Mean colony counts	Standard deviation	Number of samples	Before cleaning	After cleaning	Before cleaning	After cleaning
A	298	204	110	465	239	136	137	66
B	168	205	124	150	54	17	35	15
C	187	193	105	210	264	62	83	37
D	138	253	115	150	224	72	87	41
E	112	322	132	165	682	40	305	73
F	–	–	–	150	85	44	67	51
G	80	193	57	150	215	146	57	56
H	–	–	–	150	125	443	48	346
I	–	–	–	120	467	90	158	82
J	–	–	–	150	113	83	105	108
K	–	–	–	120	382	100	198	100
L	–	–	–	150	191	94	164	59
M	–	–	–	150	180	32	81	31
N	–	–	–	150	72	28	35	21
O	138	180	85	–	–	–	–	–
P	117	181	93	–	–	–	–	–
Q	104	158	89	–	–	–	–	–
Total	1442	210[a]	101[a]	2430	235[a]	99[a]	111[a]	78[a]

[a] Average values.

Source: A.K. Pryor et al., *Health Lab. Sci.,* 4, 153, 1967. With permission.

2.1−44   BACTERIAL COUNTS OF FLOORS

Hospital 1 with Transfer Area and Clean Zone

| | Total organisms and *Staphylococcus aureus* | | | *Clostridium welchii* | |
Site of sampling	Number of plates	Mean total organisms per 100 cm²	Mean *S. aureus* per 100 cm²	Number of plates	Mean per 100 cm²
Hospital corridor	15	483.3	1	15	36.67 ± 13.37
Protective zone	20	469	8.33	20	41.17 ± 10.57
Transfer area	15	379.3	1.67	15	65.77 ± 15.13
Clean zone	25	295.7	1.33	25	2.80 ± 2.07
Theatre	20	111	0	20	0.83

Source: G.A.J. Ayliffe et al., *J. Hyg.,* 67, 417, 1969. With permission of Cambridge University Press.

2.1—45   BACTERIAL COUNTS OF FLOORS

Hospital 2, Theatre A, No Transfer Area

Site of sampling	Total organisms and *Staphylococcus aureus*			*Clostridium welchii*	
	Number of plates	Mean total organisms per 100 cm²	Mean *S. aureus* per 100 cm²	Number of plates	Mean per 100 cm²
Hospital corridor	15	1052.3	8.7	20	50.50 ± 5.83
Protective zone	18	336.0	2.0	20	11.50 ± 1.87
Clean zone	28	206.0	1.7	20	1.33 ± 1.40
Theatre	18	283.3	1.0	20	0.5

Source: G.A.J. Ayliffe et al., *J. Hyg.,* 67, 417, 1969. With permission of Cambridge University Press.

2.1—46   BACTERIAL COUNTS OF FLOORS

Hospital 2, Theatre B, No Transfer Area or Clean Zone

Site of sampling	Total organisms and *Staphylococcus aureus*			*Clostridium welchii*	
	Number of plates	Mean total organisms per 100 cm²	Mean *S. aureus* per 100 cm²	Number of plates	Mean per 100 cm²
Hospital corridor and doorway	15	560	2.33	15	102 ± 19.4
Protective zone	15	346.67	0.33	15	15.1 ± 3.6
Theatre	10	286.67	0.33	20	20.5 ± 12.33

Source: G.A.J. Ayliffe et al., *J. Hyg.,* 67, 417, 1969. With permission of Cambridge University Press.

2.1–47   SETTLE-PLATE COUNTS

Period of study	Sex of patients in ward	Site of sampling	Number of plates	Mean per total count per plate per hour and standard error
October 1965– September 1966	Female	Window-ventilated cubicle	262	10.8 ± 0.90
		Recirculation cubicles	726	8.8 ± 0.41
		Open ward	1421	14.5 ± 0.27
		Plenum-ventilated cubicles	569	3.9 ± 0.25
October 1966– September 1967	Male	Window-ventilated cubicle	295	23.6 ± 1.45
		Recirculation cubicles	752	15.9 ± 0.64
		Open ward	1568	26.0 ± 0.50
		Plenum-ventilated cubicles	563	12.6 ± 0.59

Comparison of:	Female	Male
Plenum with open ward	$t = 28.60$	$t = 17.32$
Plenum with recirculation cubicle	$t = 10.08$	$t = 3.79$
Plenum with window– ventilated cubicle	$t = 7.34$ } $P < 0.001$	$t = 7.01$ } $P < 0.001$
Recirculation with open ward	$t = 11.49$	$t = 12.46$
Recirculation with window– ventilated cubicle	$t = 2.01$ $P < 0.05$	$t = 4.85$
Window-ventilated cubicle with open ward	$t = 6.03$ $P < 0.001$	$t = 1.57$ Not significant

Source: G.A.J. Ayliffe et al., *J. Hyg.,* 69, 511, 1971. With permission of Cambridge University Press.

2.1–48   NUMBERS OF SPORE-BEARING PARTICLES SETTLING PER SQUARE FOOT OF EXPOSED SURFACE IN THE "SENSITIVE" AREA FOR 10^8 PARTICLES DISPERSED

Equivalent Particle Diameter 13 μm

	Ventilation conditions					
	a	b	c	d	e	f
Linear air flow velocity, feet per minute	100	60	35	22	Turbulent	
Rate of air supply, cubic feet per minute	8200	4900	2900	1800	1600	130
Ventilation rate per hour	360	220	130	80	75	6
Calculated settling for turbulent ventilation	1.32×10^4	2.18×10^4	3.63×10^4	5.57×10^4	5.90×10^4	35×10^4

Number of Particles Settling

Group	Experimental conditions	a	b	c	d	e	f
1	D 0P, -C or -U	–	3, 10	–	–	–	–
	A 0P, -C or -U	–	6, 10	–	–	–	–
	D 1-P, -C or -U	5, 3, 5, 10	–	25	0	4.0×10^4	15×10^4
	D 2-P		5	13, 12,	12, 207	3.2×10^4	29×10^4
	D 2-C	2	8	–	–	–	–
	A 1-P	2	2	–	–	–	–
	A 2-P	7	0	10	2	–	–
2	D 3-P, -C or -U	40, 90	–	255, 146, 104	590, 287, 500	–	–
		87, 65		222, 62, 85	900, 845 557		
	A 2-C	100	47	57	600	–	–
	A 3-P	149	–	119	445	–	–
3	A 2-U	202		650	7200		
	A 3-U	162	380	1230, 1380	4150	4.1×10^4	34×10^4

Notes: The experimental conditions have been coded as follows: D, dispersal at the rear of the room, at point D; A, dispersal in one bed area, B1, at point A (at the front of the room); 0, no movement during dispersal; 1, movement in the rear half of the room only; 2, movement in the rear half of the room and into the nonsensitive bed area only; 3, movement over the whole area of the room; P, a solid partition separated the two bed areas; C, bed areas were separated by a curtain; U, communication between the two bed areas was unobstructed. When dispersal was at point D the "sensitive" area comprised both bed areas, B1 and B2. When dispersal took place at point A, only the area B2 formed the "sensitive" area. The contamination level as determined in control exposures without dispersal was equivalent to eight particles settling in any test.

Source: O.M. Lidwell and A.G. Towers, *J. Hyg.*, 67, 95, 1969. With permission of Cambridge University Press.

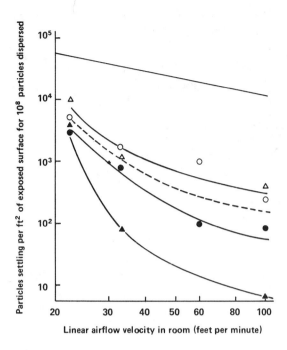

2.1–49   Transfer across the direction of airflow. Dispersal took place at point A in area B1. Triangular symbols, △ and ▲, record experiments in which movement was restricted to the rear and area B1 only (coded 2 in Table 2.1–48). Circular symbols, ○ and ●, record experiments in which movement extended over the whole of the room area (coded 3 in Table 2.1–48). Open symbols, △ and ○, record the spore-bearing particles recovered from the six plates exposed in area B2, close to the center line of the room. Filled symbols, ▲ and ●, record the spore-bearing particles recovered from the remaining six plates exposed in area B2 close to the wall of the room. The upper straight line shows the numbers of spore-bearing particles that would have been recovered if the air had been completely mixed by turbulent movements. The lower curves have been drawn in by eye to illustrate the effect of the various conditions on the number of spore-bearing particles recovered.

Source: O.M. Lidwell and A.G. Towers, *J. Hyg.*, 67, 95, 1969. With permission of Cambridge University Press.

2.1–50   NUMBERS OF SPORE-BEARING PARTICLES SETTLING PER
SQUARE FOOT OF EXPOSED SURFACE FOR 10^8 PARTICLES DISPERSED

Group	Movement	Partition or curtain	Number of particles settling in		
			Dispersal bed area, B 1	"Sensitive" bed area, B 2	Rear half of room, R
2	Rear of room	Neither	15,500	650	22,000
	and dispersal	Curtain	21,000	57	53,500
	bed area only	Partition	14,000	10	38,500
3	Over whole of	Neither	6,500	1230	36,000
	the room area	Neither[a]	12,500	1380	34,000
		Partition	15,500	119	43,500

Notes: Equivalent particle diameter, 13 μm. Linear air velocity 35 ft/min. Dispersal took place in one bed area, B 1.

[a] Although there was no partition or curtain in position during this experiment, the two subjects restricted their movements in the room as if such an obstruction had been there.

Source: O.M. Lidwell and A.G. Towers, *J. Hyg.,* 67, 95, 1969. With permission of Cambridge University Press.

2.1–51   EXTRACT FROM RECORDS (14-DAY PERIOD ONLY)
OF COLONY COUNTS OF *PSEUDOMONAS* ON CULTURES ON
ONE OF THE "GRAVITATION TREES"[a] IN A CUBICLE
OCCUPIED BY AN INFECTED PATIENT

Patient Did Not Receive Topical Cream Treatment

Height above floor, feet	Colony counts of *Pseudomonas* on day													
	3	4	5	6	7	8	9	10	11	12	13	14	15	16
9	1	1	0	13	2	11	0	1	0	13	0	5	0	8
8	3	3	1	15	3	18	1	0	1	22	1	6	0	8
7	0	5	0	11	0	14	0	18	0	11	1	7	1	11
6	1	1	1	11	0	15	0	15	4	20	0	6	0	5
5	0	1	1	9	0	12	2	19	1	16	0	5	0	2
4	3	2	0	5	0	20	1	17	1	23	1	9	0	5
3	3	3	1	5	2	14	0	19	3	25	1	7	0	12
2	0	7	1	19	0	20	0	13	1	13	1	8	4	5
1	0	6	1	16	1	13	0	15	0	17	0	3	0	7
Floor	0	2	1	15	3	11	0	14	1	17	0	5	0	13

[a] In attempting to assess the incidence of airborne *Pseudomonas* under more controlled conditions, simple structures were prepared permitting 7-hr settle plates to be placed at intervals of 1 ft between the floor and ceiling of treatment rooms.[1] These were in addition to the routine 12-hr settle plates on the floor. These structures were calld "gravitation trees" and several were used in the cubicles of infected and uninfected patients, traffic zones, and other treatment rooms in the unit.

Source: F. Dexter, *J. Hyg.,* 69, 179, 1971. With permission of Cambridge University Press.

REFERENCE

1. T. L. Barclay and F. Dexter, *Br. J. Surg.,* 55, 197, 1968. With permission.

2.1–52   EXTRACT FROM RECORDS (14-DAY PERIOD ONLY) OF COLONY COUNTS OF *PSEUDOMONAS* ON CULTURES ON ONE OF THE "GRAVITATION TREES"[a] IN A CUBICLE OCCUPIED BY AN INFECTED PATIENT

Patient Received Topical Cream Treatment

Height above floor,	Colony counts of *Pseudomonas* on day													
feet	3	4	5	6	7	8	9	10	11	12	13	14	15	16
9	0	0	0	0	0	0	0	0	0	0	0	0	0	0
8	0	0	0	0	0	0	0	0	0	0	0	0	0	0
7	0	0	0	0	0	0	0	0	1	0	0	0	0	0
6	0	0	0	0	0	0	0	0	0	0	0	0	0	0
5	0	1	0	0	0	0	0	0	0	0	0	0	1	0
4	0	0	0	0	0	0	0	0	0	0	0	0	0	0
3	0	0	0	0	0	0	0	0	0	0	0	0	0	0
2	0	0	0	0	0	0	0	0	0	0	0	0	0	0
1	1	0	0	0	0	0	0	0	0	0	0	0	0	0
Floor	0	1	0	0	0	0	0	0	0	0	0	0	0	0

[a] See Footnote a in Table 2.1–49.

Source: F. Dexter, *J. Hyg.*, 69, 179, 1971. With permission of Cambridge University Press.

2.1–53   DISINFECTION AND BACTERIOLOGICAL CONTROL OF 73 SALINE-BATH TREATMENTS

Organisms	Number of times the listed organisms were grown		
	Before bath	During bath	After disinfection[a]
Pseudomonas aeruginosa	1	44	2[b]
Proteus spp.	0	30	3
Staphylococcus aureus	1	25	0
Coliforms	0	15	1
S. albus	5	0	1
Enterococci	1	11	0
Bacillus anthracoides	0	1	0
Streptococcus pyogenes	0	0	0
Streptomyces viridans	0	0	0
Diphtheroids	0	0	0

[a] Tego® (MHG) disinfectant used.
[b] Untrained staff used incorrect disinfection procedure on these two occasions.

Source: F. Dexter, *J. Hyg.*, 69, 179, 1971. With permission of Cambridge University Press.

2.1–54   PYOCINE TYPING, 1966: CHAINLIKE INFECTIONS WITH TWO TYPES OF *PSEUDOMONAS AERUGINOSA* AFFECTING 23 PATIENTS, SUGGESTIVE OF CROSS-INFECTION

Patient	Admitted	Pyocine types of *P. aeruginosa*						
E.L.	Dec. 65	–	F	–	–	–	–	–
C.L.	Jan. 66	–	–	O3	–	–	–	–
D.E.	Feb. 66	–	–	–	B1C	–	–	–
I.B.	Feb. 66	B1A	–	–	–	–	–	–
N.E.	Feb. 66	–	–	–	B1C	–	–	–
S.M.	Mar. 66	–	–	–	B1C	–	–	–
A.F.	Mar. 66	–	–	–	B1C	–	–	–
W.I.	Apr. 66	–	–	–	–	B1B	–	–
G.O.	Apr. 66	–	–	–	B1C	–	–	–
S.I.	Apr. 66	–	–	–	B1C	–	–	–
B.E.	Apr. 66	–	–	–	B1C	–	–	–
M.I.	Apr. 66	–	–	–	–	B1B	–	–
W.R.	May 66	–	–	–	B1C	–	–	–
H.O.	June 66	–	–	–	–	B1B	–	–
W.I.A.	June 66	–	–	–	–	B1B	–	–
S.H.	July 66	–	–	–	–	B1B	–	–
C.R.	Aug. 66	–	–	–	–	B1B	–	–
W.I.	Sept. 66	–	–	–	–	–	N5	–
R.I.	Sept. 66	–	–	–	–	B1B	–	–
H.E.	Sept. 66	–	–	–	–	B1B	–	–
W.E.	Sept. 66	–	–	O3	–	–	–	–
W.H.	Oct. 66	–	–	–	–	B1B	–	–
H.O.L.	Oct. 66	–	–	–	–	B1B	–	–
B.A.	Nov. 66	–	–	–	–	B1B	–	–
S.E.	Nov. 66	–	–	–	–	B1B	–	–
L.O.	Nov. 66	–	–	–	–	B1B	–	–
H.E.	Nov. 66	–	–	–	–	–	–	NT
B.U.	Nov. 66	–	–	–	–	B1B	–	–
S.H.	Dec. 66	–	–	–	–	B1B	–	–

Source: F. Dexter, *J. Hyg.*, 69, 179, 1971. With permission of Cambridge University Press.

2.1–55   MEAN BACTERIAL COUNTS FROM COVERED AND UNCOVERED AREAS OF FLOOR AND ON SETTLE PLATES EXPOSED FOR 1 hr

	Mean total bacterial counts		
Total observations	Impression plates		Settle plates
	Covered area	Uncovered area	
20	12	164 ± 21[a]	63 ± 2.8[a]

[a] t (18 degrees of freedom) = 4.45; $P < 0.001$.

Source: G.A.J. Ayliffe et al., *J. Hyg.*, 65, 515, 1967. With permission of Cambridge University Press.

2.1−56 FLOOR BACTERIA 24 hr AFTER CLEANING WITH SUDOL OR WITH SOAP AND WATER OVER A PERIOD OF 21 DAYS

Method of cleaning	Number of samples	Number of plates	Mean bacterial count per plate
Soap and water	9	18	824
Sudol 1/100	9	18	848

Source: G.A.J. Ayliffe et al., *J. Hyg.,* 65, 515, 1967. With permission of Cambridge University Press.

2.1−57 SURVIVAL OF *STAPHYLOCOCCUS AUREUS* ON A VINYL SURFACE CONTAMINATED BY A STAPHYLOCOCCAL DISPERSER

Time of sampling after contamination	Total counts of S. aureus per sample (impression plate) in experiments					Mean percentage survivors
	1	2	3	4	5	
1 hr	15	43	29	28	15	100
1 day	6	19	17	2	16	46
2 days	−	17	5	0	1	22
3 days	10	−	3	0	0	12.5
4 days	−	6	0	0	1	6.7
7 days	0	3	0	0	0	1.9

Source: G.A.J. Ayliffe et al., *J. Hyg.,* 65, 515, 1967. With permission of Cambridge University Press.

2.1−58 THE AMOUNT OF SURFACE BACTERIA PER 24.6 cm^2 ON A WELDED PLASTIC FLOOR COVERING AND A NEEDLE-FELT CARPET[a]

Day	Plastic floor covering	Needle-felt carpet
A	241	29
B	127	27
C	233	74
D	165	20
E	183	21
F	286	39
G	228	36
H	259	49
Mean value	215	37

[a] Every test represents the mean value of 12 samples taken at the same time.

Source: R. Rylander et al., *Am. J. Public Health,* 64, 163, 1974. With permission.

2.1–59   THE AMOUNT OF BACTERIA IN ROOMS WITH TUFTED CARPETING, NEEDLE-FELT CARPETING, AND PLASTIC TILE FLOORING[a]

Type of bacteria	Tile floor	Tufted carpeting	Needle-felt carpeting
Surface bacteria	113 (11.83)	19 (17.14)	58 (5.43)
Total amount per 1.13 cm^2		52 (11.40)	52. (5.33)
Airborne, 10-min test	37 (17.40)	14 (18.8)	
Airborne, 60-min test	251 (6.146)	59 (6.33)	
Sedimenting	4(7.4)	3(7.2)	

[a] Values within parentheses refer to the number of tests and the standard deviation.

Source: R. Rylander et al., *Am. J. Public Health,* 64, 163, 1974. With permission.

2.1–60   RESULTS OF TWO EXPERIMENTS WITH SOILED CARPET SHOWING NUMBERS OF BACTERIA IN 5-mm PLUGS REMOVED WITH A CORK-BORER AND GROUND IN A WARING® BLENDER

Sample number	Total number of bacteria	
	Experiment 1	Experiment 2
1	14,639	66,000
2	33,500	17,030
3	34,333	19,660
4	15,390	12,081
5	23,270	19,300
6	10,830	26,080
Average	21,944	26,692

Note: These experiments were done on two different samples of wool carpet. Experiment 1 was done on a section placed in the corridor of an office complex and had been lightly vacuumed daily for 14 days. Experiment 2 was done on a corresponding section placed on the floor in the Medical Records Department of the hospital. It had not been vacuumed for 14 days.

Source: J.G. Shaffer, *Health Lab. Sci.,* 3, 73, 1966. With permission.

2.1–61   RESULTS OF PERIODIC CULTURE OF PLUGS FROM AN EXPERIMENTAL WOOL CARPET IN AN AREA OF HEAVY USE WITH CONDITIONS AS INDICATED

Date	Conditions	Number of samples	Bacteria per sample
5/25/65[a]	Before vacuuming	17[b]	15,851
	After vacuuming	19	13,419
6/1/65[c]	4 hr after shampoo	4	4,912
	18 hr after shampoo	4	6,060
	24 hr after shampoo	12	14,145
7/13/65	Before vacuuming	21	10,712
8/18/65	Before vacuuming	19	11,688
8/19/65	After vacuuming	22	15,850

[a] Carpet had been in use for approximately 9 months. The only cleaning done during this period was done with the vacuum cleaner.

[b] Samples in each experiment were randomly taken from a 1-yard^2 area. On removal, the upper two thirds of the pile was cut off and discarded.

[c] No disinfectant was used in the shampoo process.

Source: J.G. Shaffer, *Health Lab. Sci.,* 3, 73, 1966. With permission.

2.1–62   RODAC® COUNTS FOR ALL CARPETED AREAS BEFORE CLEANING

Magnitude of counts	Observed	Expected according to Poisson
0–49	105	103
50–99	50	52
100–149	13	13
150–199	4	2
	172	($\chi^2 \cong 2.11$, df 2)

Note: The agreement is slightly inferior for solid surfaces ($\chi^2 \approx 3.2$), and after cleaning procedures ($\chi^2 \approx 5.4$).

Source: G.J. Bonde, *Health Lab. Sci.,* 10, 308, 1973. With permission.

2.1–63   AVERAGE COUNT PER PLATE
BEFORE AND AFTER CLEANING

Type of covering	Before	After	% reduction
E.S.[a]	47	16	66
E.J.[b]	24	11	54
E.D.[c]	57	15	73
Linoleum	55	19	65
Vinyl	80	5	94

Code	Fiber	Construction	Backing
[a]E.S.	100% nylon polyprop./acryl fiber	Needle felt Staple fiber	Propylene/fiberglass
[b]E.J.	Acryl/perlon	Tufted, cut pile	Polyprop./polyester
[c]E.D.	Bri-nylon	Tufted, cut pile	p.v.c. and fiberglass

Source: G.J. Bonde, *Health Lab. Sci.,* 10, 308, 1973. With permission.

2.1–64   PERCENT DECLINE, RODAC® PLATES

Positions	Window	Bed	Basin			Door	x^2	df
	74	65	63			68	1.46	3
Covering	E.J.[a]	E.S.[b]	E.D.[c]		Linoleum	Vinyl		
	64	69	75		60	78	3.29	4
Cleaner	Z-73[d]	Pull.[e]	668+v.[f]	693+v.[g]	Hil.+v.[h]	None		
	47	67	78	75	92	58	19.65	5
Days	13 series	Min 30	Max 78				35.3	17

[a] Acryl-perlon.
[b] 100% nylon.
[c] Bri-nylon.
[d] Electrolux® Z-73 vacuum cleaner, Swedish brand "sturdy home type."
[e] Pullman filter, American commercial-grade type filter.
[f] 668, a disinfectant-quaternary (intended for antistatic treatment as well).
[g] 693, a combined alkaline disinfection-detergent quaternary applied in a concentration of 0.5% (v/v).
[h] Hilphene®, a disinfectant-detergent with o-phenylphenol and salicylic acid in a concentration of 0.25% (v/v).

Source: G.J. Bonde, *Health Lab. Sci.,* 10, 308, 1973. With permission.

2.1–65   PERCENTAGE DECLINE IN PLUGS

Position	Window	Bed		Basin		Door	x^2	df
	44	67		43		50	6.39	3
Covering	E.S.[a]	E.J.[b]		E.D.[c]				
	47	54		50		–	0.72	2
Cleaner	Z-73[d]	Pull.[e]	668+v.[f]	693+v.[g]	Hil.+v.[h]	None		
	62	46	40	53	48	36	1.46	4

[a] 100% nylon.
[b] Acryl-perlon
[c] Bri-nylon.
[d] Electrolux® Z-73 vacuum cleaner, Swedish brand "sturdy home type."
[e] Pullman filter, American commercial grade type filter.
[f] 668, a disinfectant-quaternary (intended for antistatic treatment as well).
[g] 693, a combined alkaline disinfection-detergent quaternary applied in a concentration of 0.5% (v/v).
[h] Hilphene®, a disinfectant-detergent with o-phenylphenol and salicylic acid in a concentration of 0.25% (v/v).

Source: G.J. Bonde, *Health Lab. Sci.,* 10, 308, 1973. With permission.

2.1–66   DISTRIBUTION OF ALL STRAINS

	Plugs			Rodac®					
	E.S.[a]	E.J.[b]	E.D.[c]	E.S.	E.J.	E.D.	Linoleum	Vinyl	Total
Staphylococcus aureus	24[d]	14	1	5	4	0	2	5	55
S. epidermidis	68[d]	58	36[d]	39	29	21	14	10	275
Enterobacter	5	6	3	22[d]	7	6	0	1	50
Pseudomonas	5	8	1	17[d]	10	5	1	1	48
Enterobacter agglomerans	1	1	3	6	2	4[d]	0	0	17
Xanthomonas	1	1	0	6[d]	3	0	0	0	11
Alcaligenes and *Achromobacter*	1	0	0	5	5	2	4[d]	5[d]	22
Acinetobacter	1	2	0	4	10[d]	3	0	0	20
Micrococcus	1	7	2	4	2	1	0	0	17
Coryneform	2	4	2	3	2	2	7[d]	6[d]	28
Bacillus	19	34	13	34	61[d]	26	28[d]	20	235
No growth	11	5	1	4	12	2	1	6	42
	139	140	62	149	147	72	57	54	820

[a] 100% nylon.
[b] Acryl/perlon.
[c] Bri-nylon.
[d] Significantly above expectation ($x^2 = 3.84$, df 1).

Source: G.J. Bonde, *Health Lab. Sci.,* 10, 308, 1973. With permission.

2.1–67 PERCENT DISTRIBUTION OF GROUPS OF ORGANISMS AFTER DIFFERENT CLEANING METHODS IN PLUGS AND RODAC® PLATES

	Vacuuming		Vacuuming + 668[a]		Vacuuming + 693[b]		Vacuuming + Hi[c]			Number
	Plugs	Rodac	Plugs	Rodac	Plugs	Rodac	Plugs	Rodac		
Staphylococcus aureus	62	9	0	0	24	5	0	0	100	21
S. epidermidis	37	14	4	7	7	7	13	11	100	110
Enterobacter[d]	13	48	0	0	9	4	13	13	100	23
Pseudomonas[e]	25	36	9	4	0	21	7	7	100	28
Yellow rods[f]	18	27	9	9	0	9	0	27	100	11
A.A.A.[g]	0	47	5	11	5	5	5	21	100	19
Gr. +[h]	11	23	10	8	11	21	7	9	100	131
No growth	0	0	0	33	0	33	21	13	100	24
	81	80	19	30	30	54	34	39	—	367

a 668, a disinfectant quaternary (intended for antistatic treatment as well).
b 693, a combined alkaline disinfection-detergent quaternary applied in a concentration of 0.5% (v/v).
c Hilphene®, a disinfectant-detergent with o-phenylphenol and salicylic acid in a concentration of 0.25% (v/v).
d Enterobacter: includes Escherichia coli, Enterobacter aerogenes and cloacae, Proteeae, and Aeromonas hydrophila.
e Pseudomonas includes green fluorescent and non-pigment-forming strains.
f Yellow rods: Xanthomonas and E. agglomerans.
g A.A.A.: groups of Acinetobacter, Alcaligenes, and Achromobacter.
h Gr. +: Bacillus, Coryneforms, and Micrococcus.

Source: G.J. Bonde, Health Lab. Sci., 10, 308, 1973. With permission.

2.1−68   RESULTS OF THREE TECHNIQUES TO EVALUATE BACTERIAL COUNT OF CARPETS

Sampling techniques	Bacterial colony counts		Percent reduction of colony counts
	Before vacuuming	After vacuuming	
Plugs	5342[a]	3343[a]	38
Liquid recovery	3033[a]	1923[a]	37
Beater	2.3[b]	1.4[b]	36

[a] Per mil.
[b] Per cubic foot.

Source: B.Y. Litsky, *Health Lab. Sci.*, 10, 28, 1973. With permission.

2.1−69   NUMBERS OF BACTERIA RECOVERED BY THE PROBE METHOD ON SIX CONSECUTIVE TESTS OF THE SAME CARPETED AREAS

Numbers of Bacteria per Square Centimeter from Carpet

Test	A-1	A-2	B-1	B-2
1	11,900	4,000	25,500	9,100
2	7,500	2,000	9,700	4,600
3	3,400	1,200	6,900	4,500
4	2,700	1,100	6,100	3,700
5	2,300	1,000	4,500	2,900
6	2,800	700	3,400	2,300
Percentage obtained on first test	39	40	45	34

Source: W.G. Walter and A.H. Stober, *Health Lab. Sci.*, 5, 162, 1968. With permission.

2.1−70   COMPARISON OF COUNTS OBTAINED FROM CARPET BY THE PROBE AND PLUG METHODS

Test	Numbers of organisms	
	Probe	Plug
1	18,600	163,000
2	17,700	135,000
3	13,500	101,000
4	5,500	74,000
5	6,100	57,000
6	9,000	93,000
7	9,000	130,000
8	17,200	125,000
9	13,700	121,000
10	11,300	150,000
11	4,800	86,000
12	5,300	56,000
13	8,400	97,000
14	5,900	103,000
Mean	10,000	110,000

Source: W.G. Walter and A.H. Stober, *Health Lab. Sci.*, 5, 162, 1968. With permission.

2.1–71   NUMBERS OF BACTERIA OBTAINED BY
THE PROBE METHOD FROM
THREE REPLICATES OF FIVE CARPETS
BEFORE AND AFTER VACUUMING

Mean Numbers of Bacteria per Square Centimeter

Type of carpet	Before vacuuming	Standard error of mean	After vacuuming	Standard error of mean
Veltron®	8,700	4,100	3,700	2,000
Acrilan®	21,700	8,300	10,400	4,300
Wool	23,400	7,800	11,700	3,700
Antron®	15,300	6,500	9,700	3,500
Herculon®	6,400	3,200	1,400	600

Source: W.G. Walter and A.H. Stober, *Health Lab. Sci.*, 5, 162, 1968. With permission.

2.1−72 CARPETED NURSERY COLONY COUNTS[a] BY LOCATION, CLEANING, AND REPLICATION

Nursery areas	Number of samples	Mean	Service areas	Number of samples	Mean	Corridor areas	Number of samples	Mean	Cleaning	Number of samples	Mean	Replications	Number of samples	Mean
C_1	24	4402	C_5	24	7481	C_7	24	19,849	Before	96	7207	1st	32	9,619
C_2	24	3702	C_6	24	8708	C_8	24	8,353	After	96	8923	2nd	32	6,686
C_3	24	4331	Combined		8095							3rd	32	6,671
C_4	24	7697										4th	32	11,317
Combined		5033										5th	32	6,637
												6th	32	7,462
												Combined		8,071

a Colony counts per square inch of floor area (about 13 plugs = 1 in.²).

Source: R.R. Lanese et al., *Am. J. Public Health*, 63, 174, 1973. With permission.

2.1−73 TILED NURSERY COLONY COUNTS[a] BY LOCATION, CLEANING, AND REPLICATION

Nursery area	Number of samples	Mean	Service area	Number of samples	Mean	Cleaning	Number of samples	Mean	Replications	Number of samples	Mean
T_1	96	7.2	T_5	96	9.6	Before	288	10.1	1st	96	10.7
T_2	96	8.7	T_6	95	11.7	After	288	8.3	2nd	96	6.5
T_3	96	8.6	Combined		10.7				3rd	96	7.6
T_4	96	9.3							4th	96	9.3
Combined		8.5							5th	96	8.2
									6th	96	13.0
									Combined		9.2

a Per square inch of tiled floor area. This unit of area is used for convenient reporting of counts (not to be compared directly to carpet counts obtained from plugs of the full depth of pile and backing, and subjected to different microbiological techniques).

Source: R.R. Lanese et al., *Am. J. Public Health*, 63, 174, 1973. With permission.

2.1–74 BACTERIAL CONTAMINATION OF THEATRE FOOTWEAR AND OUTDOOR SHOES FROM THREE HOSPITALS

| Type of footwear | Total organisms and Staphylococcus aureus | | | Clostridium welchii | |
	Number of samples	Mean total organisms per plate	Mean S. aureus per plate	Number of samples	Mean per plate
Theatre	40	360	1.1	44	10.45 ± 0.98
Outdoor	36	> 1000	10.6	40	74.05 ± 10.52

Source: G.A.J. Ayliffe et al., *J. Hyg.*, 67, 417, 1969. With permission of Cambridge University Press.

2.1–75 REMOVAL OF BACTERIA FROM A SHOE BY A TAKIMAT

Three Experiments

| Floor areas sampled by impression plate | Total bacterial count per plate | | | Staphylococcus aureus per plate | | |
	1	2	3	1	2	3
Site of first step (on vinyl before stepping on Takimat)	185	76	320	10	3	28
Site of second step (Takimat)	394	145	196	18	7	25
Site of third step (on vinyl after stepping on Takimat)	199	10	194	6	0	26
Controls						
Site of first step on vinyl	358	75	79	9	7	30
Site of second step on vinyl	283	35	82	9	0	3
Site of third step on vinyl	95	20	80	1	0	10
Clean Takimat	16	8	15	0	0	0

Source: G.A.J. Ayliffe et al., *J. Hyg.*, 65, 515, 1967. With permission of Cambridge University Press.

2.1–76   THE EFFECT OF A TAKIMAT AT ENTRANCE ON BACTERIAL COUNTS IN A ROOM

	Number of plates	Mean bacterial count per impression plate	
		In air lock	In cubicle
Takimat in air lock	17	120	68 ± 14.6[a]
No Takimat in air lock	13	125	105 ± 27.0[a]

[a] t (28 degrees of freedom) $= 1.29, P > 0.1$.

Source: G.A.J. Ayliffe et al., *J. Hyg.,* 65, 515, 1967. With permission of Cambridge University Press.

2.1–77   THE EFFECT OF TREADING ON A DISINFECTANT MAT ON THE CONTAMINATION OF A CLEAN FLOOR BY SHOES

	Total bacterial counts and counts of *Staphylococcus aureus* per impression plate from floor			
	Left foot		Right foot	
Floor area sampled by impression plate	Total	*S. aureus*	Total	*S. aureus*
Before stepping on mat	127	62	194	70
First step after mat	3	0	3	0
Tenth step after mat	40	7	46	16

Source: G.A.J. Ayliffe et al., *J. Hyg.,* 65, 515, 1967. With permission of Cambridge University Press.

2.1–78   THE TRANSFER OF BACTERIA FROM A CONTAMINATED TAKIMAT TO A CLEAN FLOOR

Area sampled by impression plate	Total organisms per impression plate
Floor after contact with clean shoe	2
Contaminated Takimat	1000
Floor after contact with shoe	
First step after Takimat	31
Fourth step after Takimat	12

Source: G.A.J. Ayliffe et al., *J. Hyg.,* 65, 515, 1967. With permission of Cambridge University Press.

2.1–79   COMPARISON OF *STAPHYLOCOCCUS AUREUS* ISOLATIONS FROM AIR AND SIEVE CULTURES AT SPECIFIED SITES ON CARPETED AND TILED AREAS

Date	Site	Carpet S. aureus Air[a]	Sieve[b]	Site	Tile S. aureus Air	Sieve
6/3/65	1	0	14	3	11	53
	2	0	3	4	1	15
6/10/65	1	0	0	3	0	0
	2	0	1	4	3	1
6/17/65	1	1	2	3	0	53
	2	1	1	4	4	3
6/24/65	1	0	7	3	0	24
	2	0	1	4	1	0
7/1/65	1	1	1	3	0	2
	2	0	0	4	1	0
7/8/65	1	0	4	3	0	0
	2	0	5	4	0	2
7/15/65	1	0	0	3	0	2
	2	0	0	4	0	1
7/22/65	1	0	0	3	0	1
	2	1	8	4	1	1
7/29/65	1	0	4	3	0	1
	2	0	3	4	0	0
8/5/65	1	0	2	3	0	0
	2	0	10	4	1	1
8/12/65	1	0	1	3	0	2
	2	0	2	4	0	1
8/19/65	1	0	1	3	1	2
	2	0	0	4	1	1
8/26/65	1	0	4	3	0	2
	2	0	2	4	0	1
Total *S. aureus*		4	76		25	169
Total sites with *S. aureus*		4	20		10	20

[a] 20 ft³ of air sampled at each site.
[b] Ten sieve samples were taken in an area around the site of the air sample.

Source: J.G. Shaffer, *Health Lab Sci.*, 3, 73, 1966. With permission.

2.1–80 STUDIES ON THE REDISPERSAL OF BACTERIA FROM THE FLOOR INTO THE AIR

Total counts and counts of *Staphylococcus aureus*
on four settle plates exposed for 30 min

Time of sampling	Experiment 1 (vinyl, blowing)		Experiment 2 (vinyl, blowing)		Experiment 3 (terrazzo, blowing)		Experiment 4 (vinyl, sweeping)		Experiment 5 (vinyl, exercise)		Experiment 6 (terrazzo, exercise)	
	Total	*S. aureus*	Total	*S. aureus*	Total	*S. aureus*	Total	*S. aureus*	Total	*S. aureus*	Total	*S. aureus*
During and after contamination[a] (0–30 min)	2136	1004	638	89	832	65	544	99	129	43	68	12
After contamination (30–60 min)	20	0	18	0	7	1	30	1	5	1	5	0
During and after floor disturbance (60–90 min)	41	1	39	2	56	4	222	32	2	0	10	0
After floor disturbance (90–120 min)	19	1	31	1	6	0	16	0	3	0	3	0

a   Experiment 1: contamination by staphylococcal disperser; Experiments 2–6: contamination by shaken blanket.

Source: G.A.J. Ayliffe et al., *J. Hyg.*, 65, 515, 1967. With permission of Cambridge University Press.

2.1–81   STUDIES ON THE REDISPERSAL OF BACTERIA FROM THE FLOORS INTO THE AIR[a]

Mean total counts and counts of *Staphylococcus aureus* per floor impression plate

Time of sampling	Experiment 1 (vinyl, blowing)		Experiment 2 (vinyl, blowing)		Experiment 3 (terrazzo, blowing)		Experiment 4 (vinyl, sweeping)		Experiment 5 (vinyl, exercise)		Experiment 6 (terrazzo, exercise)	
	Total	*S. aureus*	Total	*S. aureus*	Total	*S. aureus*	Total	*S. aureus*	Total	*S. aureus*	Total	*S. aureus*
Before floor disturbance	+++	1240	105	26	390[b]	14	97[c]	21	104	16	75	1
After floor disturbance	+++	1093	112	34	234[b]	8	31[c]	7	116	15	95	1

[a] Experiment 1: contamination by staphylococcal disperser; Experiments 2–6: contamination by shaken blanket.
[b] t (10 degrees of freedom) = 2.79, $P < 0.02$.
[c] t (10 degrees of freedom) = 5.26, $P < 0.001$.

Source: G.A.J. Ayliffe et al., *J. Hyg.*, 65, 515, 1967. With permission of Cambridge University Press.

2.1–82   BACTERIAL CONTAMINATION OF WALLS IN AN OPERATING THEATRE

Mean counts from ten
impression plates on area

Time of sampling (after washing)	Left intact after washing		Cleaned weekly with oiled mop	
	Total	*Staphylococcus aureus*	Total	*S. aureus*
1 day	2.8	0	3	0
1 week	5.0	0	6.4	0
2 weeks	3.4	0.6	2.8	0
3 weeks	3.4	0.2	7	0.2
4 weeks	3.2	0	1.8	0.2
5 weeks	4.6	0.2	1.6	0
12 weeks	1.2	0	1.4	0
1 day after second wash	1.0	0	0.8	0

Source: G.A.J. Ayliffe et al., *J. Hyg.,* 65, 515, 1967. With permission of Cambridge University Press.

2.1–83   TRANSFER OF *STAPHYLOCOCCUS AUREUS* FROM CONTAMINATED WALL BY FINGERS OF NONCARRIER

Area of wall sampled	Fingers	Total *S. aureus* per plate
Contaminated by disperser	Left 1	21
	Right 2	32
Contaminated by transfer on	Left 3	5
fingers of noncarrier	Right 3	2

Source: G.A.J. Ayliffe et al., *J. Hyg.,* 65, 515, 1967. With permission of Cambridge University Press.

2.1–84   STAPHYLOCOCCAL CONTAMINATION OF FINGERS OF THE DISPERSER BEFORE AND AFTER TOUCHING A WALL

Time of sampling	Fingers of disperser	Total *Staphylococcus aureus* per plate
Before contaminating the wall	Left	About 100
	Right	About 100
After contaminating the wall	Left	87
	Right	66

Source: G.A.J. Ayliffe et al., *J. Hyg.,* 65, 515, 1967. With permission of Cambridge University Press.

2.1–85   SURVIVAL OF *STAPHYLOCOCCUS AUREUS* ON A WALL CONTAMINATED BY FINGERS OF THE DISPERSER

	Total bacterial counts and counts of *S. aureus* per impression plate			
	Experiment 1 (dry fingers)		Experiment 2 (wet fingers)	
Time of sampling (after contamination)	Total	*S. aureus*	Total	*S. aureus*
1 hr	12	9	1040	560
1 day	4	0	151	55
2 days	0	0	6	0
3 days	2	0	4	0
4 days	0	0	3	0
7 days	0	0	0	0

Source: G.A.J. Ayliffe et al., *J. Hyg.,* 65, 515, 1967. With permission of Cambridge University Press.

2.1–86   QUALITATIVE RESULTS OF RODAC® SAMPLES OF ISOLATION ROOM WALLS

Isolated patient infection	Number rooms sampled	Number Rodac samples	Total Rodacs positive for *Staphylococcus aureus*	Total Rodacs positive for *Pseudomonas aeruginosa*
S. aureus	36	888	1	0
P. aeruginosa	21	524	0	0
S. aureus and *P. aeruginosa*	8	192	3	0

Source: N.J. Petersen et al., *Health Lab. Sci.,* 10, 23, 1973. With permission.

2.1–87   COMPARISON OF RODAC® PLATE COUNTS BY WALL LOCATION FOR NINE SURVEYS IN ROOM D

Wall description	Number Rodac samples	Mean number colonies per Rodac
Opposite bed	54	9
Bed	54	7
Window	36	10
Hall	36	10
Bath	36	6

Source: N.J. Petersen et al., *Health Lab. Sci.,* 10, 23, 1973. With permission.

2.1–88   COMPARISON OF RODAC® PLATE COUNTS BY VERTICAL LOCATION FOR NINE SURVEYS IN ROOM D

Inches above floor	Number of Rodac samples	Mean number of colonies per Rodac
24	54	9
52	108	9
80	54	7

Source: N.J. Petersen et al., *Health Lab Sci.,* 10, 23, 1973. With permission.

2.1–89   COMPARISON OF PLATE COUNTS ON RODAC® SAMPLES OF ISOLATION ROOM AND CONTROL ROOM WALLS

Test number	Mean number colonies per Rodac	
	Isolation room	Control room
1	36	4
2	32	7
3	1	5
4	48	10
5	17	1
6	1	20
7	27	3
8	20	3
9	10	2
10	6	1
11	3	7
Mean	18	6

Source: N.J. Petersen et al., *Health Lab. Sci.*, 10, 23, 1973. With permission.

2.1–90   RELATIONSHIPS OF HALLWAY TRAFFIC TO THE AIRBORNE BACTERIAL COUNT IN A HOSPITAL WARD

Time	Number passing station per hour				Colonies per cubic foot
	Staff	Visitors	Patients	Total	
11:00–12:00 A.M.	242	55	46	343	40–46
3:00– 4:00 P.M	135	120	11	266	15–20
9:00–10:00 A.M	113	18	8	139	7–8
10:00–11:00 P.M.	38	5	2	45	3–4
11:00–12:00 P.M.	5	0	0	5	0–3

Source: Reprinted from V.W. Greene et al., *Mod. Hosp.*, 95, 136, 1960. © 1960 by McGraw-Hill, Inc. All rights reserved.

2.1–91   MEAN NUMBERS OF AIRBORNE BACTERIA IN TILED AND CARPETED PATIENT ROOMS GROUPED BY EXTENT OF ACTIVITY IN 212 AIR SAMPLING TESTS

Activity	Group 1		Group 2		Group 3
Passes[a]	0–13		14–30		31–50
Tests	173		37		2
Floor covering	Tile	Carpet	Tile	Carpet	Carpet
Bacteria per cubic foot	8.0	9.6	12.0	14.2	18.1

Note: Any two means not underscored by the same line are significantly different (*P*<0.01).

[a] Persons and vehicles passing in 20 min.

Source: W.G. Walter et al., *Health Lab. Sci.,* 6, 140, 1969. With permission.

2.1–92   MEAN NUMBERS OF AIRBORNE BACTERIA IN TILED AND CARPETED PATIENT ROOMS IN THE SURGICAL FLOOR

Activity	Group 1	
Passes	0–13	
Floor covering	Tile	Carpet
Bacteria per cubic foot	4.7	6.0

Source: W.G. Walter et al., *Health Lab. Sci.,* 6, 140, 1969. With permission.

2.1–93   MEAN NUMBERS OF AIRBORNE BACTERIA AT CORRIDOR SITES CATEGORIZED IN ACTIVITY GROUPINGS, MEDICAL AND SURGICAL FLOORS

473 Samples in Ten Corridor Sites

Activity[a]	Group 1	Group 2	Group 4	Group 3
Bacteria per cubic foot	8.4	9.3	11.9	12.0

Note: Any two means not underscored by the same line are significantly different (*P*<0.01).

[a] Group 1, 0–13 passes; Group 2, 14–30 passes; Group 3, 31–50 passes; Group 4, >50 passes.

Source: W.G. Walter et al., *Health Lab. Sci.,* 6, 140, 1969. With permission.

2.1—94   MEAN NUMBERS OF AIRBORNE BACTERIA FOUND
IN CORRIDORS ACCORDING TO SEASON

Quarters	First	Fourth	Second	Third
Bacteria per cubic foot	9.8	11.0	11.7	13.9

Note: Any two means not underscored by the same line are significantly different (P<0.05).

Source: W.G. Walter et al., *Health Lab. Sci.,* 6, 140, 1969. With permission.

2.1—95   AIRBORNE BACTERIA PER
CUBIC METER OF AIR IN ROOMS WITH
TILE FLOORING AND TUFTED CARPETING
BEFORE AND DURING SIZABLE
WALKING ACTIVITY

Test No.	Tile floor Before	During	Carpet Before	During
1	167	224	85	342
2	160	240	67	398
3	160	146	58	128
4	133	142	73	151
5	127	151	27	124
Mean value	149	181	62	229

Source: R. Rylander et al., *Am. J. Public Health,* 64, 163, 1974. With permission.

2.1—96   AVERAGES OF TESTS EMPLOYING A HIGH-VOLUME
AIR SAMPLER AND BARRIER SUIT

5-min air sample	Tile Colonies per cubic foot of air	% reduction or increase	Carpet Colonies per cubic foot of air	% reduction or increase
No activity	0.9	—	0.9	—
Walk	3.2	+258	18	+1779
After walk	1.8	+103	6	+471

Source: B.Y. Litsky, *Health Lab. Sci.,* 10, 28, 1973. With permission.

2.1–97 INFLUENCE OF BEDMAKING ON THE AIRBORNE BACTERIAL COUNTS OF HOSPITALS

Activity	Colonies per cubic foot	
	In patient's room	Hallway near patient's room
Background	34	30
During bedmaking	140	64
10 min after bedmaking	60	40
30 min after bedmaking	36	27
Background	16	
Normal bedmaking	100	
Vigorous bedmaking	172	

Source: Reprinted from V.W. Greene et al., *Mod. Hosp.*, 95, 136, 1960. © 1960 by McGraw-Hill, Inc. All rights reserved.

2.1–98   MICROBIAL CONTAMINATION ON SURFACES AND FURNISHINGS IN LAMINAR AIRFLOW ROOM
AND CONVENTIONAL ISOLATION AND HOSPITAL ROOMS

Colonies per Rodac plate

Item	Number of Rodac® plates per experiment	Laminar airflow room				Conventional isolation rooms		Conventional hospital rooms	
		Experiment 1	Experiment 2	Experiment 3	Experiment 4	Experiment 1[a]	Experiment 2[a]	Experiment 1[a]	Experiment 2[a]
Pillows and blankets	10	10.9	4.9	4.1	10.7	33.0	37.3	57.6	72.3
Floors	15	0.1	0.5	1.0	0.4	24.5	25.7	89.0	99.3
Walls	4	0.0	0.0	0.0	0.0	2.0	3.6	8.2	12.4
Tables and shelves	15	0.2	0.3	0.2	0.8	17.7	25.4	61.4	72.3
Chair	4	1.0	0.0	0.5	0.0	21.0	39.7	91.3	93.7
Lamp	4	0.0	0.0	0.0	0.5	20.2	29.0	233.7	356.7

[a] Mean of three experiments, one in each of three different rooms.

Source: C.O. Solberg et al., *Appl. Microbiol.*, 21, 209, 1971. With permission.

2.1–99   BED FRAME CONTAMINATION WITH *KLEBSIELLA*

Status of bed frame	Patient carrier status			Bed empty	Total samples
	Positive	Negative	Unknown		
Positive	23	13	0	5	41
Negative	12	43	1	4	60
Total	35	56	1	9	101

Source: A.G. Turner and J.G. Craddock, *Hospitals,* 47, 79, 1973. With permission.

2.1–100   *KLEBSIELLA* CONTAMINATION OF FOMITES

Fomite[a]	Total cultures	Positive cultures	Percent positive
Hand sinks	16	3	19
Bedside tables	13	3	23
Irrigation saline	9	1	11
Nurse work surface	6	0	0
Bed linen	6	3	50
Scrub water–floor	5	0	0
Shelves–linen storage	5	0	0
Staphene®	4	0	0
Hand cream	3	0	0
Windowsill	3	1	33
Clean washbasins	3	0	0
Hand lotion	2	0	0
Lotocreme®	2	0	0
Scrub water–furniture	1	0	0
Draw curtains	1	0	0
Total	79	11	14

[a] Staphene was used in a 0.5% solution for patient thermometer containers. Hand cream and lotion were prepared in the hospital pharmacy. Lotocreme is a product of Abbott Laboratories, Atlanta.

Source: A.G. Turner and J.G. Craddock, *Hospitals,* 47, 79, 1973. With permission.

2.1–101   RODAC[®] PLATE BACTERIAL COLONY COUNTS OBTAINED FROM HOSPITAL OVERBED TABLES IN 1964 AND 1965

Hospital	1964 Before cleaning			1965 Mean colony counts			1965 Pooled standard deviation	
	Number of samples	Mean colony counts	Standard deviation	Number of samples	Before cleaning	After cleaning	Before cleaning	After cleaning
A	200	83	96	300	79	10	69	41
B	118	94	105	100	43	37	86	24
C	130	80	122	140	72	11	80	16
D	100	94	81	100	62	24	58	20
E	50	193	104	110	230	22	173	31
F	–	–	–	100	51	20	35	16
G	120	65	79	100	87	28	33	20
H	–	–	–	100	61	90	25	26
I	–	–	–	80	62	61	56	60
J	–	–	–	100	48	24	64	39
K	–	–	–	80	109	36	20	30
L	–	–	–	100	46	49	42	87
M	–	–	–	90	87	24	63	36
N	–	–	–	100	22	14	20	17
O	97	135	108	–	–	–	–	–
P	80	44	35	–	–	–	–	–
Q	70	118	126	–	–	–	–	–
Total	965	100[a]	95[a]	1600	75[a]	32[a]	59[a]	33[a]

[a] Mean value.

Source: A.K. Pryor et al., *Health Lab. Sci.*, 4, 153, 1967. With permission.

2.1–102   WARD ENVIRONMENT EXAMINATIONS. BACTERIAL CONTAMINATION OF WARD FIXTURES AND FITTINGS, AND BACTERIAL CONTENT OF AIR

	Number of samples	Number positive for *Staphylococcus pyogenes*	Number positive for Gram-negative bacilli	Total counts Maximum	Total counts Minimum	Total counts Average
Sellotape[®] transfers	131	41 (31)[a]	7 (5)			
Broth-moistened swabs	103	6 (6)	4 (4)			
Air						
Slit sampler	30	16 (53)	NR[d]	320[b]	6[b]	77[b]
Sieve sampler	77	51 (66)	NR	320[c]	11[c]	112[c]
Settle plates	67	27 (40)	NR	176	3	61

[a] Figures in parentheses indicate percentages.
[b] Counts per 5 ft³.
[c] Counts per 12.5 ft³.
[d] NR = not recorded.

Source: F.W. Winton and A.J. Keay, *J. Hyg.*, 66, 325, 1968. With permission of Cambridge University Press.

2.1–103   BACTERIAL CONTAMINATION OF TROLLEY WHEELS

Hospital	Number of samples	Mean total organisms per plate	Mean *Staphylococcus aureus* per plate	% of plates showing *S. aureus*	Mean *Clostridium welchii* per plate
1 Theatre trolleys	24	550	18.6	12.5	3.13 ± 0.47
1 Hospital trolleys	24	834.5	18.6	41.7	67.90 ± 7.68
2 Hospital and theatre trolleys	20	287.6	6.0	35	45.10 ± 3.90

Source: G.A.J. Ayliffe et al., *J. Hyg.,* 67, 417, 1969. With permission of Cambridge University Press.

2.1–104   BACTERIAL CONTAMINATION OF TROLLEYS

Hospital	Site of sampling	Number of samples	Mean total organisms per plate	Mean *Staphylococcus aureus* per plate	Mean *Clostridium welchii* per plate
1 Theatre trolleys	Top	12	48.2	0.42	0.75
	Bars	12	89.3	2	–
	Handles	6	12.3	0	–
1 Hospital trolleys	Top	12	36.8	5.6	0.92
	Bars	12	88.7	0.17	–
	Handles	6	12.6	0	–
2 Hospital and theatre trolleys	Top	10	35.2	0.33	0.5
	Bars	10	9	0.1	–
	Handles	5	11.4	0	–

Source: G.A.J. Ayliffe et al., *J. Hyg.,* 67, 417, 1969. With permission of Cambridge University Press.

2.1–105   GEOMETRIC MEANS (LOG$_{10}$) OF SURVIVORS PER SQUARE CENTIMETER ON POLYESTER-COTTON SHEETING BEFORE WASHING, AFTER WASHING, AND AFTER WASHING AND DRYING[a]

Time of count	Wash temperature, °C	T3 phage	Serratia marcescens	Staphylococcus aureus	Bacillus stearothermophilus
Before washing	–	44.1	5.19	5.52	4.68
After washing	24	1.83	3.84	3.77	3.95
	35	1.69	2.41	3.96	4.01
	46	1.34	1.01	2.45	3.34
	57	0.69	0.0	1.03	3.17
	68	0.0	0.0	1.20	2.97
After drying	24	0.24	0.0	0.54	3.17
	35	0.0	0.0	1.12	3.14
	46	0.0	0.0	0.84	2.74
	57	0.0	0.0	0.45	2.73
	68	0.0	0.0	0.63	2.45

[a] A regular detergent was used with the permanent-press wash cycle, and a low (40°C) drying temperature was used.

Source: J.C. Wiksell et al., *Appl. Microbiol.*, 25, 431, 1973. With permission.

2.1–106   ARITHMETIC MEANS (LOG$_{10}$) OF MICROORGANISMS RECOVERED PER MILLILITER OF WASH WATER AT THE END OF THE WASH CYCLE AND OF RINSE WATER AT THE END OF THE RINSE CYCLE[a]

Sample	Wash temperature, °C	T3 phage	Serratia marcescens	Staphylococcus aureus	Bacillus stearothermophilus
Wash water	24	3.66	2.32	1.95	3.62
	35	3.08	1.48	1.54	3.36
	46	3.34	1.36	0.0	3.43
	57	1.30	0.0	0.70	3.54
	68	0.0	0.0	0.70	3.08
Rinse water	24	1.60	0.48	1.00	2.45
	35	0.90	0.0	1.36	2.46
	46	0.0	0.0	1.30	2.40
	57	0.0	0.0	1.36	2.43
	68	0.0	0.0	0.0	2.20

[a] Experimental conditions were the same as those listed for Table 2.1–105.

Source: J.C. Wiksell et al., *Appl. Microbiol.*, 25, 431, 1973. With permission.

2.1–107   GEOMETRIC MEANS (LOG$_{10}$) OF SURVIVORS PER SQUARE CENTIMETER ON POLYESTER-COTTON SHEETING WASHED IN COLD WATER WITH REGULAR AND COLD-WATER DETERGENTS[a]

Time of count	Detergent	T3 phage	Serratia marcescens	Staphylococcus aureus	Bacillus stearothermophilus
After washing	Regular	1.76	3.13	3.86	3.98
	Cold water	1.13	3.56	3.45	4.08
After drying	Regular	0.12	0.0	0.83	3.16
	Cold water	0.0	0.0	0.27	3.45

[a] Data for two wash temperatures, 24 and 35°C, did not differ statistically and were combined. The permanent-press wash cycle was used. The pH values of the wash waters for the regular and cold-water detergents varied between 8.8 and 8.9.

Source: J.C. Wiksell et al., *Appl. Microbiol.*, 25, 431, 1973. With permission.

2.1–108   GEOMETRIC MEANS (LOG$_{10}$) OF SURVIVORS PER SQUARE CENTIMETER ON POLYESTER-COTTON SHEETING WASHED IN COLD WATER WITH THE PERMANENT-PRESS AND REGULAR WASH CYCLES[a]

Time of count	Wash cycle	T3 phage	Serratia marcescens	Staphylococcus aureus	Bacillus stearothermophilus
After washing	Permanent press	1.44	3.34	3.66	4.03
	Regular	0.43	2.84	2.65	3.32
After drying	Permanent press	0.06	0.0	0.55	3.31
	Regular	0.0	0.0	0.58	3.00

[a] Data for two wash temperatures, 24 and 35°C, did not differ statistically and were combined. A regular detergent was used.

Source: J.C. Wiksell et al., *Appl. Microbiol.*, 25, 431, 1973. With permission.

2.1–109   GEOMETRIC MEANS (LOG$_{10}$) OF MICROORGANISMS RECOVERED FROM STERILIZED SAMPLES OF POLYESTER-COTTON SHEETING THAT WERE WASHED WITH INOCULATED SAMPLES[a]

Wash temperature, °C	T3 phage	Serratia marcescens	Staphylococcus aureus	Bacillus stearothermophilus
24	0.24	1.12	1.48	1.45
35	0.0	0.0	1.10	1.13
46	0.56	0.81	1.23	0.56
57	0.0	0.0	1.28	0.63
68	0.0	0.0	0.54	0.45

[a] The permanent press cycle and a regular detergent were used.

Source: J.C. Wiksell et al., *Appl. Microbiol.*, 25, 431, 1973. With permission.

2.1–110   AIRBORNE MICROBIAL COUNTS ACCORDING TO TYPE OF DISHWASHING FACILITY

Type of facility	"Clean" area			"Soiled" area		
	Number of samples	Mean number of colonies per cubic foot	Median of means	Number of samples	Mean number of colonies per cubic foot	Median of means
Open alcove	456	12.4	8.0	457	15.9	15.6
Separate room	464	13.9	13.7	465	11.7	12.7
Physical separation	260	16.2	6.7	263	70.8[a]	20.2

[a] One hospital had an extremely high count.

Source: W.H. Jopke and D.R. Hass, *Hospitals,* 44, 126, 1970. With permission.

2.1–111   CHARACTERIZATION OF AIRBORNE CONTAMINATION ACCORDING TO TYPE OF DISHWASHING FACILITY

	Open alcove	Separate room	Physical separation	Total
Number of isolates	5727	5973	2998	14,698
Gram-positive cocci	33.8%	34.2%	29.1%	33.0%
Gram-positive rods	4.1%	4.7%	2.5%	4.0%
Gram-positive sporeforming rods	0.5%	0.7%	0.5%	0.6%
Gram-negative rods	2.1%	2.7%	3.1%	2.6%
Diphtheroids	22.6%	23.4%	37.2%	25.9%
Actinomycetes	1.1%	1.8%	1.3%	1.4%
Yeasts	3.5%	4.8%	2.0%	3.7%
Molds	32.3%	27.8%	24.2%	28.8%

Source: W.H. Jopke and D.R. Hass, *Hospitals,* 44, 126, 1970. With permission.

2.1—112   AIRBORNE MICROBIAL COUNTS IN DISHWASHING FACILITIES IN OPERATION AND NOT IN OPERATION

Type of facility	In operation			Not in operation		
	Number of samples	Mean number of colonies per cubic foot	Median of means	Number of samples	Mean number of colonies per cubic foot	Median of means
All types	2365	20.1	12.8	937	14.2	11.0
Open alcove	913	14.2	12.8	363	13.7	10.5
Separate room	929	12.6	13.7	318	14.6	13.5
Physical separation	523	43.6	9.6	256	14.4	8.1

Source: W.H. Jopke and D.R. Hass, *Hospitals*, 44, 126, 1970. With permission.

2.1—113   CHARACTERIZATION OF AIRBORNE CONTAMINATION IN DISHWASHING FACILITIES IN OPERATION AND NOT IN OPERATION

	In operation	Not in operation
Number of isolates	10,268	4386
Gram-positive cocci	36.7%	24.6%
Gram-positive rods	3.8%	4.4%
Gram-positive sporeforming rods	0.7%	0.6%
Gram-negative rods	2.4%	2.6%
Diphtheroids	26.9%	21.0%
Actinomycetes	1.8%	3.1%
Yeasts	3.1%	4.7%
Molds	24.6%	38.9%

Source: W.H. Jopke and D.R. Hass, *Hospitals,* 44, 126, 1970. With permission.

2.1–114   AIRBORNE MICROBIAL COUNTS IN DISHWASHING FACILITIES WITH AND WITHOUT AIR CLEANING SYSTEMS

Air cleaning[a]	Number of samples	Mean number of colonies per cubic foot	Median of means	Range of means
Present	1075	10.9	8.2	2.4–57.1
Absent	555	33.7	18.5	7.0–443.6

[a] Includes mechanical and electrical filters.

Source: W.H. Jopke and D.R. Hass, *Hospitals,* 44, 126, 1970. With permission.

2.1–115   CHARACTERIZATION OF AIRBORNE CONTAMINATION IN DISHWASHING FACILITIES WITH AND WITHOUT AIR CLEANING SYSTEMS

	With air cleaning	Without air cleaning
Number of isolates	7095	7603
Gram-positive cocci	44.4%	22.3%
Gram-positive rods	3.2%	4.8%
Gram-positive sporeforming rods	0.7%	0.5%
Gram-negative rods	2.3%	2.8 %
Diphtheroids	21.2%	30.3%
Actinomycetes	1.9%	1.0%
Yeasts	2.5%	4.8%
Molds	23.8%	33.5%

Source: W.H. Jopke and D.R. Hass, *Hospitals,* 44, 126, 1970. With permission.

2.1–116 MICROBIAL CONTAMINATION ON HOSPITAL TABLEWARE

Type of tableware[a]	Immediately after washing					During storage					Before use				
	Number of samples	Mean (average) count	Percentage distribution of microbial counts			Number of samples	Mean (average) counts	Percentage distribution of microbial counts			Number of samples	Mean (average) count	Percentage distribution of microbial counts		
			0	1–50	>50			0	1–50	>50			0	1–50	>50
Plates	627	13.9	71	25	4	630	5.5	64	34	2	628	3.4	77	22	1
Trays	627	24.2	65	25	10	629	10.4	60	35	5	629	11.2	54	42	4
Cups	315	7.4	51	46	3	315	15.2	34	59	7	315	14.6	24	71	5
Glasses	313	3.9	65	34	1	314	15.8	38	55	7	313	10.3	36	60	4
Spoons	105	17.5	73	19	8	104	30.3	59	31	10	105	109.5	53	27	20
Forks	105	11.6	84	10	6	105	35.4	57	32	11	105	72.6	55	30	15
Knives	105	7.6	72	21	7	105	42.4	55	36	9	105	34.1	49	39	12

[a] Expressed as colonies per utensil for the flatware and colonies per Rodac® plate for the other types of tableware (spoons, forks, knives).

Source: W.H. Jopke et al, *Hosp. Prog.*, 53, 31, 1972. Reprinted with permission. Copyright 1972 by The Catholic Hospital Association.

2.1–117   PERCENT OF MEAN COUNTS[a] EXCEEDING VARIOUS RECOMMENDED LEVELS

	Stage of handling, %											
	After washing				During storage				Before use			
Type of tableware	10[b]	30	50	100	10	30	50	100	10	30	50	100
Plates	12	5	5	5	14	2	2	0	10	2	2	0
Trays	33	21	17	5	14	10	7	0	24	14	5	5
Cups	15	5	0	0	33	9	9	5	38	9	5	0
Glasses	5	0	0	0	24	14	14	9	19	5	5	0
Spoons	24	9	9	5	29	19	14	14	38	38	33	19
Forks	14	9	5	5	43	29	9	9	42	33	24	14
Knives	14	9	5	5	29	9	9	9	52	19	19	9

[a] Forty-two mean counts were obtained for dinner plates and service trays; 21 mean counts were obtained for the other types of tableware.

[b] Colonies per utensil for flatware and colonies per Rodac® plate for other types of tableware.

Source: W.H. Jopke et al., *Hosp. Prog.,* 53, 31, 1972. Reprinted with permission. Copyright 1972 by The Catholic Hospital Association.

2.1–118 AVERAGE TEST RESULTS. CHUTES RANKED IN ORDER OF BACTERIAL SURFACE CONTAMINATION (CLEANLINESS)

Hospital	Chute No.	Number of floors served[a]	Department served	Approximate number of patients	Method of disposal	Chute vented[b]	Average counts on the chute surfaces			Average floor count	% of Staphylococcus aureus isolated from chute
							Test 1	Test 2	Mean		
G	11	1	Operating theatres	–	Linen bags	Yes	3.3	1.0	2.1	5.3	0
	10	1	Operating theatres	–	Linen bags	Yes	5.1	1.8	3.4	12.3	2.0
	9	3	Neurosurgical wards	60	Linen bags	Yes	6.3	2.7	4.5	107	0.4
C	4	4	Children's wards and operating theatres	100	Linen bags	Yes	3.8	11.7	7.7		
F	8	6	General wards and operating theatres	300	Polythene bags	Yes	6.8	10.6	8.7	125	0.6
E	7	1	Children's ward (infectious diseases)	25	Linen and polythene bags	No	7.2	11.8	9.5	88	0
D	6	5	Medical wards	270	Linen bags	Yes	8.1	19.2	13.6	93	1.1
B	2	2	Wards	95	Linen bags	No	13.7	21.4	17.5	109	0
A	1	6	Maternity wards and operating theatres	180	Loose	No	28.6 (9.6)[d]	26.9 (14.3)	27.7 (12.4)	–	0.1
B	3	2	Wards	80	Linen bags	No	12.3	47.8	30.1	169	0.2
D	5	5	Surgical wards	300	Linen bags	Yes	39.2	32.9	36.1	164	4.0
H[c]	12	3	General wards	} 66	Loose	Yes	41.5 (19.7)	–	41.5 (19.7)	422 / 163[e]	0.5
	13	3	General wards		Loose	Yes	53.8 (41.2)	–	53.8 (41.2)	385 / 126[e]	0.5

Note: All counts are given as average numbers of bacterial colonies per Rodac® plate.

a The number of floors served does not include the basement or exit floor.

b Vented by an opening at the top, either to outside or into roof space; none of the chutes had mechanical extract ventilation.

c Both chutes at Hospital H served the same area, No. 12 being used for pre-rinsed soiled linen, No. 13 for dry dirty linen.

d Figures in parentheses give the average chute counts with the results from the sloping entry connections omitted.

e Average floor counts with the results from the basement floor omitted.

Source: W. Whyte et al., *J. Hyg.,* 67, 427, 1969. With permission of Cambridge University Press.

2.1–119 THE BUILD-UP OF BACTERIAL CONTAMINATION ON STERILE SURFACES OF CHUTES NOS. 12 AND 13[a]

Elapsed time, (days)	Chute No. 12		Chute No. 13	
	Basement	First floor	Basement	First floor
0	1.3	2.3	0.5	2.5
½	3.0	3.8	21.8	20.3
1	8.7	14.8	52.3	20.5
2	102.2	19.7	70.6	26.8
3	25.0	8.8	54.7	15.2
4	17.0	3.5	104.7	29.7
5	9.2	18.5	95.2	17.5
6	9.7	12.7	72.8	14.5
8	12.8	–	45.8	8.7
10	11.8	–	67.4	11.3
14	7.2	–	34.0	9.3
21	23.8	–	83.5	6.8
30	19.7	17.3	28.8	10.5
35	6.2	7.3	35.2	12.3
42	6.0	5.5	28.0	11.5

Note: Counts given as number of bacteria per Rodac® plate.

[a] Hospital H. General wards. Both chutes served the same area, No. 12 being used for pre-rinsed soiled linen, No. 13 for dry dirty linen.

Source: W. Whyte et al., *J. Hyg.,* 67, 427, 1969. With permission of Cambridge University Press.

2.1–120 THE BUILD-UP OF BACTERIAL CONTAMINATION ON STERILE SURFACES OF CHUTE NO. 1[a]

Elapsed time, hours	Basement	First floor
0	0	0.2
4	71.0	0.3
14	18.2	0.2
22	91.0	0.5
39	54.8	1.0
46	35.2	3.0

Note: Counts given as number of bacteria per Rodac® plate

[a] Hospital A. Maternity wards and operating theatres.

Source: W. Whyte et al., *J. Hyg.,* 67, 427, 1969. With permission of Cambridge University Press.

2.1–121   NUMBER AND TYPE OF *PSEUDOMONAS* ISOLATES RECOVERED DURING AND AFTER A PSEUDOMONAS CASE[a]

Areas sampled	Number and type of isolates recovered on dates sampled					
	May 14	May 19	May 22	May 26	May 28	May 30
Surfaces						
Floor D, Room 1	1E	2E	43E	0	0	0
Bed table D, Room 1	0	0	8E	0	0	0
Bedspread B, Room 1	10E	9E	2E	0	0	0
Bedpan, Room 1	0	12E	3E	–	–	–
Sink basin, Room 1	0	1D	1E	5E	0	0
Sink handle, Room 1	0	0	3E	0	0	0
Water reservoirs						
Sink drain, Room 1	3E	7E	6E	TN[b]E	TNE	TNE
Inhalation therapy equipment						
Exhale tubing	TNE	TNE	TNE	0	–	–
Inhale tubing	TNE	TNE	TNE	0	–	–
Reflux trap	TNE	TNE	TNE	0	–	–
Tubing from trap to nebulizer	36E	29E	13E	0	–	–
Nebulizer kettle	16E	0	0	0	–	–
Inlet tubing	0	0	0	0	–	–

Note: Patient expired on May 24.

[a] *Pseudomonas* was isolated from the patient's tracheostomy, blood specimens, urine specimens, and lung mucous fluid on May 10, May 20, and May 24.
[b] Too numerous to count.

Source: D. Drollette et al., *Hospitals,* 45, 91, 1971. With permission.

2.1–122   NUMBER OF OPERATING THEATRE SITES FOUND CONTAMINATED BY GRAM-NEGATIVE BACTERIA

Number of sites	April	May	June	July	August	September	October
Swabbed	13	36	38	66	13	63	48
Contaminated by Gram-negative organisms	13 (100)	24 (67)	17 (45)	11 (17)	2 (15)	22 (35)	10 (21)
Contaminated by *Pseudomonas aeruginosa*	13 (100)	12 (33)	13 (34)	6 (9)	1 (8)	12 (19)	0 –

Note: Figures in parentheses are percentages.

Source: M.E.M. Thomas et al., *J. Hyg.,* 70, 63, 1972. With permission of Cambridge University Press.

2.1–123   TYPES OF HOSPITAL-ACQUIRED INFECTIONS MOST FREQUENTLY ENCOUNTERED AND PREDOMINANT ETIOLOGIC AGENTS RESPONSIBLE

Type of infection	Implicated etiologic agents
Urinary tract infections	*Escherichia coli: Proteus* sp.; *Pseudomonas* sp.; *Streptococcus* sp.; *Staphylococcus albus; Aerobacter-Klebsiella* group
Postoperative wound infections	*S. aureus; E. coli; Proteus* sp.; *Streptococcus* sp.; *Aerobacter-Klebsiella* group
Respiratory infections	*S. aureus; Diplococcus pneumoniae; Aerobacter-Klebsiella* group; *Pseudomonas* sp.; *Proteus* sp.; *Streptococcus* sp.
Skin and subcutaneous infections (including septic phlebitis)	*S. aureus; Pseudomonas* sp.; *Staphylococcus albus; Aerobacter-Klebsiella* group
Endometritis, conjunctivitis, serum hepatitis, otitis, sinusitis, chickenpox, diarrhea, pharyngitis, and others	*S. aureus; E. coli; Proteus* sp.; *Pseudomonas* sp.; "viruses"
Bacteremia (subsequent to another infection listed above)	*Aerobacter-Klebsiella* group; *S. aureus; E. coli*

Source: R.G. Bond et al., Eds., *Environmental Health and Safety in Health-Care Facilities,* Macmillan, New York, © 1973, 35. With permission.

REFERENCES

L.D. Edwards, *Public Health Rep.,* 84, 451 1969.
T.C. Eickhoff and P.S. Brachman, *J. Hosp. Res.,* 6, 9, 1968.
R. Thoburn et al., *Arch. Intern. Med.,* 121, 1, 1968.

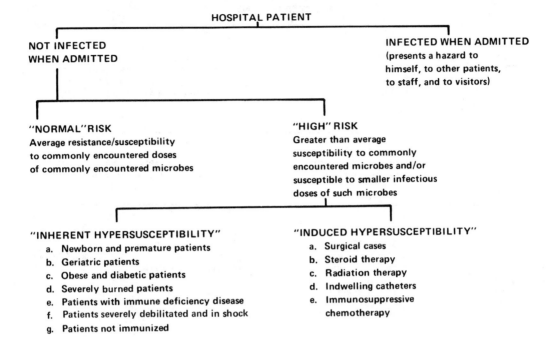

HOSPITAL PATIENT

NOT INFECTED WHEN ADMITTED

INFECTED WHEN ADMITTED (presents a hazard to himself, to other patients, to staff, and to visitors)

"NORMAL" RISK Average resistance/susceptibility to commonly encountered doses of commonly encountered microbes

"HIGH" RISK Greater than average susceptibility to commonly encountered microbes and/or susceptible to smaller infectious doses of such microbes

"INHERENT HYPERSUSCEPTIBILITY"
a. Newborn and premature patients
b. Geriatric patients
c. Obese and diabetic patients
d. Severely burned patients
e. Patients with immune deficiency disease
f. Patients severely debilitated and in shock
g. Patients not immunized

"INDUCED HYPERSUSCEPTIBILITY"
a. Surgical cases
b. Steroid therapy
c. Radiation therapy
d. Indwelling catheters
e. Immunosuppressive chemotherapy

2.1–124   Classification of infection "risks" in the hospital.

Source: R.G. Bond et al., Eds., *Environmental Health and Safety in Health-Care Facilities,* Macmillan, New York, © 1973, 36. With permission.

REFERENCE

V.W. Greene, *Hospitals,* 44, 124, 1970.

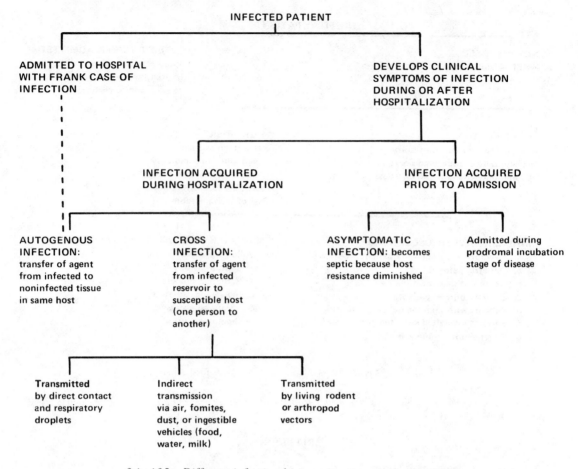

2.1–125   Different infection-history situations in the hospital.

Source: R.G. Bond et al., Eds., *Environmental Health and Safety in Health-Care Facilities,* Macmillan, New York, © 1973, 37. With permission.

REFERENCE

V.W. Greene, *Hospitals,* 44, 124, 1970.

2.1–126   ILLUSTRATIVE CONTAMINATION CONTROL GUIDELINES

Criticality	Examples	Microbial standards
Environments		

critical	Surgical suite
	Obstetric suite
	Nursery
	Burns unit
	Immunosuppressive
	  patients
		Floor counts[a]: $<5/cm^2$
		Air counts:$5-10/ft^3$
		  (maximum activity)

Domiciliary	Medical wards Treatment rooms Physiotherapy	Floor counts[a]: $5-10/cm^2$ Air counts:$10-20/ft^3$
Administrative	Offices Shops Conference rooms Teaching areas	Aesthetic considerations
Instruments and materials		
Critical	Introduced beneath body surfaces	Sterility
Semicritical	Contact mucous membranes	Must be free of common vegetative cells, tubercle bacilli, and viruses; may be contaminated with some spores
Noncritical	Touch only unbroken skin	Must be free of common vegetative pathogens such as *Pseudomonas*, staphylococci, tubercle bacilli, and viruses; may be contaminated with normal environmental organisms

[a] Floor counts are suggested maximum permissible levels during normal activity.

Source: R.G. Bond et al., Eds., *Environmental Health and Safety in Health- Care Facilities*, Macmillan, New York, © 1973, 41. With permission.

REFERENCES

S.D. Rubbo and J.T. Gardner, *A Review of Sterilization and Disinfection*, Year Book Medical Publishers, Chicago, 1965.
E.H. Spaulding, Chemical disinfection of medical and surgical materials, in C.A. Lawrence and S.S. Block, Eds., *Disinfection, Sterilization, and Preservation*, Lea & Febiger, Philadelphia, 1968, 517.

2.1−127   RELATIONSHIP OF THE CARRIER STATE TO WOUND COLONIZATION AND SEPSIS

Carriage pre-op.	Number of patients	Wounds colonized		Wounds septic	
		Number	%	Number	%
Noncarriers	173	27	16	16	9
Carriers					
Nose only	51	11	22	6	12
Skin	45[a]	25[b]	56	10[c]	22
Total	96	36	37	16	17
All patients	269	63	23	32	12

[a] Ten were carriers on skin only.
[b] Three were carriers on skin only.
[c] One was a carrier on skin only.

Source: F. Colia et al., *J. Hyg.,* 67, 49, 1969. With permission of Cambridge University Press.

2.1−128   RELATIONSHIP OF CARRIER STRAIN TO WOUND COLONIZATION AND SEPSIS

	Patients with wound colonization	Patients with wound sepsis
Total	63	32
Noncarriers	27	16
Carriers	36	16
With carrier strain only	27	13
With carrier plus exogenous strain	6	3
With exogenous strain only	3	0

Source: F. Colia et al., *J. Hyg.,* 67, 49, 1969. With permission of Cambridge University Press.

2.1–129   WOUND COLONIZATION AND SEPSIS RELATED TO CARRIAGE ON NORMAL VS. BROKEN SKIN[a]

	Number of skin carriers	Number of wounds colonized	Number of wounds septic
Total	45	25 (56%)	10 (22%)
On normal skin	28	14 (50%)	5 (18%)
On broken skin	17	11 (65%)	5 (29%)

[a] For example, vascular leg ulcers, burns, dermatitis, etc.

Source: F. Colia et al., *J. Hyg.,* 67, 49, 1969. With permission of Cambridge University Press.

2.1–130   QUANTITY OF PREOPERATIVE NASAL CARRIAGE RELATED TO SKIN CARRIAGE, WOUND COLONIZATION, AND SEPSIS

Nasal carriage	Number of patients	Skin carriers		Wounds colonized[a]		Wounds septic[a]	
		Number	%	Number	%	Number	%
Heavy	36	17 (5)[b]	47	16	44	11	31
Moderate	31	8 (2)	26	9	29	3	10
Light or none	29	20 (10)	69	8	30	2	7

[a] With carrier strain.
[b] Numbers in parentheses represent patients with carriage on broken skin, some of whom carried on intact skin as well.

Source: F. Colia et al., *J. Hyg.,* 67, 49, 1969. With permission of Cambridge University Press.

2.1–131   RESULTS OF CULTURES OF OPERATING ROOM SPECIMENS OBTAINED FROM 74 PATIENTS

	Number	Sterile	*Staphylococcus aureus*
Wound swabs	48	13	0
Skin strips	60	14	9[a]

[a] Four patients were preoperative carriers; none had same strain in skin strip. Eight patients had subsequent cultures, and none were colonized later with skin-strip strain.

Source: F. Colia et al., *J. Hyg.,* 67, 49, 1969. With permission of Cambridge University Press.

2.1–132   CARRIAGE BY STAFF OF STAPHYLOCOCCUS PYOGENES AND GRAM-NEGATIVE BACILLI

	Total number	Number positive for	
		S. pyogenes	Gram-negative bacilli
Staff investigated	51	28 (55)[a]	8 (16)
Nasal swabs taken	116	18 (16)	0
Hand swabs taken	116	35 (30)	10 (9)
Nasal carriers		9 (18)	0
Hand carriers		27 (53)	8 (16)

[a] Figures in parentheses indicate percentages.

Source: F.W. Winton and A.J. Keay, *J. Hyg.*, 66, 34, 1968. With permission of Cambridge University Press.

2.1–133   RATE OF NASAL COLONIZATION WITH COAGULASE-POSITIVE STAPHYLOCOCCI IN THE NURSERY

Month	Number of infants	Number colonized	Rate of colonization, %
October	246	11	4.5
November	263	10	3.8
December	202	5	2.4
January	201	4	1.9
February	211	12	5.6
March	218	6	2.8
Total	1341	48	3.6

Source: R. Thoburn et al., *Arch Intern. Med.*, 121, 1, 1968. With permission. Copyright 1968, American Medical Association.

2.1–134   EFFECT OF THE INTRODUCTION OF THE USE OF ANTISEPTICS ON THE PREVALENCE AND ACQUISITION OF MeRS[a] IN THE NOSE IN STAFF MEMBERS

Period	n[b]	Prevalence[c]	n^{acq} [d]	Acquisition
October 1969	80	7(8.7%)	—	—
November 1969	133	7(5.3%)	126	2(1.5%)
December 1969	149	5(3.4%)	144	1(0.7%)
January 1970	145	4(2.8%)	141	4(2.8%)
February 1970	150	6(4.0%)	144	5(3.3%)
March 1970	127	8(6.3%)	119	4(3.2%)
Total	704	37(5.3%)	674	16(2.4%)
April 1970	136	6(4.4%)	130	1(0.7%)
May 1970	133	7(5.3%)	126	0
June 1970	121	3(2.5%)	118	0
August 1970	136	4(2.9%)	132	0
September 1970	132	3(2.3%)	129	2(1.5%)
Total	658	23(3.5%)	635	3(0.5%)

Note: The use of antiseptics was started in March 1970.

[a] MeRS = methicillin-resistant staphylococci.
[b] n = number of nasal cultures.
[c] Prevalence $\chi^2_{(1)}$ = 2.10 (0.10<P<0.20); acquisition $\chi^2_{(1)}$ = 6.99 (0.005<P<0.01).
[d] n^{acq} = number of nasal cultures diminished by the number of carriers.

Source: M.F. Michel and C.C. Priem, *J. Hyg.*, 69, 453, 1971. With permission of Cambridge University Press.

2.1–135   PREVALENCE OF MeRS[a] IN THE NOSE IN CHILDREN

Unit	Period	n	Prevalence	$x^2_{(1)}$
Neonatal	Before	173	48(27.7%)	37.58
	After	148	3(2%)	$P<0.0005$
Infant	Before	206	21(10.1%)	1.77
	After	115	6(5.25%)	$0.10<P<0.20$
Surgical	Before	427	68(15.9%)	24.61
	After	314	13(4.1%)	$P<0.005$

Note: The prevalence is compared in each unit for a period of 6 months before and 6 months after the introduction of the use of antiseptics.

[a] MeRS = methicillin-resistant staphylococci.

Source: M.F. Michel and C.C. Priem. *J. Hyg.,* 69, 453, 1971. With permission of Cambridge University Press.

2.1–136   MONTHLY ISOLATIONS OF MeRS[a] FROM INFECTIONS IN CLINICAL UNITS AND THE OUTPATIENT DEPARTMENT

	Unit						
	Neonatal	Infant	Surgical	Quarantine	Medical	O.D.[b]	Total
1969							
October	0	1	6	0	2	0	9
November	0	1	1	1	0	0	7
December	5	0	2	2	1	1	11
1970							
January	1	1	4	2	0	3	12
February	1	2	4	3	1	5	17
March	0	1	4	2	0	0	8
April	0	0	0	1	0	0	3
May	0	1	0	2	0	0	4
June	0	1	0	2	0	1	5
July	1	0	0	2	2	1	6
August	0	0	0	1	1	0	2
September	0	0	0	2	1	0	5

Notes: In March 1970 the use of antiseptics was introduced in three out of five clinical units, i.e., the neonatal, infant, and surgical unit.

$x^2_{(1)} = 12.4$ $(P < 0.0005)$.

[a] MeRS = methicillin-resistant staphylococci.
[b] O.D. = outpatients' department.

Source: M.F. Michel and C.C. Priem, *J. Hyg.,* 69, 453, 1971. With permission of Cambridge University Press.

2.1–137   TYPE AND SOURCE OF MeRS[a] COLLECTED DURING FEBRUARY AND JUNE 1970

		Strains					
			Source				
Period	Number	Nose	Sputum	Pus	Blood	Urine	Phage type
February 1970	37	27	2	7	1	–	85/+
	12	8	2	2	–	–	6/47/77/84/85/+
	2	1	–	1	–	–	6/7/47/53/54/88/83A/85/+
	1	–	–	–	–	1	53/88/85
	1	1	–	–	–	–	29/52/+
June 1970	5	3	1	1	–	–	85/+
	3	–	2	1	–	–	6/42/77/84/85/+
	1	–	–	–	–	–	6/7/47/53/54/88/83A/85/+
	1	–	–	1	–	–	7/47/54/77/+

[a] MeRS = methicillin-resistant staphylococci.

Source: M.F. Michel and C.C. Priem, *J. Hyg.,* 69, 453, 1971. With permission of Cambridge University Press.

2.1–138   NOSOCOMIAL INFECTION RATE, 1971–72 – OTTAWA GENERAL HOSPITAL

		Percent of all patients	Percent of infections
Patients discharged	18,315	–	
Patients infected	2,469	13.5	–
Nosocomial infections	1,443	7.9	58.5

Source: J.C. Westwood et al., *Can. Med. Assoc. J.,* 110, 769, 1974. With permission.

2.1–139   EFFECT OF PREEXISTING INFECTION ON NOSOCOMIAL INFECTION RATE – OTTAWA GENERAL HOSPITAL

		Percent of all patients	Percent of patients infected on admission
Total patients	18,315	–	
Patients infected on admission	2,359	12.9	–
Superinfected in hospital	337	–	14.3
Patients not infected on admission	15,956	87.1	–
Infected in hospital	1,106	–	6.9

Source: J.C. Westwood et al., *Can. Med. Assoc. J.,* 110, 769, 1974. With permission.

2.1–140   NOSOCOMIAL INFECTION RATES BY SERVICE AND TYPE – OTTAWA GENERAL HOSPITAL

	A Medical		B Surgical		C Postop. (all services)		D Surgical (no op.)		E Gynecology (no op.)		F Maternity (no op.)		G Nursery (no op.)		H Pediatrics (no op.)	
	No.	Infection rate, %	No.	Infection rate, %	No.	Infection rate, %	No.	Infection rate, %	No.	Infection rate, %	No.	Infection rate, %	No.	Infection rate, %	No.	Infection rate, %
Total patients	6713	–	6847	–	6804	–	2268	–	273	–	1167	–	1090	–	2466	–
Nosocomial infections																
Sepsis	129[a]	1.9	170	2.5	262	3.9	15	0.7	1	0.4	3	0.3	4	0.4	10	0.4
Respiratory	122	1.8	57	0.8	97	1.4	8	0.4	–	–	–	–	–	–	8	0.3
Urinary tract	242	3.6	160	2.3	329	4.8	44	1.9	9	3.3	14	1.2	9	0.8	62	2.5
Other	19	0.3	15	0.2	25	0.4	1	–	–	–	2	0.2	3	0.3	7	0.3
Fever >101°F only	109	1.6	172	2.5	205	3.0	29	1.3	2	0.8	5	0.4	4	0.4	78	3.2
Total	621	9.3	574	8.4	918	13.5	97	4.3	12	4.4	24	2.1	20	1.8	165	6.7

[a] Includes 70 cases of cannula sepsis among artificial kidney unit (AKU) patients.

Source: J.C. Westwood et al., *Can. Med. Assoc. J.*, 110, 769, 1974. With permission.

2.1−141   NOSOCOMIAL URINARY TRACT INFECTIONS, 1971−72 − OTTAWA GENERAL HOSPITAL

Hospital "area" or service	Number of patients	Urinary tract infections Number	Percent of Total patients	Percent of Nosocomial infections
Medical	6,713	242	3.6	43.9
Postop.	6,804	329	4.8	35.8
Surgical, no op.	2,268	44	1.9	45.4
Gynecology, no op.	273	9	3.3	75.0
Maternity, no op.	1,167	14	1.2	58.3
Nursery	1,090	9	0.8	45.0
Total	18,315	647	3.5[a]	44.8[a]

Note: Patients on surgical and pediatric services are included in "Medical," "Postop.," and "Surgical, no op." categories. In Table 2.1−140 these categories are duplicated.

[a] Average value.

Source: J.C. Westwood et al., *Can. Med. Assoc. J.*, 110, 769, 1974. With permission.

2.1−142   RATE OF HOSPITAL-ACQUIRED URINARY TRACT INFECTIONS BY SERVICE − THE JOHNS HOPKINS HOSPITAL

Service	Number of infections	Percent of urinary tract infections	Infection rate, %
Gynecology	62	25.8	6.3
Neurosurgery	16	6.7	4.2
Orthopedics	17	7.1	3.6
Urology	21	8.7	3.3
General surgery	36	15	2.3
Pediatrics	25	10.4	1.9
Ward medicine	28	11.7	1.8
Cardiac surgery	2	0.8	1.4
Obstetrics	15	6.2	1.3
Private medicine	15	6.2	1.2
General pediatric surgery	2	0.8	0.2
Otolaryngology	1	0.4	0.1
Plastic surgery	0	0	0
Ophthalmology	0	0	0
Total	240	100	1.9[a]

[a] Average value.

Source: R. Thoburn et al., *Arch. Intern. Med.*, 121, 1, 1968. With permission. Copyright 1968, American Medical Association.

2.1–143   COMPARISON OF OVERALL INFECTION RATES AT VARIOUS HOSPITALS

Hospitals	Number of beds	Infections expressed as percent of patients discharged			Hospital-acquired infections as percent of total infections	Urinary tract infections as percent of hospital-acquired infections
		Total	Community acquired	Hospital acquired		
Ottawa General	600	13.5	5.6	7.9	58.5	44.8
Presbyterian-St. Luke's, Chicago[1]	850	15.7	8.5	7.2	45.9	30
University, Illinois[2]	640	17.5	10.4	7.1	40.6	14[a]
Staten Island[3]	800	25.7	17.9	7.8	30.4	–
Atlanta (Community Hospitals)[4]	200–500	–	–	3.5	–	36
Boston City[5]		41.8	26.3	15.5	37.1	33
Latter-day Saints[6]	642	19.1	10.6	8.5	44.5	40
Bronx-Lebanon[7]		7.9	4.3	3.6	45.6	–
Johns Hopkins[8]		–	–	4.0	–	41

[a] Urinary tract infection rate was reduced from 32 to 14% by measures taken.[2]

Source: J.C. Westwood et al., *Can. Med. Assoc. J.*, 110, 769, 1974. 1974. With permission.

REFERENCES

1.   L.D. Edwards et al., *Am. J. Public Health*, 62, 1053, 1972.
2.   S. Streeter et al., *Am. J. Nurs.*, 67, 526, 1967.
3.   L.D. Edwards, *Public Health Rep.*, 84, 45, 1969.
4.   J.V. Bennett et al., *Can. Hosp.*, 47, 42, 1970.
5.   F.F. Barrett et al., *N. Engl. J. Med.*, 278, 5, 1968.
6.   M.L. Moody and J.P. Burke, *Arch. Intern. Med.*, 130, 261, 1972.
7.   V.L. Lorian and B. Topf, *Arch. Intern. Med.*, 130, 104, 1972.
8.   R. Thoburn et al., *Arch. Intern. Med.*, 121, 1, 1968.

2.1–144   HOSPITAL-ACQUIRED URINARY TRACT INFECTION
AND ITS ASSOCIATION WITH BACTEREMIA –
THE JOHNS HOPKINS HOSPITAL

Organism	Number of infections	Percent of total	Patients with bacteremia Number	Patients with bacteremia Percent
Escherichia coli	83	34.6	2	2.4
Proteus	40	16.7	0	
Klebsiella Aerobacter	31	12.9	3	9.7
Streptococcus fecalis	27	11.2	0	
Paracolon	15	6.2	0	
Pseudomonas	8	3.3	1	12.5
Staphylococcus aureus	3	1.2	0	
Mixed culture	10	4.2	0	
Other organisms, or not cultured	23	9.6	0	
Total	240	100	6	2.5

Source: R. Thoburn et al., *Arch. Intern. Med.,* 121, 1, 1968. With permission. Copyright 1968, American Medical Association.

2.1–145   VARIOUS PROTEUS STRAINS ISOLATED FROM
URINARY TRACT AND WOUND INFECTIONS[a] –
THE JOHNS HOPKINS HOSPITAL

Strain	Urinary tract infections Number	Urinary tract infections Percent of total	Wound infections Number	Wound infections Percent of total
Proteus mirabilis	31	74	15	62.5
P. morgani	3	7.1	6	25
P. rettgeri	6	14.1	2	8.3
P. vulgaris	2	4.8	1	4.2
Total	42	100	24	100

[a] Relative frequency.

Source: R. Thoburn et al., *Arch. Intern. Med.,* 121, 1, 1968. With permission. Copyright 1968, American Medical Association.

**2.1–146 RATE OF HOSPITAL-ACQUIRED
URINARY TRACT INFECTIONS[a] –
THE JOHNS HOPKINS HOSPITAL**

Patient category	Number of infections	Number of patients	Attack rate, %
White male	42	3717	1.1
White female	93	4879	1.9
Negro male	17	1825	0.9
Negro female	50	2837	1.8
Private	41	1801	2.3
Semiprivate	24	4780	1.5
Ward	102	6677	1.5

[a] Relation of race, sex, and pay status.

Source: R. Thoburn et al., *Arch. Intern. Med.,* 121, 1, 1968. With permission. Copyright 1968, American Medical Association.

**2.1–147 COMPARISON OF BACTERIA ISOLATED FROM COMMUNITY-
AND HOSPITAL-ACQUIRED INFECTIONS,
SEPTEMBER 1971 TO AUGUST 1972**

	Community-acquired infections		Hospital-acquired infections	
	Number of isolates	Percent of total isolates	Number of isolates	Percent of total isolates
Staphylococcus aureus	283	13.6	256	12.5
Streptococcus group A	32	1.5	14	0.7
Streptococcus faecalis	99	4.7	150	7.3
Diplococcus pneumoniae	94	4.5	56	2.7
Total Gram-positive cocci	508	24.4	476	23.2
Escherichia coli	582	27.9	519	25.3
Enterobacter and *Klebsiella*	297	14.2	345	16.8
Proteus	142	6.8	162	7.9
Pseudomonas aeruginosa	110	5.3	139	6.8
Total Gram-negative (enterobacteriaceae)	1131	54.2	1165	56.8
Anaerobes	38	1.8	45	2.2
Candida albicans	61	2.9	114	5.6
Other	348	16.7	251	12.2
Average totals	2086	100	2051	100.0

Source: J.C. Westwood et al., *Can. Med. Assn. J.,* 110, 769, 1974. With permission.

2.1–148   ORGANISMS ISOLATED FROM CASES OF CLINICAL INFECTION

Organism	Total	Eye	Skin	Umbilicus	Mouth	Nose and throat	Urine	Stool	Gastric washing
Staphylococcus pyogenes	16	8	4	2	–	–	–	2	–
S. albus	5	4	–	–	–	1	–	–	–
Escherichia coli	2	–	1	–	–	–	1	–	–
Proteus spp.	1	–	–	–	–	–	1	–	–
Pseudomonas pyocyanea	1	–	–	1	–	–	–	–	–
Paracolon bacilli	3	2	–	–	–	1	–	–	–
Streptococcus pyogenes	3	–	–	2	–	1	–	–	–
Pneumococcus	1	1	–	–	–	–	–	–	–
Candida spp.	4	–	–	–	3	–	–	–	1
None	18	3	–	1	–	1	1	12	–
Total	54	18	5	6	3	4	3	14	1

Source: F.W. Winton and A.J. Keay, *J. Hyg.*, 66, 325, 1968. With permission of Cambridge University Press.

2.1–149   BACTERIA RESPONSIBLE FOR INFECTIONS

Causative organism	Number of infections	Site of infection			
		Sputum	Tracheostomy	Wound swab	Urine
Coliforms	39	18	6	5	10
Pseudomonas aeruginosa	21	11	7	2	1
Proteus spp.	9	5	1	1	2
Haemophilus influenzae	6	5	1	0	0
Streptococcus pneumoniae	3	3	0	0	0
Staphylococcus aureus	21	10	4	7	0
Streptococcus faecalis	4	0	1	3	0
S. pyogenes	3	0	0	3	0
Candida spp.	14	3	11	0	0
Clostridium welchii	1	0	0	1	0
Total	121	55	31	22	13

Source: D.M. Harris et al., *J. Hyg.*, 67, 525, 1969. With permission of Cambridge University Press.

2.1–150   BACTERIA ISOLATED AS PREDOMINANT PATHOGEN FROM POSTOPERATIVE WOUND INFECTIONS – THE JOHNS HOPKINS HOSPITAL

Organism	Isolations	Percent of total isolations	Patients receiving antibiotics at time of culture — Number	Patients receiving antibiotics at time of culture — Percent of total	Patients not receiving antibiotics at time of culture — Number	Patients not receiving antibiotics at time of culture — Percent of total	Percent receiving antibiotics
Staphylococcus aureus	28	23	6	15.8	22	27.2	21.4
S. albus	15	12.3	9	23.6	6	7.5	60
β-Streptococcus (Group A)	9	7.4	0	0	9	10.6	0
Streptococcus fecalis	9	7.4	2	5.3	7	8.3	22.2
Pneumococci	1	0.8	0	0	1	1.2	0
Clostridium welchii	1	0.8	0	0	1	1.2	0
Escherichia coli	22	18	9	23.6	13	15.4	41
Klebsiella-Aerobacter	12	9.8	2	5.3	10	11.7	16
Proteus	12	9.8	5	13.2	7	8.3	41.5
Paracolon	7	5.8	3	7.9	4	4.8	43
Pseudomonas	6	4.9	2	5.3	4	4.8	25
Total	122[a]	100	38	100	84	100	31.1

[a] An additional 56 wound infections were associated with more than one pathogen, no pathogen, or no cultural information.

Source: R. Thoburn et al., *Arch. Intern. Med.,* 121, 1, 1968. With permission. Copyright 1968, American Medical Association.

2.1–151   COMPARISON OF ANTIBIOTIC SENSITIVITIES OF *STAPHYLOCOCCUS AUREUS* ISOLATED FROM VARIOUS STRAINS

	Hospital-acquired infections (256 isolates)	Community-acquired infections (283 isolates)	New staff, 1972 (411 isolates)
Sensitive to penicillin G	19	25	31.1
Resistant to penicillin G	76.6	69	64.2
Resistant to penicillin G + tetracycline	3.6	4.9	4.4
Resistant to Penicillin G + tetracycline + lincomycin	0.8	1.1	0.3
Total	100	100	100

Note: Figures represent percentages in each class.

Source: J.C. Westwood et al., *Can. Med. Assn. J.,* 110, 769, 1974. With permission.

**2.1–152   THE INCIDENCE OF INFECTION AMONG PATIENTS ADMITTED FROM THE
CASUALTY DEPARTMENT, OR FROM THE WARDS OR OTHER HOSPITALS,
IN RELATION TO THE TYPE OF TREATMENT GIVEN**

| | Total patients | | Patients admitted from casualty[a] | | | Patients admitted from wards or other hospitals | | |
	Number	Died	Number	Infected on admission	Infection acquired	Number	Infected on admission	Infection acquired
I	85	44	34	1	10 (29)[b]	51	15 (29)	24 (47)
II	27	6	12	0	3 (25)	15	5 (33)	6 (40)
III	95	13	48	0	1	47	4 (8)	9 (19)
Total	207	63	94	1	14 (15)	113	24 (21)	39 (34)

Note: Group I: patients requiring Intermittent Positive Pressure Ventilation (I.P.P.V.). Group II: patients other than those in Group I, requiring endotracheal intubation or tracheostomy. Group III: patients admitted for continuous monitoring.

[a] From outside a government hospital.
[b] Figures in parentheses indicate percentages.

Source: D.M. Harris et al., *J. Hyg.,* 67, 525, 1969. With permission of Cambridge University Press.

2.1–153   INFLUENCE OF INTERMITTENT POSITIVE PRESSURE VENTILATION (I.P.P.V.) ON ACQUISITION OF INFECTION

Diagnosis	Patients requiring I.P.P.V.						Patients requiring tracheostomy or endotracheal tube, but not I.P.P.V.					
	Total number	Number with E.T.T.[a]	Number with T.[b]	Number with infection			Total number	Number with E.T.T.[a]	Number with T.[b]	Number with infection		
				On admission	Acquired	Total				On admission	Acquired	Total
Drug overdose	18	18	0	1	3	4	8	8	0	0	0	0
Intracranial hemorrhage, etc.	12	12	0	0	3	3	4	3	1	0	3	3
Respiratory inadequacy and post-op. complications	19	13	6	7	10	17	7	4	3	4	2	9
Total	49	43	6	8	16	24 (48.7%)	19	15	4	1	6	9 (47.4%)

[a] E.T.T. = endotracheal tube.
[b] T. = tracheostomy.

Source: D.M. Harris et al., *J. Hyg.*, 67, 525, 1969. With permission of Cambridge University Press.

2.1–154 INFECTIONS DEVELOPING AFTER ADMISSION TO THE JOHNS HOPKINS HOSPITAL

Type	Number	Total infections, %	Incidence of infection,[a] %	Prevalence of infection,[b] %
Urinary tract infection	240	40.5	1.9	1.8
Postoperative wound infection	178	30.1	2.3	1.4
Pneumonia	85	14.5	0.8	1.3
Phlebitis, presumably septic	62	10.5	–	–
Endometritis	9	1.5	–	–
Conjunctivitis	8	1.3	–	–
Hepatitis (serum)	2	0.3	–	–
Otitis, furunculosis, sinusitis, chickenpox, diarrhea, skin infection, pharyngitis, and rectal abscess (one of each)	8	1.3	–	–
Total	592	100	4	4.7

[a] October 1, 1965 to March 31, 1966.
[b] March 13, 1967.

Source: R. Thoburn et al., *Arch. Intern. Med.*, 121, 1, 1968. With permission. Copyright 1968, American Medical Association.

2.1–155   FREQUENCY OF WOUND INFECTION
FOLLOWING SELECTED OPERATIONS –
THE JOHNS HOPKINS HOSPITAL

Surgical procedure	Number of patients infected	Number of operations performed	Infection rate, %
Gastrointestinal			
Abdominal-perineal resection	4	11	36.3
Choledochostomy	2	13	15.3
Partial resection of small bowel	2	21	9.5
Partial colectomy	4	47	8.5
Cholecystectomy	4	58	6.9
Cardiovascular			
Portacaval shunt	2	9	22.1
Cardiac valve replacement	3	16	18.7
Arterial bypass graft	4	24	16.7
Genitourinary			
Oophorectomy	2	11	18.2
Ureterolithotomy	2	15	13.3
Abdominal hysterectomy	11	107	10.3
Vaginal hysterectomy	6	108	5.5
Abdominal hysterectomy with bilateral salpingo-oophorectomy	4	85	4.7
Miscellaneous			
Ventriculoperitoneal shunt	6	16	37.5
Radical neck dissection	4	20	20
Craniotomy	2	20	10
Ventral and incisional herniorrhaphy	3	35	8.4

Source: R. Thoburn et al., *Arch. Intern. Med.,* 121, 1, 1968. With permission.
Copyright 1968, American Medical Association.

2.1–156   POSTOPERATIVE WOUND INFECTION RATE BY SURGICAL SPECIALTY – THE JOHNS HOPKINS HOSPITAL

Service	Number of infections	Number of operations	Infection rate, %
Cardiac	12	121	9.9
Gynecology	59	1451	4.1
General surgery	55	1686	3.2
Neurosurgery	12	430	2.8
Orthopedics	10	418	2.4
Urology	9	494	1.8
Plastic	6	601	1
General pediatric	4	404	1
Ophthalmology	8	1460	0.5
Otolaryngology	3	819	0.4
Total	178	7884	2.3

Source: R. Thoburn et al., *Arch. Intern. Med.*, 121, 1, 1968. With permission. Copyright 1968, American Medical Association.

2.1–157   INCIDENCE OF HOSPITAL-ACQUIRED PNEUMONIA BY SERVICE – THE JOHNS HOPKINS HOSPITAL

Service	Number with pneumonia	Percent of total patients with pneumonia	Attack rate, %
Cardiac surgery	4	4.7	3.1
Neurosurgery	10	11.8	3
General surgery	17	20	1.2
Pediatrics	10	11.8	0.8
Ward medicine	17	20	0.8
Gynecology	11	13	0.8
Orthopedic surgery	4	4.7	0.6
Plastic surgery	1	1.2	0.3
Private medicine	9	10.4	0.3
Urology	1	1.2	0.2
Obstetrics	1	1.2	0.1
Otolaryngology	0	0	0
Ophthalmology	0	0	0
General pediatric surgery	0	0	0
Total	85	100	0.8

Source: R. Thoburn et al., *Arch. Intern. Med.*, 121, 1, 1968. With permission. Copyright 1968, American Medical Association.

2.1–158 FREQUENCY OF VARIOUS ORGANISMS IN SPUTUM OF PATIENTS WITH HOSPITAL-ACQUIRED PNEUMONIA – THE JOHNS HOPKINS HOSPITAL

Organism	Number of isolations	Percent of total isolations
Pneumococci	11	13
Klebsiella-Aerobacter	11	13
Staphylococcus aureus	11	13
Pseudomonas aeruginosa	8	9.4
Paracolon	6	7
Escherichia coli	3	3.5
Haemophilus influenzae	2	2.3
Streptococcus fecalis	2	2.3
Proteus mirabilis	1	1.2
Mixed	3	3.5
Uncertain	27	31.8
Total	85	100

Source: R. Thoburn et al., *Arch. Intern. Med.,* 121, 1, 1968. With permission. Copyright 1968, American Medical Association.

2.1–159 ORGANISMS CULTURED FROM THE SITE OF INTRAVENOUS CATHETERIZATION IN PATIENTS WITH PHLEBITIS – THE JOHNS HOPKINS HOSPITAL

Organism	Number	Percent of total
Staphylococcus aureus	6	33.4
Klebsiella-Aerobacter	5	27.8
S. albus	2	11.3
Proteus morgani	1	5.5
Escherichia coli	1	5.5
Paracolon	1	5.5
Streptococcus fecalis	1	5.5
Flavobacterium	1	5.5
Total	18	100

Source: R. Thoburn et al., *Arch. Intern. Med.,* 121, 1, 1968. With permission. Copyright 1968, American Medical Association.

2.1–160 BACTEREMIA ASSOCIATED WITH HOSPITAL-ACQUIRED INFECTION – THE JOHNS HOPKINS HOSPITAL

Site of infection	Number of infections	Number with bacteremia	Rate, %
Urinary tract	240	6	2.5
Surgical wound	178	4	2.2
Pneumonia	85	6	7
Phlebitis	62	4	6.5
Other[a]	2	2	–
Total	567	22	3.9

[a] One patient with dysgammaglobulinemia and reticulum cell sarcoma had *Escherichia coli* bacteremia following spontaneous rupture of the colon. Another with lymphosarcoma developed repeated bacteremia due to *Bacillus subtilis* from an unknown portal of entry.

Source: R. Thoburn et al., *Arch. Intern. Med.,* 121, 1, 1968. With permission. Copyright 1968, American Medical Association.

2.1–161 ORGANISMS ISOLATED FROM THE BLOOD OF PATIENTS WITH NOSOCOMIAL INFECTIONS – THE JOHNS HOPKINS HOSPITAL

Organism	Number	Percent of total
Klebsiella-Aerobacter	7	31.8
Staphylococcus aureus	5	22.8
Escherichia coli	4	18.2
Paracolon	2	9.2
Pneumococci	1	4.5
Anaerobic streptococci	1	4.5
Pseudomonas aeruginosa	1	4.5
Bacillus subtilis	1	4.5
Total	22	100

Source: R. Thoburn et al., *Arch. Intern. Med.,* 121, 1, 1968. With permission. Copyright 1968, American Medical Association.

2.1–162 DISTRIBUTION OF DEFINITE INFECTIONS IN PRELIMINARY SURVEY BY SITE OF CULTURE

	Hospital origin		Indeterminate origin		
	Gram +	Gram −	Gram +	Gram −	Mixed
Wound-soft tissue	11	1	6	8	2
Nervous system	–	–	–	1	–
Respiratory tract	–	–	8	–	3
Gastrointestinal	–	–	–	3	–
Genital	–	–	–	1	–
Urinary	–	6	5	31	1
Eye, ear, nose, and throat	–	–	2	–	–
Blood	1	–	1	3	–
Multiple sites	–	1	–	5	4
Other	–	–	–	–	2

Source: D.M. Kessner and M.H. Lepper, *Am. J. Epidemiol.*, 85, 45, 1967. With permission.

2.1–163 DISTRIBUTION OF DIAGNOSTIC EVALUATIONS IN HOSPITAL AND COMMUNITY SUBJECTS[a]

Evaluation	Hospital	Community
Hospital acquired		
Definite infection	61 (4)[b]	
Questionable infection	13 (1)	
Asymptomatic carrier	24 (1)	
Indeterminate status		
Definite infection	620 (37)	43 (9)
Questionable infection	246 (15)	
Asymptomatic carrier	714 (42)	65 (13)
Community acquired		
Definite infection	1	26 (5)
Questionable infection		
Asymptomatic carrier		365 (72)
Unknown	8	2
Total number evaluations	1687 (100)	501 (99)
Total number patients	1391	471

[a] These tabulations include single and multiple admissions for each patient and reflect the fact that individual patients were counted once if all culture sites were given the same diagnostic evaluation; patients with multiple types at different sites were tallied once in each category to which they were assigned.
[b] Percentages of total number in parentheses.

Source: D.M. Kessner and M.H. Lepper, *Am. J. Epidemiol.*, 85, 45, 1967. With permission.

2.1–164 SITE OF CULTURE[a]

Site	Hospital	Community
Bone and soft tissue	20 (1.1)[b]	1 (0.2)
Skin	10 (0.5)	65 (13.0)
Clean wound	61 (3.3)	105 (21.0)
Contaminated wound	123 (6.6)	
Abscess	26 (1.4)	4 (0.8)
Central nervous system	7 (0.4)	
Respiratory tract	441 (23.7)	48 (9.6)
Gastrointestinal tract	42 (2.3)	150 (30.1)
Biliary tract	16 (0.9)	
Genital tract	161 (8.6)	56 (11.2)
Urinary tract	764 (41.0)	67 (13.4)
Eye	14 (0.8)	1 (0.3)
Ear	11 (0.6)	
Blood	114 (6.1)	
Unknown	52 (2.8)	2 (0.4)
Total	1862 (100.1)	499 (99.9)

[a] Multiple cultures at one site in a respondent were counted only once.
[b] Percentages of the total number in parentheses.

Source: D.M. Kessner and M.H. Lepper, *Am. J. Epidemiol.*, 85, 45, 1967. With permission.

2.1–165   SITES OF "HOSPITAL" AND "COMMUNITY" INFECTIONS

Site	Hospital infections	Community infections
Urinary tract	45 (72.6)[a]	22 (100)
Clean wound	9 (14.5)	
Blood	2 (3.2)	
Ear	1 (1.6)	
Eye	1 (1.6)	
Contaminated wound	1 (1.6)	
Respiratory tract	1 (1.6)	
Gastrointestinal tract	1 (1.6)	
Genital tract	1 (1.6)	
Total	62[b] (99.9)	22 (100)
Total number of patients	61	22

[a] Percentages of total in parentheses.
[b] One patient had a hospital-acquired infection at two sites.

Source: D.M. Kessner and M.H. Lepper, *Am. J. Epidemiol.*, 85, 45, 1967. With permission.

2.1–166   ESTIMATED MINIMUM ATTACK RATES FOR "HOSPITAL INFECTIONS" BY SERVICE TO WHICH PATIENT WAS ADMITTED

Service	Number of infections	Rate per 1000 patients	Rate per 10,000 days patient care
Neurosurgery	10 (16.4)[a]	23.1	11.0
Urology	8 (13.3)	20.6	13.1
Gynecology	11 (18.0)	15.0	13.5
Medicine	10 (16.4)	8.7	4.6
General surgery	10 (16.4)	5.5	4.4
Orthopedics	6 (9.8)	5.3	2.4
Plastic surgery	1 (1.6)	4.4	3.5
Pediatrics	2 (3.3)	2.1	1.3
Obstetrics	3 (4.9)	1.2	3.4
Total	61 (100)	6.4[b]	4.8[b]

[a] Percentages of total numbers of infections in parentheses.
[b] Mean rate.

Source: D.M. Kessner and M.H. Lepper, *Am. J. Epidemiol.*, 85, 45, 1967. With permission.

2.1–167   ESTIMATED MINIMUM ATTACK RATES FOR "INDETERMINATE INFECTIONS" (HOSPITAL SOURCE) BY SERVICE TO WHICH PATIENT WAS ADMITTED

Service	Number of infections	Rate per 1000 patients	Rate per 10,000 days patient care
Urology	93 (15.6)[a]	239.7	152.1
Medicine	137 (22.9)	118.7	63.0
Gynecology	62 (10.4)	84.4	76.2
Pediatrics	79 (13.2)	81.4	50.6
Neurosurgery	35 (5.9)	80.7	38.4
General surgery	103 (17.2)	57.0	43.8
Dermatology	2 (0.3)	40.0	15.2
Orthopedics	38 (6.4)	33.4	15.3
Obstetrics	39 (6.5)	16.4	44.2
Psychiatry	1 (0.2)	9.9	1.3
Ear, nose, throat	7 (1.2)	9.0	12.0
Plastic surgery	2 (0.3)	8.7	7.1
Total	598 (100.1)	58.0[b]	44.2[b]

[a] Percentages of total number in parentheses.
[b] Mean rate.

Source: D.M. Kessner and M.H. Lepper, *Am. J. Epidemiol.,* 85, 45, 1967. With permission.

2.1–168   DISTRIBUTION OF GRAM-NEGATIVE ORGANISMS FROM HOSPITAL POPULATION BY CLINICAL EVALUATION

	Hospital origin			Indeterminate origin					
	Carrier	Probable infection	Definite infection	Carrier	Probable infection	Definite infection	No evaluation made	Total	Percent
Escherichia coli	8	2	25	387	131	354	12	919	31.5
Klebsiella	6	1	28	197	76	190	5	503	17.3
Pseudomonas	4	4	24	162	66	196	8	464	15.9
Proteus mirabilis	6	4	20	188	75	88	11	392	13.4
Aerobacter	5	1	10	110	39	71	1	237	8.1
Herella	1		6	39	5	30		81	2.8
Paracolon			2	25	12	31	1	71	2.4
Citrobacter (E. freundi)			4	21	10	30		65	2.2
Serratia	1		1	30	7	12		51	1.7
Proteus morganii				15	9	11		35	1.2
P. rettgeri				7		25		32	1.1
Alkaligenes				12	1	5	1	19	0.7
Proteus vulgaris				7	1	7		15	0.5
Providence			1	4	1	7	1	14	0.5
Mima				3	3	2		8	0.3
Salmonella						3		4	0.1
Shigella					1	2		2	0.1
Arizona						1		2	0.1
Aeromonas						1		1	0.1
Total number organisms[a]	31 (1.1)[b]	12 (0.4)	121 (4.2)	1207 (41.4)	438 (15.0)	1066 (36.6)	40 (1.4)	2915 (100.1)	100
Total number patients[a]	24 (1.4)	10 (0.6)	61 (3.6)	710 (42.2)	245 (14.5)	619 (36.8)	15 (0.9)	1684 (100.0)	

[a]These figures reflect the fact that individual patients had different evaluations at varying culture sites.
[b]Percents of total in parentheses.

Source: D.M. Kessner and M. H. Lepper, *Am. J. Epidemiol.*, 85, 45, 1967. With permission.

2.1–169 DISTRIBUTION OF ORGANISMS BY SITE OF CULTURE IN HOSPITAL- AND COMMUNITY-ACQUIRED DEFINITE INFECTIONS AND CARRIER STATES

	Hospital					Community					Grand total
	Definite infection		Carrier		Total	Definite infection		Carrier		Total	
	Number	%	Number	%	Total	Number	%	Number	%	Total	
Stool											
Escherichia coli			28	71.8	28			125	71.5	125	153
Proteus mirabilis			3	7.7	3			5	2.9	5	8
Pseudomonas			1	2.6	1			4	2.3	4	5
Klebsiella			1	2.6	1			14	8.0	14	15
Aerobacter			6	15.4	6			16	9.1	16	22
Other			0		0			11	6.3	11	11
Total			39	100.1	39			175	100.1	175	214
Urine											
E. coli	24	24.5	6	33.3	30	15	30.6	20	74	35	65
P. mirabilis	19	19.4	2	11.1	21	8	16.4	1	3.7	9	30
Pseudomonas	17	17.3	1	5.6	18	14	28.6	2	7.4	16	34
Klebsiella	19	19.4	2	11.1	21	1	2.0	0		1	22
Aerobacter	8	8.2	5	27.8	13	3	6.1	2	7.4	5	18
Other	11	11.2	2	11.1	13	8	16.3	2	7.4	10	23
Total	98	100	18	100	116	49	100.0	27	99.9	76	192
Nongenitourinary-gastrointestinal											
E. coli	1	4.0	41	30.1	42			46	34.1	46	88
P. mirabilis	2	8.0	30	22.0	32			22	16.3	22	54
Pseudomonas	7	28.0	8	5.9	15			34	25.2	34	49
Klebsiella	8	32.0	16	11.8	24			1	0.7	1	25
Aerobacter	3	12.0	23	16.9	26			21	15.5	21	47
Other	4	16.0	18	13.2	22			11	8.2	11	33
Total	25	100	136	100	161			135	100.0	135	296

Source: D.M. Kessner and M.H. Lepper, *Am. J. Epidemiol.*, 85, 45, 1967. With permission.

2.1–170   THE EFFECT OF ANTIBIOTIC TREATMENT ON THE DISTRIBUTION OF GRAM-NEGATIVE BACILLI IN "INDETERMINATE INFECTIONS" AND "CARRIER" STATES OF HOSPITAL ORIGIN

| | Indeterminate infection | | | | | Carrier | | | | |
| | Antibiotic therapy | | No antibiotic therapy | | | Antibiotic therapy | | No antibiotic therapy | | |
	Number	%	Number	%	Total	Number	%	Number	%	Total
Urinary										
Escherichia coli	247	29.4	65	34.9	312	69	45.4	58	41.7	127
Proteus mirabilis	181	21.5	30	16.1	211	25	16.4	33	23.7	58
Pseudomonas	138	16.4	11	5.9	149	10	6.6	9	6.5	19
Klebsiella	127	15.1	29	15.6	156	24	15.8	19	13.7	43
Aerobacter	48	5.7	12	6.5	60	8	5.3	8	5.8	16
Other	100	11.9	39	21.0	139	16	10.5	12	8.6	28
Total	841	100	186	100	1027	152	100	139	100	291
Nongenitourinary-gastrointestinal										
E. coli	224	28.2	27	23.5	251	120	26.3	55	27.9	175
P. mirabilis	113	14.2	23	20.0	136	80	17.5	26	13.2	106
Pseudomonas	146	18.4	16	14.0	162	85	18.6	31	15.7	116
Klebsiella	123	15.5	22	19.1	145	82	17.9	34	17.3	116
Aerobacter	83	10.4	13	11.3	96	37	8.1	20	10.2	57
Other	106	13.3	14	12.2	120	53	11.6	31	15.7	84
Total	795	100	115	100.1	910	457	100	197	100	654

Source: D.M. Kessner and M.H. Lepper, *Am. J. Epidemiol.*, 85, 45, 1967. With permission.

2.1–171 INCIDENCE OF INFECTIOUS MONONUCLEOSIS AT 19 COLLEGES AND UNIVERSITIES DURING THE ACADEMIC YEAR 1969–1970

School	Undergraduate population	Number of cases	Rate per 100,000 per school year
Alaska	1,502	9	599
Bryn Mawr	743	6	808
California	13,848	255	1841
Clemson	4,503	40	888
Emory	2,321	50	2154
Harvard	6,012	107	1780
Hawaii	13,613	15	110
Illinois	23,064	303	1314
Indiana	21,752	386	1775
Minnesota	33,277	248	745
Nebraska	16,463	209	1270
Oklahoma State	14,951	154	1030
Oregon	11,037	140	1268
Princeton	3,400	76	2235
Purdue	19,016	251	1320
San Diego	17,858	126	706
Texas	25,705	217	844
Washington	22,908	185	808
Yale	4,490	74	1722
Total	256,463	2851	1112

Source: A.L. Brodsky and C.W. Heath, *Am. J. Epidemiol.*, 96, 87, 1972. With permission.

2.1–172   INCIDENCE OF INFECTIOUS MONONUCLEOSIS AT SEVEN COLLEGES AND UNIVERSITIES BY RACE DURING THE ACADEMIC YEAR 1969–1970

School	Number of cases			Undergraduate population		Rates per 100,000 per school year	
	White	Nonwhite	Unknown	White	Nonwhite	White	Nonwhite
Alaska	9			1,288	214	699	
Bryn Mawr	5	1		689	54	726	1852
Emory	48		2	2,275	46	2110	
Harvard	103	3	1	5,642	370	1826	811
Oklahoma State	151	3		14,660	291	1030	1031
Purdue	248	3		14,612	4404	1697	68
Yale	67		7	4,295	195	1257	
Total	631	10	10	43,461	5574	1452	179

Source: A.L. Brodsky and C.W. Heath, *Am. J. Epidemiol.*, 96, 87, 1972. With permission.

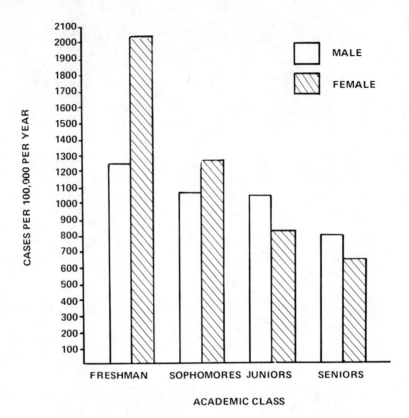

2.1–173   Infectious mononucleosis attack rates, by sex and academic class, at 14 colleges and universities, academic year 1969–1970.

Source: A.L. Brodsky and C.W. Heath, *Am. J. Epidemiol.*, 96, 87, 1972. With permission.

2.1–174   CONTACT HISTORY FOR PATIENTS WITH INFECTIOUS MONONUCLEOSIS

School	School year	Total cases	Contact with known case, %			Weeks prior to onset, %					Relationship of contact, %				
			Yes	No	Unknown	≤1	2–4	5–6	≥7	Unknown	Roommate	Friend	Spouse	Date	Other or unknown
Georgia Tech	1968–69	70	4	96		14	45	11	23	13	16	1		26	3
Harvard	1969–70	107	18	47	36	11	47	5	21	16	27	58		53	5
	1968–69	137	11	88	1		53	13	7	27		27			13
Nebraska	1969–70	209	26	54	20	9	52	7	28	4	7	42	2	30	19
Oregon	1968–69	149	13	85	3	21	21	26	11	21	11	42	5	31	16
Purdue	1969–70	251	16	75	9	15	25	13	43	5	8	60		18	15
	1968–69	288	10	89	1	24	34	10	7	24	21	4	14	10	48
Total		1211	15	76	9	13	39	11	23	12	14	42	3	28	20

Source: A.L. Brodsky and C.W. Heath, *Am. J. Epidemiol.*, 96, 87, 1972. With permission.

2.1–175   OCCURRENCE OF ACUTE RESPIRATORY ILLNESS BY COTTAGE IN A BOYS' HOME, JUNE 21, 1964–DECEMBER 20, 1964

Cottage	Number of boys	Average age	Clinical diagnosis					Total illnesses	Illness rate per 100 boys per 6-month period
			Pneumonitis	Bronchitis	Pharyngitis	Coryza	Others		
A	16	9	2	5	1	0	2	10	63
B	16	11.5	3	2	4	0	0	9	56
C	16	13.5	2	7	2	0	0	11	69
D	14	14.5	1	4	4	1	0	10	71
E	16	16	0	1	1	2	0	4	25
F	12	16	2	2	0	1	0	5	42
Total	90	13.3	10	21	12	4	2	49	54
Percent of totals			20	43	25	8	4	100	

Source: G.S. Saliba et al., *Am. J. Epidemiol.*, 86, 408, 1976. With permission.

2.1–176   OCCURRENCE OF ACUTE RESPIRATORY ILLNESS IN A BOYS' HOME, JUNE 21, 1964–DECEMBER 20, 1964

Number of cases – month ending

Diseases	July 21	Aug. 20	Sept. 20	Oct. 20	Nov. 20	Dec. 20
Coryza			1	2		1
Pharyngitis	2	1		5	2	2
Bronchitis		1	6	10	1	3
Pneumonitis			2	8		
Others[a]		2				

[a] One case of fever of unknown origin and one case of acute conjunctivitis and gastroenteritis.

Source: G.S. Saliba et al., *Am. J. Epidemiol.*, 86, 408, 1967. With permission.

2.1–177   DISTRIBUTION OF *MYCOPLASMA PNEUMONIAE* INFECTION BY COTTAGE IN A BOYS' HOME, SEPTEMBER 1963–MAY 1965

Cottage	Number of boys	Age group in years	M. pneumoniae infections Number	M. pneumoniae infections Percent
A	16	5–10	11	69
B	16	10–13	14	88
C	16	13–14	4	25
D	14	14–15	6	43
E	16	15–18	1	6
F	12	15–18	2	17
Total	90	5–18	38	42

Source: G.S. Saliba et al., *Am. J. Epidemiol.*, 86, 408, 1967. With permission.

**2.1–178   *MYCOPLASMA PNEUMONIAE* INFECTION AS
RELATED TO AGE AND PREEXISTING GROWTH
INHIBITION ANTIBODY TITER IN 82 CHILDREN IN
A BOYS' HOME, SEPTEMBER 1963–MAY 1965**

Age in years	Number of boys	Preexisting growth inhibition antibody titer				Totals rises	
		0–2		Greater than 2			
		Number	Rises	Number	Rises	Number	Percent
5–12	25	18	16	7	3	19	76
13–14	25	11	5	14	1	6	24
15–19	32	18	5	14	1	6	19
Total	82	47	26	35	5	31	38

Source: G.S. Saliba et al., *Am. J. Epidemiol.*, 86, 408, 1967. With permission.

**2.1–179   COMPARISON OF GROWTH
INHIBITION AND
COMPLEMENT-FIXATION TESTS
IN 82 CHILDREN FROM WHOM
SERA WERE COLLECTED AT
4-MONTH INTERVALS,
SEPTEMBER 1963–MAY 1965**

Complement-fixation test	Growth-inhibition test		Total
	Positive	Negative	
Positive	13	2	15
Negative	18	49	67
Total	31	51	82

Source: G.S. Saliba et al., *Am. J. Epidemiol.*, 86, 408, 1967. With permission.

2.1–180   COMPARATIVE RESULTS OF COMPLEMENT FIXATION AND GROWTH INHIBITION ANTIBODY TITERS IN SIX SUBJECTS FOLLOWED AN AVERAGE OF 30 WEEKS POSTINFECTION

	Complement fixation[a]	Growth inhibition
Geometric mean antibody titer at		
a. Preinfection	2.3	0.8
b. Approximately 4 weeks postinfection	5.8	8.3
c. Approximately 17 weeks postinfection	3.7	6.0
d. Approximately 30 weeks postinfection	2.5	4.8
Change in geometric mean titers		
e. Mean rise (b-a)	3.5	7.5
f. 13-week fall (b-c)	2.1	2.3
g. 26-week fall (b-d)	3.3	3.5
Percent change in geometric mean titer		
h. Mean 13-week fall/mean rise (f/e)	60%	31%
i. Mean 26-week fall/mean rise (g/e)	94%	47%

[a] Titers expressed as reciprocal of $\log_2$, e.g., titer of less than 1:8 designated as 2.

Source: G.S. Saliba et al., *Am. J. Epidemiol.*, 86, 408, 1967. With permission.

2.1–181   EFFECT OF PRETREATMENT ON THE ISOLATION OF VACCINE ORGANISMS BY RECTAL SWAB FROM ADULT MALES IN PHASE 3 VACCINATED WITH 40 × 10⁹ STREPTOMYCIN-DEPENDENT *SHIGELLA FLEXNERI* 2a

Pretreatment	Isolation of organisms by rectal swab
Nothing	6/20
2 g NaHCO₃	17/20
8 oz milk with 0.8 g NaHCO₃	19/20

Source: M.M. Levine et al., *Am. J. Epidemiol.*, 96, 40, 1972. With permission.

2.1–182 CLINICAL RESPONSE OF INSTITUTIONALIZED CHILDREN TO ORAL ATTENUATED *SHIGELLA FLEXNERI* 2a VACCINES AND PLACEBO IN PHASE 4 IN COMPARISON WITH HEALTHY YUGOSLAV CHILDREN

Vaccine	Number of children	Vomiting	Diarrhea	Fever	Cramps	Isolation of vaccine per rectal swabs
Streptomycin-dependent	95	6 (6.3%)	2 (2.1%)	10 (10.5%)	3 (3.2%)	92 (97%)
Mutant-hybrid	102	8 (7.8%)	3 (2.9%)	15 (14.7%)	1 (1%)	54 (53%)
Placebo	95	5 (5.3%)	1 (1.1%)	8 (8.4%)	1 (1.1%)	
Streptomycin-dependent: Yugoslav children[1]	1307	48 (3.7%)	38 (2.9%)			

Source: M.M. Levine et al., *Am. J. Epidemiol.*, 96, 40, 1972. With permission.

REFERENCE

1. D.M. Mel et al., unpublished data.

2.1–183 REACTION RATES ENCOUNTERED IN POPULATION GROUPS 1 AND 2 IN COMPARISON WITH GROUP 3

Vaccine	Number of children	% vomiting	% diarrhea	% fever
	Groups 1 and 2 combined	**Groups 1 and 2 combined**		
SmD[a]	50	2	2	6
MH[b]	53	1.9	3.8	7.6
Placebo	35	0	0	8.9
	Group 3	**Group 3**		
SmD	45	11.1	2.2	15.5
MH	49	14.3	2	22.5
Placebo	60	8.3	1.7	8.3
	Yugoslav children[1]	**Yugoslav children[1]**		
SmD	1307	3.7	2.9	

[a] SmD: streptomycin dependent.
[b] MH: mutant hybrid.

Source: M.M. Levine et al., *Am. J. Epidemiol.*, 96, 40, 1972. With permission.

REFERENCE

1. D.M. Mel et al., unpublished data.

2.1–184 VACCINE REACTION RATES ACCORDING TO DOSE

Adverse reaction	First dose			Second dose			Third dose			Fourth dose			Total doses		
	No. of reactions	% reactions	No. of doses	No. of reactions	% reactions	No. of doses	No. of reactions	% reactions	No. of doses	No. of reactions	% reactions	No. of doses	No. of reactions	% reactions	No. of doses
Streptomycin-dependent Vaccine															
Vomiting	4	4.2	95	1	1.1	95	1	1.1	95	0		95	6	1.6	380
Diarrhea	1	1.1	95	1	1.1	95	0		95	0		95	2	0.5	380
Fever	5	5.3	95	5	5.3	95	3	3.2	95	1	1.1	95	14	3.7	380
Mutant-hybrid Strain Vaccine															
Vomiting	5	5.4	92	3	3.3	92	2	2.2	91	0		92	10	2.7	367
Diarrhea	2	2.2	92	1	1.1	92	0		91	0		92	3	0.8	367
Fever	5	5.4	92	7	7.6	92	7	7.7	91	3	3.3	92	22	6.0	367

Source: M.M. Levine et al., *Am. J. Epidemiol.*, 96, 40, 1972. With permission.

2.1–185   COMPARISON OF CENSUS AND
BACTERIOLOGICALLY CONFIRMED CASES OF SHIGELLOSIS
AT TWO STATE INSTITUTIONS

Institution	Average census, 1967–1968	Number of children with shigellosis, 1967–1968
Willowbrook State School	5310	541
Rosewood State Hospital	2685	152

Source: H.L. Dupont et. al., *Am. J. Epidemiol.*, 92, 172, 1970. With permission.

2.1–186   SHIGELLOSIS IN CHILDREN WITHIN THE
FIRST YEAR FOLLOWING ADMISSION

Institution	Total admissions[a] in 1967	Number with shigellosis	Attack rate
Willowbrook State School	237	75	32%
Rosewood State Hospital	324	33	10%

[a] Includes only those children remaining at institution for at least 12 months.

Source: H.L. Dupont et al., *Am. J. Epidemiol.*, 92, 172, 1970. With permission.

2.1–187   SHIGELLOSIS IN CHILDREN WITHIN THE
FIRST YEAR FOLLOWING ADMISSION TO
SELECTED COTTAGES

Institution	Number of cottages where shigellosis occurred	New admissions to these cottages in 1967	Number with shigellosis	Attack rate
Willowbrook State School	7	203 (86%)[a]	75	37%
Rosewood State Hospital	10	139 (43%)[a]	33	24%

[a] Percent of total 1967 admissions.

Source: H.L. Dupont et al., *Am. J. Epidemiol.*, 92, 172, 1970. With permission.

2.1–188 SPECIFIC CLINICAL FEATURES IN CHILDREN WITH SHIGELLOSIS AT ROSEWOOD STATE HOSPITAL, APRIL 1966–MARCH 1968

Serogroup[a]	Total number	Fever		Diarrhea		Stool with blood or mucus	
			%		%		%
B	93	46	(50)	82	(88)	25	(27)
D	69	44	(64)	62	(90)	25	(36)
B + D	5	3	(60)	5	(100)	0	
A	1	1		1		0	
Total	168[b]	94	(56)	150	(89)	50	(30)

[a] B: *Shigella flexneri;* D: *S. sonnei;* A: dysentery.
[b] Clinical features unknown in 16 patients.

Source: H.L. Dupont et al., *Am. J. Epidemiol.,* 92, 172, 1970. With permission.

2.1–189 A SURVEY OF SHIGELLA PREVALENCE AMONG INSTITUTIONALIZED CHILDREN IN SELECTED COTTAGES

	Children surveyed at random		Children with diarrhea cultured[a]		
	Number	Positive		Number	Positive
			%		%
Willowbrook State School (October 1967)	156	11	(7)	86	8 (9)
Rosewood State Hospital (April 1968)	198	11	(6)	21	5 (24)

[a] At Willowbrook all "ill" children were included.

Source: H.L. Dupont et al., *Am. J. Epidemiol.,* 92, 172, 1970. With permission.

2.1–190   ASSOCIATION OF VARIOUS FACTORS WITH THE MEAN NASAL-CARRIAGE RATE

Factor	Reference group	Relative carriage rate[b] All strains	T strains
Age			
Under 40	40–60	1.1	1.2
Over 60	40–60	1.1	2.5
Sex, female	Males	1.0	0.9
Diagnosis[a]			
A	All patients	1.1–1.8	1.2–3.0
B	All patients	0.9–1.2	0.8–1.3
C	All patients	0.2	0.8
Received penicillin	Received no antibiotics	0.9	2.1
Received antibiotic other than penicillin	Received no antibiotics	0.9	2.5
In single room	In four-bed bay	1.2	1.4
Week before death	All patients	1.2	2.3

Note: The figures are the ratio of the carriage rates in the "factor" groups to those in the reference groups.

[a] Diagnostic set A includes skin conditions (41), diabetes (48), disorders of the urogenital system (108), and diseases of the cerebrovascular system (535). Set B includes respiratory diseases (510), diseases of the stomach and duodenum (124), neoplasm (286), and diseases of the heart and circulatory system (885), together with conditions not specifically classified (1181). Set C was rheumatoid arthritis (132). The numbers in parentheses give the patient-weeks experience for each group. The diagnoses are here listed in order of decreasing carriage rate of T (tetracycline-resistant) strains.
[b] The range of values given for the relative carriage rate is the spread for the several diagnoses included in the particular set.

Source: O.M. Lidwell et al., *J. Hyg.*, 69, 113, 1971. With permission of Cambridge University Press.

2.1–191   ASSOCIATION OF VARIOUS FACTORS WITH THE NASAL-ACQUISITION RATE

Factor	Reference group	Relative acquisition rate[a] All strains	T strains[b]
Age, over 60	Under 60	1.4	1.6
Received penicillin	Received no antibiotics	1.0 (0.6)	2.6
Received other antibiotic	Received no antibiotics	1.3 (0.5)	3.9
In single room	In four-bed bay	1.1	1.3
Confined to bed	All patients	0.9	–
Fully ambulant	All patients	1.0	–
Week before death	All patients	0.8	1.0

Note: The figures are the ratio of the nasal acquisition rates in the "factor" groups to those in the reference groups.

[a] The figures in parentheses are the relative acquisition rates for strains sensitive to all antibiotics or resistant to penicillin only.
[b] T strains: tetracycline resistant.

Source: O.M. Lidwell et al., *J. Hyg.*, 69, 113, 1971. With permission of Cambridge University Press.

2.1−192   FACTORS INFLUENCING THE RATE OF NASAL ACQUISITION OF STAPHYLOCOCCUS AUREUS: COEFFICIENTS OF THE MULTIREGRESSION ANALYSIS

| | | Probable sources[a] | | | | | |
| | | All known sources | | Other patient carriers | | No known source | |
Factor	Proportion, %	S + P strains	T strains	S + P strains	T strains	S + P strains	T strains
Age							
Under 60	57	25	42	6	3	(−1)	6
Over 60	43	11	32	(4)	(2)	(−2)	(4)
Diagnostic group							
Heart and circulation	25	−	−	(−6)	−	−	(−5)
Respiratory	15	−	−	−	12	−	−
Cerebrovascular	16	(12)	−	−	−	17	−
Neoplasm	8	34	−	−	−	−	−
Others (except skin conditions and unclassified)	12	−	(−14)	(−7)	−	−	−
In single room	13	−	−	12	−	(11)	−
Noncarrier	68	−	−	−	−	−	−
Received no antibiotics	43	(10)	(−12)	10	−	19	−
Received antibiotics (other than penicillin)	15	−	20	−	5	−	12
Carrier	32	−	−	−	−	−	−
Received no antibiotics	21	−	−27	−	−	−	−
Received antibiotics (other than penicillin)	7	(−18)	−	−	−	−	13
First week in ward	35	−	−14	−	−	29	−
Mean acquisition rate	−	27	23	9	5	21	7
Multiple correlation coefficient	−	0.18	0.19	0.12	0.10	0.19	0.11
Number of acquisitions	[257]	88	77	28	17	70	22

Note: The figures in the table, apart from the last three rows and the first two columns, are the coefficients of the linear regression equation giving the nasal acquisition rate per 1000 patient-weeks. The values in bold figures exceed three times their standard errors while the values in parentheses lie between one and two times their standard error. A dash indicates that the regression analysis terminated without involving the factor concerned, none of those omitted attaining a regression coefficient equal to or greater than its standard error if included in the analysis.

[a] S + P strains are those sensitive to all antibiotics or resistant to penicillin only. T strains are those resistant to tetracycline. The strains were only tested against penicillin and tetracycline.

Source: O.M. Lidwell et al., *J. Hyg.*, 69, 113, 1971. With permission of Cambridge University Press.

2.1–193   COLONIZATION OF WOUNDS WITH
STAPHYLOCOCCUS AUREUS, 1965–1969

Site	Incidence of colonization with multiple-resistant strains	Incidence of colonization with sensitive or penicillin-resistant-only strains
Open ward	16/722 (2.2%)	20/722 (2.8%)
Recirculation cubicles	2/61 (3.3%)	1/61 (1.6%)
Plenum cubicles	0/78	3/78 (3.8%)
Window-ventilated cubicle	1/26 (3.8%)	0/26

Source: G.A.J. Ayliffe et al., *J. Hyg.,* 69, 511, 1971. With permission of Cambridge University Press.

2.1–194   ACQUISITION OF MULTIPLE-RESISTANT
STAPHYLOCOCCUS AUREUS IN NOSES, 1965–69

Site	Total patients	Number of acquisitions
Open ward	1325	51 (3.8%)
Recirculation cubicles	134	8 (5.9%)
Plenum-ventilated cubicles	143	1 (0.7%)
Window-ventilated side-ward	72	3 (4.2%)

Significant differences: comparison of
  Plenum with open ward $t = 1.94$ $P \cong 0.05$
  Plenum with recirculation cubicle $t = 2.47$ $P < 0.05$

Source: G.A.J. Ayliffe et al., *J. Hyg.,* 69, 511, 1971. With permission of Cambridge University Press.

2.1–195   CLEAN PROCEDURES IN AN OPERATING THEATRE DURING 10 MONTHS WITH THE ORGANISMS ISOLATED FROM SUBSEQUENT WOUND INFECTIONS

	January	February	March	April	May	June	July	August	September	October
Total procedures	52	50	40	50	36	35	30	20	30	41
Wound swabs sent to laboratory	15	17	10	16	14	19	8	7	10	10
No bacteria found on culture	4	1	1	4	2	5	4	2	4	2
Gram-positive bacteria										
Penicillin resistant *Staphylococcus aureus*	4	7	3	2	1	3	1	3	0	2
Penicillin sensitive *S. aureus*	0	1	0	2	3	3	0	1	3	3
Other Gram-positives, streptococci, and skin commensals	2	2	1	3	3	5	1	0	1	1
Total Gram-positives and percentage clean procedures	6 (12%)	10 (20%)	4 (10%)	7 (14%)	7 (19%)	11 (31%)	2 (7%)	4 (20%)	4 (13%)	6 (15%)
Gram-negative bacteria										
Pseudomonas aeruginosa	0	1[a]	2[a,b,c]	3[a,cc]	0	0	1	0	0	0
Klebsiella species	0	1	0	2	1[a]	1	0	1	0	0
Common gut species, *Escherichia coli*, *Proteus* species, etc.	5	5	5	1	5	4[aa]	1	0	2	3
Total Gram-negatives and percentage clean procedures	5 (10%)	7 (14%)	7 (18%)	6 (12%)	6 (17%)	5 (14%)	2 (7%)	1 (5%)	2 (7%)	2 (5%)

[a] Staphylococcus also present in one swab.
[aa] Multiple strains of the type indicated were found in two of the swabs taken.
[b] Hemolytic streptococcus also present in one swab.
[c] Endemic strain of *P. aeruginosa* also present in one swab.
[cc] Multiple strains of the type indicated were found in two of the swabs taken.

Source: M.E.M. Thomas et al., *J. Hyg.*, 70, 63, 1972. With permission of Cambridge University Press.

2.1–196 WOUND SWABS EXAMINED BEFORE AND AFTER IMPLEMENTATION OF HYGIENIC MEASURES IN OPERATING THEATRE ENVIRONMENT

	January to June	July to October
Clean incisional operations	263	121
Wound swabs sent to laboratory	91	35
Gram-positive organisms found	45 (17)	16 (13)
Staphylococcus aureus found	29 (11)	13 (11)
Gram-negative organisms found	36 (14)	7 (6)
Pseudomonas aeruginosa found	6 (2)	1 (1)

Note: Figures in parentheses are percentages of the total operations in each group.

Source: M.E.M. Thomas et al., *J. Hyg.*, 70, 63, 1972. With permission of Cambridge University Press.

2.1–197 PYOCINE TYPING 1967: RANDOM SCATTER OF PYOCINE TYPES RESULTING FROM CROSS-INFECTION COUNTER MEASURES

Patient	Admitted	Pyocine types of *Pseudomonas aeruginosa*						
G.E.	Jan. 67	—	—	—	—	—	Q22	—
H.U.	Jan. 67	—	—	—	—	—	—	O25
W.E.	Jan. 67	—	—	—	—	—	—	O25
B.R.	Feb. 67	—	—	—	—	—	—	O25
M.I.	Feb. 67	—	—	—	B1B	—	—	—
F.I.	Feb. 67	—	—	—	—	—	Q22	—
C.L.	Feb. 67	—	—	—	—	NT	—	—
H.O.	Apr. 67	—	—	—	B1B	—	—	—
S.I.	May 67	—	P30	—	—	—	—	—
H.U.	May 67	—	—	—	—	NT	—	—
H.U.	June 67	B1A	—	—	—	—	—	—
A.P.	July 67	—	—	—	—	NT	—	—
C.O.	Aug. 67	—	—	—	—	NT	—	—
W.I.	Aug. 67	B1A	—	—	—	—	—	—
A.C.	Sept. 67	—	—	F	—	—	—	—
A.P.	Oct. 67	—	—	—	—	NT	—	—
T.H.	Nov. 67	—	—	—	—	NT	—	—
H.A.	Nov. 67	B1A	—	—	—	—	—	—
W.A.	Dec. 67	—	—	—	B1B	—	—	—
O.L.	Dec. 67	—	—	—	B1B	—	—	—

Source: F. Dexter, *J. Hyg.*, 69, 179, 1971. With permission of Cambridge University Press.

2.1–198   TOTAL NUMBER OF ISOLATES OF EACH PYOCINE TYPE OF *PSEUDOMONAS AERUGINOSA* RECOVERED FROM THE ENVIRONMENT

Number of isolates recovered from
areas samples

Pyocine type	Surfaces	Water reservoirs	Inhalation therapy equipment	Solutions and medications	Personnel
A	0	1	0	0	0
B	0	0	0	0	1
C	0	2	0	0	0
D	2	2	0	0	0
E[a]	12	8	14	0	2
Untypable	2	3	0	0	0

[a] Pyocine type isolated from the infected patient.

Source: D. Drollette et al., *Hospitals*, 45, 91, 1971, With permission.

2.1–199   INFECTION RATES (*STAPHYLOCOCCUS AUREUS*) OF EIGHT DEPARTMENTS 1960–62

Annual infection rates, %

Departments	1960	1961	1962
Long-term diseases	15.78	5.96	3.80
Internal chest diseases	11.71	8.67	2.15
Internal medicine	1.39	0.62	0.98
Obstetrics and gynecology	0.48	0.26	0.29
General surgery	1.41	1.54	1.11
Chest surgery	10.49	10.76	4.07
Plastic surgery	7.04	10.12	6.53
Orthopedics	4.69	4.08	3.99
All eight departments	2.27	1.76	1.46

Source: C. Ericson and I. Juhlin, *J. Hyg.*, 63, 25, 1965. With permission of Cambridge University Press.

2.1−200   RELATIVE INFECTION NUMBERS OF EIGHT DEPARTMENTS DURING TWO CONSECUTIVE 18-MONTH PERIODS, AND DIFFERENCES BETWEEN THE RESULTS FROM THE TWO PERIODS

Departments	Classes of *Staphylococcus aureus* strains	Relative infection numbers		
		1. i. 60− 30. vi. 61	1. vii. 61− 31. xii. 62	Difference, %
Long-term diseases	Virginal	63.6	23.5	−63.1
	Multiple-R + "80/81"	60.6	23.4	−61.4
	All	124.2	46.9	−62.2
Internal chest diseases	Virginal	16.6	9.9	−40.4
	Multiple-R + "80/81"	73.1	22.0	−69.9
	All	89.7	31.9	−64.4
Internal medicine	Virginal	24.1	40.8	+69.3
	Multiple-R + "80/81"	52.8	30.7	−41.9
	All	76.9	71.5	−7.0
Obstetrics and gynecology	Virginal	36.3	24.5	−32.5
	Multiple-R + "80/81"	19.7	5.9	−70.1
	All	56.0	30.4	−45.7
General surgery	Virginal	86.4	53.5	−38.3
	Multiple-R + "80/81"	75.9	58.5	−22.9
	All	162.3	111.8	−31.1
Chest surgery	Virginal	31.2	16.3	−47.8
	Multiple-R + "80/81"	54.2	34.8	−35.8
	All	85.4	51.1	−40.2
Plastic surgery	Virginal	48.8	64.8	+32.8
	Multiple-R + "80/81"	77.8	43.4	−43.9
	All	126.2	108.2	−14.3
Orthopedics	Virginal	55.2	71.6	+29.7
	Multiple-R + "80/81"	73.8	50.5	−31.6
	All	129.0	122.1	−5.3

Source: C. Ericson and I. Juhlin, *J. Hyg.*, 63, 25, 1965. With permission of Cambridge University Press.

2.1−201   RELATIVE INFECTION NUMBERS OF EIGHT DEPARTMENTS TAKEN TOGETHER DURING TWO CONSECUTIVE 18-MONTH PERIODS, AND DIFFERENCES BETWEEN THE RESULTS FROM THE TWO PERIODS

Classes of *Staphylococcus aureus* strains	Relative infection numbers				Difference, %
	1. i. 60−30. vi. 61		1. vii. 61−31. xii. 62		
	Number	%	Number	%	
Virginal	362.2	42.6	304.7	53.1	−15.9
Multiple-R "80/81"	318.7	37.5	212.4	37.0	−33.3
	168.8	19.9	56.8	9.9	−66.4
All	849.7	−	573.9	−	−32.5

Source: C. Ericson and I. Juhlin, *J. Hyg.*, 63, 25, 1965. With permission of Cambridge University Press.

2.1–202   THE DISTRIBUTION OF *STAPHYLOCOCCUS AUREUS* STRAINS FROM VARIOUS SOURCES INTO DIFFERENT CLASSES

Material Collected in Malmo, 1960–62

Origin	Virginal		Multiple-R		"80/81"		Total number
	Number	%	Number	%	Number	%	
Infections in in-patients	509	47.6	390	36.4	171	16.0	1070
Infections in out-patients	169	76.8	3	1.4	48	21.8	220
Carriers among hospital staff	167	81.9	30	14.7	7	3.4	204
Carriers outside the hospital	57	100.0	0	–	0	–	57

Source: C. Ericson and I. Juhlin, *J. Hyg.*, 63, 25, 1965. With permission of Cambridge University Press.

2.1–203   DISTRIBUTION OF DIAGNOSES BETWEEN WARDS

Ward	Distribution, %					Total patients
	Eczema	Psoriasis	Light sensitivity	Ulcer	Other	
A (male)	40	20	2	3	35	569
B (male)	30	27	9	1	33	276
C (female)	30	35	2	4	29	236
D (female)	21	28	8	6	36	467

Source: P.E. Wilson et al., *J. Hyg.*, 69, 125, 1971. With permission of Cambridge University Press.

2.1–204 CARRIAGE OF STAPHYLOCOCCUS AUREUS ON ADMISSION

Staphylococcus[a]	Carriage on admission, %					
	Eczema	Psoriasis	Light sensitivity	Leg ulcer	Other	Total
Nose						
S/PT	38	32	14	17	23	28
R/P	13	13	15	7	10	11
R/T	9	3	4	12	5	5
R/PT	6	3	3	7	5	4
Chest						
S/PT	32	15	11	2	12	16
R/P	11	7	6	0	7	6
R/T	5	2	3	14	3	3
R/PT	5	1	3	9	4	3
Groin						
S/PT	21	12	8	14	12	12
R/P	9	5	4	0	4	5
R/T	5	2	0	5	4	3
R/PT	4	2	3	10	4	3
Total patients	457	389	80	57	509	1492[b]

[a] S/PT, sensitive to penicillin and tetracycline; R/P, resistant to penicillin; R/T, resistant to tetracycline; R/PT, resistant to penicillin and tetracycline.

[b] Because of occasional weekend admissions, some patients were not swabbed on admission or within 48 hr of admission and have been excluded from this analysis.

Source: P.E. Wilson et al., *J. Hyg.*, 69, 125, 1971. With permission of Cambridge University Press.

2.1–205 ACQUISITIONS IN RELATION TO ANTIBIOTIC TREATMENT

Acquisition site	Antibiotic therapy	Acquisitions per 100 patient-weeks occurring after admission		
		1–14 days	15–28 days	29 + days
Nose	Topical	24	15	3
	Systemic	21 [a]	6 [a]	6 [b]
	None	16	8	3
Chest	Topical	17	12	4
	Systemic	15 [a]	8 [a]	4 [b]
	None	10	5	3
Groin	Topical	17	8	5
	Systemic	12 [a]	8 [c]	5 [d]
	None	8	6	2
Total patient-week contributing	Topical	304	256	170
	Systemic	446	358	170
	None	2322	1416	410

[a] Statistical significance: $P < 0.01\%$.
[b] Not significant at the 5% level.
[c] Statistical significance: $2\% < P < 5\%$.
[d] Statistical significance: $0.1\% < P < 1\%$.

Source: P.E. Wilson et al., *J. Hyg.*, 69, 125, 1971. With permission of Cambridge University Press.

2.1–206   ACQUISITION OF TETRACYCLINE-RESISTANT STAPHYLOCOCCI IN RELATION TO ANTIBIOTIC THERAPY

	Strain acquired[a]	Antibiotic therapy, % acquisitions		
		Topical	Systemic	None
Nose	S/T	50	46	65
	R/T	50	54	35
Chest	S/T	39	47	62
	R/T	61	53	38
Groin	S/T	37	36	51
	R/T	63	64	49
Total acquisitions		282	404	1038

[a] S/T, strain sensitive to tetracycline; R/T, strain resistant to tetracycline.

Source: P.E. Wilson et al., *J. Hyg.,* 69, 125, 1971. With permission of Cambridge University Press.

2.1–207   DISTRIBUTION OF "INFECTIONS"

Percent patients with lesions yielding	1st period	2nd period
Staphylococcus aureus[a]		
S/PT	22	10
R/P	4	7
R/T	11	3
R/PT	3	5
β-Hemolytic streptococcus	9	7
Pseudomonas aeruginosa	9	2
Proteus	3	2
Coliform-type organisms	2	3
Total patients	531	1100

[a] S/PT, sensitive to penicillin and tetracycline; R/P, resistant to penicillin only; R/T, resistant to tetracycline only; R/PT, resistant to penicillin and tetracycline.

Source: P.E. Wilson et al., *J. Hyg.,* 69, 125, 1971. With permission of Cambridge University Press.

2.1–208   DISTRIBUTIONS OF ANTIBIOTIC PRESCRIPTIONS

Route	Antibiotic	Number of prescriptions	
		1st period	2nd period
Systemic	Penicillin	16 (12%)	68 (30%)
	Orbenin + ampicillin	28 (22%)	41 (18%)
	Tetracycline	60 (46%)	100 (44%)
	Other	25 (19%)	18 (8%)
	Total	129	227
Topical	Tetracycline	49 (70%)	42 (45%)
	Neomycin	17 (24%)	1 (1%)
	Gentamicin	0	38 (40)
	Other	4 (6%)	13 (14%)
	Total	70	94
Total patients		531	1017

Source: P.E. Wilson et al., *J. Hyg.*, 69, 125, 1971. With permission of Cambridge University Press.

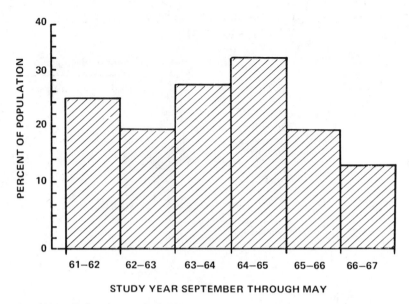

2.1–209   Percentage of children in a children's home shedding herpes simplex virus, 1961–1967.

Source: T.C. Cesario et al., *Am. J. Epidemiol.,* 90, 416, 1969. With permission.

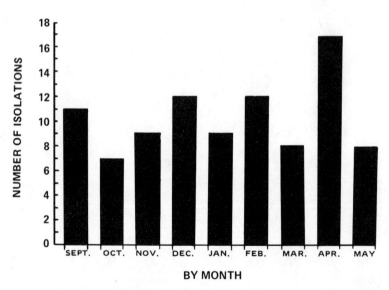

2.1–210   Isolations of herpes simplex virus in throat washings of 46 children by month, 1961–1967.

Source: T.C. Cesario et al., *Am. J. Epidemiol.,* 90, 416, 1969. With permission.

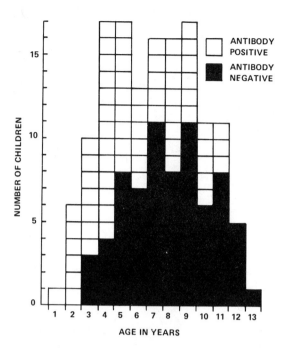

2.1–211   Herpes simplex antibody status of 143 children by age at the time of entry into a children's home, 1961–1967.

Source: T.C. Cesario et al., *Am. J. Epidemiol.,* 90, 416, 1969. With permission.

2.1–212   VIRUS ISOLATIONS BY MONTH IN A CHILDREN'S HOME, KANSAS CITY, KANSAS, 1961–1962

Agents	September	October	November	December	January	February	March	April	May	Total
Herpes simplex	2	0	0	3	1	1	1	4	3	15
Adenovirus type 3	0	0	1	5	0	0	0	0	0	6
Parainfluenza virus type 2	0	0	1	1	0	0	0	0	3	5
Influenza virus type B[a]	0	0	0	0	1	0	0	0	0	1
Mumps virus	0	0	0	0	0	2	1	1	1	5
Coxsackievirus type B5	0	0	1	1	0	0	0	0	0	2
Echovirus type 14	0	0	1	4	1	0	1	1	4	12
Echovirus 25	1	0	0	0	0	0	0	0	0	1
Reovirus type 1	0	0	0	0	1	0	0	0	0	1
Vaccinia virus	0	0	0	0	1	0	0	0	1	2
Unidentified agent	0	0	0	1	0	0	0	0	0	1
Total	3	0	4	15	5	3	3	6	12	51

[a] An outbreak of influenza B occurred in January and February 1962. Twenty-one children had fourfold antibody rise for influenza B. Type B influenza virus was also recovered from two adult employees.

Source: J.D. Poland et al., *Am. J. Epidemiol.*, 84, 92, 1966. With permission.

2.1–213   DISTRIBUTION OF HEMAGGLUTINATION-INHIBITION ANTIBODIES TO TWO INFLUENZA VIRUSES BY AGE GROUPS, KANSAS CITY, KANSAS, 1961–1962

| | | Influenza virus type | | | |
| | | A | | B | |
Age group in years	Number of children tested	Number positive	Percent positive	Number positive	Percent positive
2–4	4	3	75	2	50
5–8	21[a]	9	43	8	40[a]
9–12	16	11	69	10	63
Total	41[a]	23	56	20	50[a]

[a] One child in the 5–8 years age group was not tested to influenza B due to depletion of serum.

Source: J.D. Poland et al., *Am. J. Epidemiol.*, 84, 92, 1966. With permission.

2.1–214   FREQUENCY OF ANTIBODY RESPONSE AND FEBRILE ILLNESS IN RELATION TO PREEXISTING INFLUENZA B ANTIBODY TITERS, KANSAS CITY, KANSAS, 1961–1962

| | | Fourfold rise | | Febrile illness | |
Initial antibody titer[a]	Number of children	Number	Percent	Number	Percent
<10	20	15	75	9	45
10	4	3	75	1	25
20	9	2	22	1	11
≥40	7	1	14	0	0
Total	40	21	52	11	28

[a] Reciprocal of serum dilution.

Source: J.D. Poland et al., *Am. J. Epidemiol.*, 84, 92, 1966. With permission.

2.1–215 DISTRIBUTION OF PARAINFLUENZA VIRUS NEUTRALIZING ANTIBODIES IN INITIAL SERA BY AGE GROUPS, KANSAS CITY, KANSAS, 1961–1962

Age group in years	Number of children tested	Parainfluenza virus type							
		1		2		3		4	
		Number positive	Percent positive	Number positive	Percent positive	Number positive	Percent positive	Number positive	Percent positive
2–4	5[a]	0	0	0	0	3	50[a]	3	60[b]
5–8	22[b]	7	32	1	5	20	91	14	66[b]
9–12	17	7	41	4	24	14	82	10	59
Total	44[a,b]	14	32	5	11	37	82[a]	27	63[b]

[a] Six sera were tested to type 3 in the 2–4 years age group.
[b] Only 21 sera were tested to type 4 in the 5–8 years age group.

Source: J.D. Poland et al., *Am. J. Epidemiol.*, 84, 92, 1966. With permission.

2.1–216 CHILDREN WITH A FOURFOLD OR GREATER RISE IN NEUTRALIZING ANTIBODY TITER TO THE PARAINFLUENZA VIRUSES BY AGE GROUPS, KANSAS CITY, KANSAS, 1961–1962

Age group in years	Number of children tested	Fourfold or greater rise to parainfluenza virus type							
		1		2		3		4	
		Number	Percent	Number	Percent	Number	Percent	Number	Percent
2–4	5[a]	0	0	0	0	1	17[a]	2	40
5–8	22[b]	3	14	2	9	6	27	5	25[b]
9–12	17	1	6	4	23	2	12	1	6
Total	44[a,b]	4	9	6	14	9	20[a]	8	19[b]

[a] Six paired sera were tested to type 3 in the 2–4 years age group.
[b] Only 20 paired sera were tested to type 4 in the 5–8 years age group.

Source: J.D. Poland et al., *Am. J. Epidemiol.*, 84, 92, 1966. With permission.

2.1–217   CHILDREN WITH A FOURFOLD ANTIBODY RISE FOR PARAINFLUENZA VIRUSES IN RELATION TO CLINICAL DISEASE, KANSAS CITY, KANSAS, 1961–1962

Antibody rise for parainfluenza virus, by type	Number of children (N = 18)	Number of antibody rises (N = 27)	Relation of antibody rise to respiratory illness[a] (kind of illness)
2	3	3	0
3	4	4	2 (bronchitis, pharyngitis)
4	4	4	2 (pharyngitis)
1, 3	1	2	0
1, 4	1	2	0
2, 3	1	2	0
2, 4	1	2	1 (common cold)
3, 4	1	2	1 (common cold)
1, 2, 3	1	3	1 (influenza[b])
1, 3, 4	1	3	1 (otitis media)

[a] Illnesses occurred during interim between paired sera demonstrating fourfold antibody rise.

[b] Also antibody rise for influenza B virus.

Source: J.D. Poland et al., *Am. J. Epidemiol.,* 84, 92, 1966. With permission.

2.1–218   FREQUENCY OF HERPES SIMPLEX VIRUS ISOLATION AND NEUTRALIZING ANTIBODY RISE BY INITIAL ANTIBODY STATUS, KANSAS CITY, KANSAS, 1961–1962

Initial antibody titer	Number of children tested	Number with isolations	Number with fourfold rise
<1:8	21	3 (14%)	3 (14%)
≥1:8	26	8 (31%)	3 (12%)
Total	47	11 (23%)	6 (13%)

Source: J.D. Poland et al., *Am. J. Epidemiol.,* 84, 92, 1966. With permission.

2.1–219   DETAILS OF PATIENTS ADMITTED TO TWO MALE WARDS DURING THE SURVEY PERIOD

	Ward A	Ward B
Number admitted	117	169
	%	%
Age		
< 15 years	3	3
15–39 years	27	39
>39 years	70	58
Diagnosis		
Eczema	42	34
Psoriasis	35	31
Others	23	34
Length of stay		
1–7 days	9	17
8–14 days	28	24
15–21 days	23	22
22–29 days	11	14
29–56 days	20	16
>56 days	9	7
Antibiotic therapy		
Topical	23	17
Systemic	25	17
Total	40	25

Source: D.A. Somerville, *J. Hyg.,* 70, 667, 1972. With permission of Cambridge University Press.

2.1–220   INCIDENCE OF YEASTS AMONG PATIENTS ON ADMISSION

	Number of patients	Number carrying on				
		Gums	Nose	Chest	Groin	Total
Candida albicans Patients with						
Eczema	107	39 (36)	2 (2)	2 (2)	3 (3)	39 (36)
Psoriasis	94	24 (26)	1 (1)	2 (2)	6 (6)	27 (29)
Other disease	85	15 (18)	1 (1)	1 (1)	1 (1)	17 (20)
Total	286	78 (27)	4 (1)	5 (2)	10 (4)	83 (29)
Other yeasts Patients with						
Eczema	107	15 (14)	4 (4)	4 (4)	9 (8)	24 (22)
Psoriasis	94	5 (5)	3 (3)	6 (6)	6 (6)	14 (15)
Other disease	85	5 (6)	3 (4)	4 (5)	10 (12)	19 (22)
Total	286	25 (9)	10 (3)	14 (5)	25 (9)	55 (19)

Note: Figures in parentheses are percentages.

Source: D.A. Somerville, *J. Hyg.,* 70, 667, 1972. With permission of Cambridge University Press.

2.1–221   ACQUISITION OF *CANDIDA ALBICANS* IN THE HOSPITAL

	Number of patients	Number acquiring on						
		Gums[a]		Nose	Chest	Groin	Lesions	Total
		++	+					
Age								
<40 years	106	2 (2)	14 (13)	5 (5)	12 (11)	14 (13)	1 (1)	30 (28)
≥40 years	180	17 (9)	19 (11)	15 (8)	18 (10)	32 (18)	12 (7)	67 (37)
Disease								
Eczema	107	12 (11)	10 (9)	6 (6)	10 (9)	18 (17)	6 (6)	43 (40)
Psoriasis	94	1 (1)	14 (15)	9 (10)	11 (12)	18 (19)	5 (5)	29 (31)
Others	85	6 (7)	9 (11)	5 (6)	9 (11)	10 (12)	2 (2)	25 (29)
Length of hospital stay								
<14 days	114	2 (2)	8 (7)	3 (3)	3 (3)	6 (5)	1 (1)	20 (18)
14–29 days	101	9 (9)	12 (12)	3 (3)	12 (12)	12 (12)	1 (1)	34 (34)
>29 days	71	8 (11)	13 (18)	14 (20)	15 (21)	28 (39)	11 (15)	43 (61)
Total	286	19 (7)	33 (12)	20 (7)	30 (10)	46 (16)	13 (5)	97 (34)

Note: Figures in parentheses are percentages.

[a] ++: ≥20 colonies isolated on primary culture; +: <20 colonies isolated on primary culture.

Source: D.A. Somerville, *J. Hyg.,* 70, 667, 1972. With permission of Cambridge University Press.

2.1–222   TOTAL CARRIAGE OF *CANDIDA ALBICANS*

	Number of patients	Number carrying on						
		Gums[a]		Nose	Chest	Groin	Lesions	Total
		++	+					
Age								
<40 years	106	17 (16)	32 (30)	7 (7)	14 (13)	17 (16)	1 (1)	56 (53)
≥40 years	180	45 (25)	36 (20)	17 (9)	21 (12)	39 (22)	12 (7)	94 (52)
Diseases								
Eczema	107	34 (32)	27 (25)	8 (7)	12 (11)	20 (19)	6 (6)	69 (64)
Psoriasis	94	12 (13)	27 (29)	10 (11)	13 (14)	25 (27)	5 (5)	45 (48)
Others	85	16 (19)	14 (16)	6 (7)	10 (12)	11 (13)	2 (2)	36 (42)
Total	286	62 (21)	68 (24)	24 (8)	35 (12)	56 (20)	13 (5)	150 (52)

Note: Figures in parentheses are percentages.

[a] ++: ≥ 20 colonies isolated on primary culture; +: <20 colonies isolated on primary culture.

Source: D.A. Somerville, *J. Hyg.,* 70, 667, 1972. With permission of Cambridge University Press.

2.1–223   EFFECT OF TREATMENT WITH ANTIBIOTICS ON CARRIAGE OF YEASTS

	Number of patients	*Candida albicans*			Other yeasts		
		Number carrying on			Number carrying on		
		Gums	Skin	Total	Gums	Skin	Total
Topical antibiotics	55	33 (60)	20 (36)	39 (71)	16 (29)	25 (45)	35 (64)
Systemic antibiotics	58	37 (64)	31 (53)	49 (84)	24 (41)	34 (59)	48 (83)
No antibiotics	195	77 (39)	35 (18)	88 (45)	28 (14)	52 (27)	78 (48)

Note: Figures in parentheses are percentages.

Source: D.A. Somerville, *J. Hyg.,* 70, 667, 1972. With permission of Cambridge University Press.

2.1–224   ACQUISITION OF YEASTS OTHER THAN *CANDIDA ALBICANS*

	Number of patients	Number acquiring on					
		Gums	Nose	Chest	Groin	Lesions	Total
Age							
<40 years	106	10 (9)	14 (13)	10 (9)	14 (13)	3 (3)	34 (32)
≥40 years	180	37 (21)	37 (21)	43 (24)	44 (24)	14 (8)	89 (49)
Disease							
Eczema	107	19 (18)	20 (19)	20 (19)	26 (24)	6 (6)	50 (47)
Psoriasis	94	13 (14)	18 (19)	21 (22)	20 (21)	6 (6)	43 (46)
Others	85	15 (18)	13 (15)	12 (14)	12 (14)	5 (6)	30 (35)
Length of hospital stay							
<14 days	114	6 (5)	8 (7)	5 (4)	8 (7)	2 (2)	23 (20)
14–29 days	101	14 (14)	16 (16)	17 (17)	21 (21)	5 (5)	47 (47)
>29 days	71	27 (38)	27 (38)	31 (44)	29 (41)	10 (14)	53 (75)
Total	286	47 (16)	51 (18)	53 (19)	58 (20)	17 (6)	123 (43)

Note: Figures in parentheses are percentages.

Source: D.A. Somerville, *J. Hyg.,* 70, 667, 1972. With permission of Cambridge University Press.

2.1–225   YEASTS OTHER THAN *CANDIDA ALBICANS* ISOLATED FROM PATIENTS (ONLY ONE ISOLATE FROM EACH PATIENT IS INCLUDED)

	Number of isolations				
	Gums	Chest	Groin	Lesions	Total
Candida parapsilosis	19	48	54	10	106
C. tropicalis	4	–	3	–	5
C. pseudotropicalis	3	–	–	1	4
C. krusei	4	–	1	–	5
C. melinii	6	1	1	1	11
C. zeylanoides	1	–	1	–	3
Other Candida spp.	2	1	2	3	6
Torulopsis glabrata	30	7	11	2	43
T. famata	7	9	7	3	28
T. inconspicua	13	7	2	–	18
Pityrosporum pachydermatis	13	23	27	5	47

Source: D.A. Somerville, *J. Hyg.,* 70, 667, 1972. With permission of Cambridge University Press.

2.1−226   INCIDENCE OF YEASTS IN INFECTED LESIONS AND OTHER SPECIMENS SENT TO LABORATORIES

Specimens	Number of specimens	Number carrying		
		Candida albicans	Other *Candida*	Other yeasts
Swabs from				
Ulcers	78	4 (5)	12 (15)	1 (1)
Eyes/ears	78	8 (10)	5 (6)	4 (5)
Axillae/groin	59	11 (19)	5 (3)	1 (2)
Toewebs	29	0	8 (28)	3 (10)
Other areas	809	49 (6)	49 (6)	40 (5)
Sputum	131	62 (47)	7 (5)	20 (15)
Throat swab	306	110 (36)	16 (5)	28 (9)
Nose swab	155	9 (6)	5 (3)	4 (3)

Note: Figures in parentheses are percentages.

Source: D.A. Somerville, *J. Hyg.,* 70, 667, 1972. With permission of Cambridge University Press.

2.1−227   FATALITIES ASSOCIATED WITH HOSPITAL-ACQUIRED INFECTION − THE JOHNS HOPKINS HOSPITAL

Infection	Organism isolated	Predisposing factors
Lung abscess	*Klebsiella-Aerobacter*	Lymphosarcoma
Pneumonia	*Klebsiella-Aerobacter*	Leukemia
Pneumonia	*Klebsiella-Aerobacter, Proteus*	Uremia
Pneumonia	*Proteus*	Metastic carcinoma
Wound infection	*Klebsiella-Aerobacter, Proteus*	Metastic carcinoma, choledochostomy
Wound infection	*Escherichia coli*	Cirrhosis, portacaval shunt
Wound infection	*Staphylococcus aureus*	Intestinal obstruction, surgical fixation of hip joint
Peritonitis, bacteremia	*E. coli*	Dysgammaglobulinemia, reticulum cell sarcoma, perforated colon
Bacteremia (unknown portal of entry)	*Bacillus subtilis*	Lymphosarcoma

Source: R. Thoburn et al., *Arch. Intern. Med.,* 121, 1, 1968. With permission. Copyright 1968, American Medical Association.

2.1—228   AIRBORNE BACTERIA INSIDE AND OUTSIDE PLASTIC VENTILATED ISOLATORS

	Settle plate counts[a]		Andersen sampler counts (total per cubic foot of air)		
	Mean counts per plate	Number of observations	Experiment 1 (quiet ward)	Experiment 2 (busy ward)	Experiment 3 (quiet ward)
Isolator with filters	0.1 (range 0—0.4)	5	0.2	<0.01	0.03
Isolator with coarse filter only	—	—	—	0.5	0.22
Isolator with no filter	1.0 (range 0—2.2)	10	1.2	—	0.13
Open ward	46.7 (range 11.0—82.8)	15	2.4	7.3	2.3

[a] Mean counts of colonies on 3½-in. (8.8-cm) plates exposed for 6 hr. Each observation represents a sampling with a number of settle plates on 1 day.

Source: E.J.L. Lowbury et al., *J. Hyg.*, 69, 529, 1971. With permission of Cambridge University Press.

2.1–229 PARTICLE SIZE DISTRIBUTION OF BACTERIA INSIDE AND OUTSIDE PLASTIC VENTILATED ISOLATOR

Estimated particle size	Experiment 1 (quiet ward) (viable counts per 30 ft³ of air)				Experiment 2 (busy ward) (viable counts per 60 ft³ of air)			
	Isolator with filters	Isolator with coarse filter only	Isolator with no filter	Open ward	Isolator with filters	Isolator with coarse filter only	Isolator with no filter	Open ward
9.2 μm and above	1	—	1	15ᵃ (1 Staphylococcus aureus)	0	0	—	145
5.5–9.2 μm	2	—	3 (1 S. aureus)	15 (1 S. aureus)	0	1	—	109
3.3–5.5 μm	0	—	5	19	0	0	—	79
2.0–3.3 μm	2	—	6	11	0	7	—	63
1.0–2.0 μm	0	—	19 (2 S. aureus)	11	0	22	—	42
Less than 1.0 μm	1	—	1	0	0	0	—	0

ᵃ S. aureus were counted only in Experiment 1.

Source: E.J.L. Lowbury et al., J. Hyg., 69, 529, 1971. With permission of Cambridge University Press.

2.1–230   SETTLE PLATE COUNTS INSIDE
AND OUTSIDE ISOLATORS

Mean settle plate counts (total)[a]

Isolator	Inside isolator	Outside isolator (near bed)	Open ward (remote from bed)
Air curtains ("Sterair" unit)	9.2 (range 2.5–16.8)	29.0 (range 20.5–42.9)	90.6 (range 50.5–179.8)
Open-topped plastic isolator	16.7 (range 9.3–21.5)	23.6 (range 17.2–31.5)	98.3 (range 79.0–153.3)

[a]Five tests were made in each isolator, with five or six plates exposed for 6 hr in each test.

Source: E.J.L. Lowbury et al., *J. Hyg.,* 69, 529, 1971. With permission of Cambridge University Press.

2.1–231   CONTROLLED TRIALS OF ISOLATORS: COMPARABILITY
OF GROUPS

	Trials 1 and 2			Trial 3		
	Group a	Group b	Group c	Group a	Group b	Group c
Number of patients	20	17	17	10	10	10
Number in age groups						
<5	5	10	7	5	5	9
5–10	6	4	4	2	3	1
10–20	7	2	4	2	2	0
20–30	1	0	2	0	0	0
>30	1	1	0	1	0	0
Mean area of burn, %	13.5	17	14	11	9	11
Range %	(4–30)	(7–19)	(5–30)	(6–20)	(5–13)	(8–20)
Number treated by covered method	15	12	15	6	5	8
Number treated by exposure method	4	2	1	2	5	1
Number treated by mixed covered and exposure methods	1	3	1	2	0	1

Source: E.J.L. Lowbury et al., *J. Hyg.,* 69, 529, 1971. With permission of Cambridge University Press.

2.1–232 CONTROLLED TRIALS OF PLASTIC ISOLATORS: BACTERIAL INFECTION OF BURNS

Patients in	Staphylococcus (multi resistant)			Pseudomonas aeruginosa			Proteus spp.			Coliform bacilli			Total patients
	+[a]	CM[b]	% + or CM	+	CM	% + or CM	+	CM	% + or CM	+	CM	% + or CM	
Isolators with filters	12	1	65	2	0	10	9	3	60	17	1	90	20
Isolators with coarse filters	4	1	55.5	1	1	22.2	3	1	44.4	7	1	88.8	9
Isolators with no filters	3	2	62.5	0	0	–	1	2	37.5	7	0	87.5	8
Isolators (all types)	19	4	62.2	3	1	10.8[c]	13	6	51.1	31	2	89.2	37
No isolators (control)	10	4	82.3	11	0	64.7[c]	3	2	29.3	16	0	94.1	17

[a]+ = growth on solid medium.
[b]CM = growth only in fluid medium (cooked meat).
[c]$\chi^2 = 14.5$, $P < 0.001$.

Source: E.J.L. Lowbury et al., *J. Hyg.*, 69, 529, 1971. With permission of Cambridge University Press.

2.1–233   CONTROLLED TRIAL OF PLASTIC ISOLATORS: ACQUISITION OF *STAPHYLOCOCCUS AUREUS* (TRIALS 1 AND 2)

Patients in		Multiresistant *S. aureus* (+[a] and CM[b])				Total patients
		In burns		In nares		
		Patients	%	Patients	%	
Isolators	Whole period	23	62[c]	20	54[c]	37
	1st week	15	40[c]	9	24[d]	
Control series	Whole period	14	82[c]	14	82[c]	17
	1st week	11	64[c]	10	59[d]	

[a]+ = growth on solid medium.
[b]CM = growth only in fluid medium (cooked meat).
[c]Not significant.
[d]$\chi^2 = 4.4, P < 0.05$.

Source: E.J.L. Lowbury et al., *J. Hyg.*, 69, 529, 1971. With permission of Cambridge University Press.

2.1–234   CONTROLLED TRIAL OF ISOLATORS: AIR CURTAINS AND OPEN-TOPPED PLASTIC ISOLATOR

Numbers of patients who acquired

| Isolator | In burns | | | | | | | | | | | | In nares | | | Number of patients |
| | Pseudomonas aeruginosa | | | Staphylococcus aureus RR[a] | | | Proteus spp. | | | Coliform bacilli | | | S. aureus | | | |
	+[b]	CM[c]	% + or CM	+	CM	% + or CM	+	CM	% + or CM	+	CM	% + or CM	+	CM	% + or CM	
"Sterair" isolator (air curtains)	5	0	50[d]	4	0	40	3	3	60	7	3	100	3	2	50	10
Open-topped plastic isolator	0	0	0[d]	6	1	70	4	0	40	8	0	80	3	0	30	10
Controls (open ward)	5	0	50[d]	3	1	40	2	1	30	6	1	70	3	0	30	10

[a]RR = multiresistant.
[b]+ = growth on solid medium.
[c]CM = growth only in fluid medium (cooked meat).
[d]$\chi_c^2 = 2.06$, $P < 0.025$ (see reference below).

Source: E.I.L. Lowbury et al., *J. Hyg.*, 69, 529, 1971. With permission of Cambridge University Press.

REFERENCE

R.A. Fisher and F. Yates, *Statistical Tables for Biological, Agricultural, and Medical Research*, 3rd ed., Oliver and Boyd, London, 1948, 47.

2.1–235   CARRIAGE OF BACTERIA BY PATIENT ON ADMISSION AND SUBSEQUENT INFECTION OF BURNS

Bacteria	Site of carriage on admission	Bacteria carried on admission		Bacteria not carried on admission but in burns later	Total	
		Not in burns later	In burns later		Sampled for carriage on admission	Patients
Staphylococcus aureus (RR)[a]	Nose	0	3	40	73	
Pseudomonas aeruginosa	Rectum	2	1	18	54	84
Proteus spp.	Rectum	0	4	23	54	
Coliform bacilli	Rectum	4	41	4	54	

[a]RR = multiresistant.

Source: E.J.L. Lowbury et al., *J. Hyg.,* 69, 529, 1971. With permission of Cambridge University Press.

2.1–236 TYPES OF *PSEUDOMONAS AERUGINOSA* IN FECES OR RECTAL SWAB AND ON BURNS OF PATIENTS IN PLASTIC ISOLATORS

P. aeruginosa isolated

Patient	Date of admission	Isolator group	From burns			From feces or rectal swab			Comments
			Serotype	Phage type	Date	Serotype	Phage type	Date	
1	4/14/68	Prefilter and filter	3	16/31/F8/109/119X/352/M6/Col 11	4/21	–	Not typed	4/21	Burns Unit strain; rectal swab on admission negative (i.e., no *P. aeruginosa*)
2	12/10/68	Prefilter and filter	–	Not typed (one isolate only)	12/20	–	None isolated	–	–
3	4/9/68	Pre filter only	(a) 5c (b) NT	7/F7/119X 119X	4/14 4/27	5c –	7/31/F7/119X Not isolated	4/17 –	(a) Recent Burns Unit strain (b) Current Burns Unit strain. First four rectal swabs negative
4	8/1/68	Prefilter only	NT	119X	8/14	NT	119X	8/11	Burns Unit Strain. First four rectal swabs negative
5	5/19/68	Prefilter only	–	None isolated	–	NT	119X	5/27	Burns Unit strain. First five rectal swabs negative
6	7/29/69	No filter or prefilter	–	None isolated	–	3	21/31/44/68/F7/F8/109&119X	–	Not Burns Unit strain. First rectal swab negative

Source: E.J.L. Lowbury et al., *J. Hyg.,* 69, 529, 1971. With permission of Cambridge University Press.

2.1–237   CONTAMINATION OF VARIOUS
ITEMS ISSUED TO PATIENTS

Items	Number of samples	Number of samples contaminated with			
		Staphylo- coccus aureus[a]		Gram- negative bacilli	
		+[b]	CM[c]	+	CM
Books, papers, etc.	20	–	–	2	1[d]
Washing bowls	16	1	3	–	3
Crockery, glassware	55	3	–	1	3
Cutlery	8	–	2	–	–
Clean pillows	7	1	–	–	1
Disposable bedpan supports	8	6	–	2	1[e]
Urine bottles	2	1	–	–	1
Toys	8	1	–	–	–
Receiving bowls	3	–	1	–	1
Foods (various)	45	1	1	4	5
Total	172	14	7	9	16

[a] All strains were found resistant to two or more antibiotics except those from food, which were not tested.
[b] + = growth on solid medium.
[c] CM = growth only in fluid medium (cooked meat).
[d] *Proteus* sp.
[e] *Pseudomonas aeruginosa.*

Source: E.J.L. Lowbury et al., *J. Hyg.,* 69, 529, 1971. With permission of Cambridge University Press.

2.1–238   EFFECT OF ISOLATOR CONFINEMENT CARE ON MICROFLORA OF THE PHARYNGEAL AND NASOPHARYNGEAL REGIONS

Bacterial isolations and identifications	Pharyngeal regions			Nasopharyngeal regions		
	Average rate of recovery[b]	Frequency of recovery[b]	Normal values[c]	Average rate of recovery	Frequency of recovery	Normal values
α Streptococci	<300	100%	>100	>100	100%	<300
β Streptococci	< 10	25%	> 10	> 10	75%	< 10
γ Streptococci	10–100	100%	10 100	10–100	100%	10–100
Staphylococcus epidermidis	<100	100%	10–100	>100	80%	<100
S. aureus	< 10	Single occurrence	< 10	< 10	Single occurrence	NR
Pneumococci	< 10	Single occurrence	< 10	–	–	NR
Diphtheroids	<100	75%	10–100	< 10	50%	10–100
Neisseria	< 10	50%	< 10	< 10	Single occurrence	NR
Hemophilus	< 10	Single occurrence	NR[d]	–	–	NR
Escherichia coli	< 10	Single occurrence	NR	–	–	NR
Klebsiella pneumoniae, 3	< 10	35%	< 10	< 10	90%	10–100
Total	>300	100%	>300	>300	100%	>300

[a] Average recovery rate: approximate number of colonies recovered after primary dilution and plating.

[b] Frequency of recovery: number of individual specimens cultured from which positive isolations were reported.

[c] Normal recovery values: based on results of cultured specimens taken from same patient during hospitalization on open ward under conditions of normal environmental exposure.

[d] NR: not reported.

Source: S. Shadomy et al., *Arch. Environ. Health,* 11, 191, 1965. With permission. Copyright 1965, American Medical Association.

2.1–239   EFFECT OF ISOLATOR CONFINEMENT CARE ON THE MICROFLORA OF SELECTED REGIONAL PATIENT SKIN SURFACES

Abdominal Surfaces

Pubic, umbilical and lateral regions

Bacterial isolations and identifications	Average recovery rate (CFU range)[a]		Frequency of recovery[b]		Normal recovery values[d]
	First period	Second period	First period[c]	Second period	
α Streptococci	CRN[e]	CRN	–	–	10–100
β Streptococci	CRN	CRN	–	–	CRN
γ Streptococci	< 10	CRN	1	–	10–100
Staphylococcus epidermidis	10–100	10–100	5	2	>100
S. aureus	< 10	< 10	2	2	< 10
Streptococcus faecalis	< 10	NR[f]	1	–	10–100
Escherichia coli	< 10	< 10	3	1	10–100
Klebsiella-Aerobacter	CRN	CRN	–	–	< 10
Others	NR	NR	–	–	10–100
Total (positive cultures)	<100	10–100	6:11 (55%)	2:7 (29%)	>300

Rectal Surfaces

Bacterial isolations and identifications	Anal and perineal regions				
α Streptococci	< 10	< 10	4	4	10–100
β Streptococci	CRN	CRN	–	–	CRN
γ Streptococci	10–100	< 10	9	3	>100
S. epidermidis	>100	<100	11	7	>100
S. aureus	< 10	< 10	6	5	10–100
S. faecalis	< 10	< 10	9	3	10–100
E. coli	>100	10–100	10	7	>100
Klebsiella-Aerobacter	< 10	< 10	10	5	10–100
Others	10–100	NR	2	–	>100
Total (% positive cultures)	>300	>100	11 (100%)	7 (100%)	>300

Back Surfaces

Bacterial isolations and identifications	Lower median, sacral, and lumbar regions				
α Streptococci	CRN	CRN	–	–	< 10
β Streptococci	CRN	CRN	–	–	CRN
γ Streptococci	CRN	CRN	–	–	10–100
S. epidermidis	10–100	< 10	10	2	>300
S. aureus	< 10	< 10	6	2	10–100
S. faecalis	< 10	CRN	2	–	10–100

2.1–239   EFFECT OF ISOLATOR CONFINEMENT CARE ON THE MICROFLORA OF SELECTED REGIONAL PATIENT SKIN SURFACES (continued)

Back Surfaces

Lower median, sacral, and lumbar regions

Bacterial isolations and identifications	Average recovery rate (CFU range)[a]		Frequency of recovery[b]		Normal recovery values[d]
	First period	Second period	First period[c]	Second period	
E. coli	< 10	CRN	2	–	10–100
Klebsiella-Aerobacter	< 10	CRN	2	–	< 10
Others	NR	< 10	–	1	10–100
Total (% positive cultures)	<100	< 10	10:11 (91%)	3:7 (43%)	>300

Forearm Surfaces

Bacterial isolations and identifications	Volar and ulnar antibrachial regions				
α Streptococci	< 10	CRN	1	–	10–100
β Streptococci	CRN	CRN	--	–	< 10
γ Streptococci	CRN	CRN	–	–	10–100
S. epidermidis	< 10	< 10	6	–	10–100
S. aureus	< 10	< 10	3	1	< 10
S. faecalis	NR	NR	–	1	NR
E. coli	CRN	CRN	–	–	CRN
Klebsiella-Aerobacter	CRN	CRN	–	–	CRN
Others	NR	NR	–	–	NR
Total (% positive cultures)	<100	< 10	6:11 (55%)	1:7 (14%)	>300

Thigh Surfaces

Bacterial isolations and identifications	Anterior and interior regions				
α Streptococci	CRN	CRN	–	–	10–100
β Streptococci	CRN	CRN	–	–	CRN
γ Streptococci	< 10	CRN	1	–	< 10
S. epidermidis	10–100	< 10	4	3	>300
S. aureus	< 10	< 10	2	2	< 10
S. faecalis	NR	NR	–	–	10–100
E. coli	< 10	CRN	1	–	10–100
Klebsiella-Aerobacter	CRN	CRN	–	–	< 10
Others	< 10	NR	1	–	10–100
Totals (positive cultures)	<100	< 10	5:11 (45%)	3:7 (43%)	>300

2.1–239   EFFECT OF ISOLATOR CONFINEMENT CARE ON THE MICROFLORA OF SELECTED REGIONAL PATIENT SKIN SURFACES (continued)

Note: Average rates and frequencies of recovery are derived from bacteriological data obtained during two consecutive periods of 7 to 5 days, respectively, during which two different skin cleansing agents were employed in the patient bathing procedures.

[a] Average recovery rate: approximate number of colony forming units (CFU) recovered on primary dilution and plating.

[b] Frequency of recovery: number of cultures examined in which positive isolations were reported.

[c] First period: through morning of 7th day, Dial® soap used in patient bathing procedures, 11 cultures taken; second period: from afternoon of 7th day through last day of isolation, pHisoHex® used in patient bathing procedures.

[d] Normal recovery values: based on cultures taken from the same patient on an open ward in a normal hospital environment.

[e] CRN: culture reports negative.

[f] NR: not reported.

Source: S. Shadomy et al., *Arch. Environ. Health,* 11, 191, 1965. With permission. Copyright 1965, American Medical Association.

2.1–240   EVALUATION OF ULTRAHIGH EFFICIENCY FILTRATION SYSTEM BY CHALLENGE WITH AEROSOL SUSPENSIONS OF *SERRATIA MARCESCENS*

Sampling points	Sample repli-cates	CFU[a] counts per exposed plate following 10-min exposure periods	
		Experiment series	
		1	2
1. Efferent air stream from roughing filter	A	30[b]	>500[b]
2. Efferent air stream from UHEF unit	A	None	None
	B	None	None
	C	None	None
3. Air settling plates at test site, external to UHEF system	A	43[b]	115[b]
	B	15[b]	120[b]

Note: The effectiveness of the filtration was monitored by locating recovery plates (blood agar) at various points in the efferent air stream.

[a] CFU: colony forming units recovered on blood agar plates following exposure with subsequent incubation at 25–27°C for 48 hr.

[b] Mixture of *S. marcescens* and various bacterial and mycotic components of room air microflora of the test site.

Source: S. Shadomy et al., *Arch. Environ. Health,* 11, 183, 1965. With permission. Copyright 1965, American Medical Association.

2.1–241 EVALUATION OF ULTRAHIGH EFFICIENCY FILTRATION SYSTEM BY CHALLENGE WITH THE NATURALLY OCCURRING ROOM AIR MICROFLORA

Sampling points	CFU[a] counts per exposed plate following 1- to 2-hr exposure periods		
	Experimental Series		
	1	2	3
1. Efferent air stream from roughing filter	12	5	50
2. Efferent air stream from UHEF filter	0	0	0
3. Head of mattress within isolator enclosure, distal to air inlet	0	1	0
4. Head of mattress within isolator enclosure, under air inlet	0	0	0
5. Foot of mattress within isolator enclosure, distal to air inlet	0	0	0

Note: The UHEF units were subjected to challenge by exposure through prolonged operation to the naturally occurring room air microflora of the test site. The effectiveness of the filtration was monitored by locating recovery plates (blood agar) at various points in the efferent air streams within the isolator enclosure.

[a] CFU: colony forming units recovered on blood agar plates following exposure with subsequent incubation at 37°C for 48 hr.

Source: S. Shadomy et al., *Arch. Environ. Health*, 11, 183, 1965. With permission. Copyright 1965, American Medical Association.

2.1–242   REACTION OF POLYVINYL CHLORIDE PLASTIC TENT MATERIAL WITH CONCENTRATED GERMICIDE SOLUTIONS

Agent	Concentration[b]	Germicide		Plastic[a]	
		Color	Appearance	Color	Appearance
Wescodyne G®[c]	5.0%	Yellow	Clear	Yellow	Clear
Terigsy®[d]	10.0%	White	Cloudy (slowly separating emulsion)	None	Thin film of agent adhering to strip
Benzalkonium chloride[e]	1:75	None	Clear	None	Clear
Amphyl®[d]	5.0%	Pink	Clear	None	Heavy film of agent adhering to strip
SBT[f]	5.0%	White	Cloudy	Light gray	Heavy agent residue adhering to strip
SBT-24[f]	5.0%	White	Cloudy (slowly separating emulsion)	None	Heavy film of agent adhering to strip
Ves-phene®[g]	10.0%	Gray	Clear	None	Clear
Saline	9.0%	None	Clear	None	Clear
Water	–	None	Clear	None	Clear

Note: Evaluation of germicides for use in concurrent sanitization of the occupied patient isolator. Candidate germicidal agents were subjected to an evaluation of their chemical and physical compatibilities with the polyvinyl chloride plastic used in the fabrication of the isolator enclosure. Testing was accomplished by prolonged exposure of plastic strips to concentrated solutions of germicide for 24 hr with subsequent examination for either chemical or physical alterations of the treated strips.

[a] After 24-hr immersion in germicide with subsequent 2-min rinse under running, deionized water.
[b] 10 X recommended working concentration for surface disinfection.
[c] West Chemical Products, Inc.
[d] Lehn and Fink Products Corp.
[e] Certified Laboratories, Inc.
[f] Lever Bros. Co., Industrial Division.
[g] Vestal Laboratories.

Source: S. Shadomy et al., *Arch. Environ. Health,* 11, 183, 1965. With permission. Copyright 1965, American Medical Association.

2.1-243   PERSISTENCE OF R-FACTOR-MEDIATED RESISTANCE TO KANAMYCIN IN THE INTESTINAL FLORA OF HIGH-RISK INFANTS AFTER DISCHARGE FROM THE INTENSIVE CARE NURSERY

Intervals after discharge

R-factor-carrying organism	Number of infants	1–2 months	3–4 months	5–11 months	≥12 months
Klebsiella pneumoniae	24	8/10 (80)[a] *K. pneumoniae*	4/7 (57) *K. pneumoniae* 1/7 (14) *E. coli*	3/12 (25) *K. pneumoniae* 2/12 (17) *E. coli* 1/12 (8) *Enterobacter* 1/12 (8) *Citrobacter*	1/11 (9) *K. pneumoniae* 2/11 (18) *E. coli*
Escherichia coli	20	1/1 (100) *E. coli*	1/1 (100) *E. coli*	1/5 (20) *K. pneumoniae* 2/5 (40) *E. coli*	1/17 (6) *K. pneumoniae* 8/17 (47) *E. coli*
K. pneumoniae and *E. coli*	11		1/1 (100) *E. coli*	1/1 (100) *K. pneumoniae* and *E. coli*	3/10 (30) *E. coli* 1/10 (10) *Enterobacter* 1/10 (10) *K. pneumoniae* and *E. coli*
Enterobacter	1		1/1 (100) *Enterobacter*	0	1/1 (100) *Enterobacter*
All organisms	56	9/11 (82)	8/10 (80)	11/19 (58)	18/39 (46)

[a] Number of cultures positive for R-factor-carrying, kanamycin-resistant organisms per number of cultures performed. Numbers in parentheses indicate percentages.

Source: J.J. Domato et al., *J. Infect. Dis.*, 129, 205, 1974. With permission of the University of Chicago Press.

2.1–244   POINT-PREVALENCE OF HB Ag AND HB Ab IN PATIENTS AND STAFF OF 15 DIALYSIS CENTERS[a]

Subjects	Patients		Staff	
	Number	%	Number	%
Tested	583	100.0	451	100.0
With HB Ag	71[b]	12.2	10[b]	2.2
With HB Ab	198	34.0	141	31.3
With HB Ag and HB Ab	26	4.6	1	0.2
With HB Ag or HB Ab	295	50.6	152	33.7

[a] HB Ag indicates hepatitis B antigen; HB Ab indicates hepatitis B antibody.

[b] Seventeen patients were positive by radioimmunoassay only.

Source: W. Szmuness et al., *JAMA*, 227, 901, 1974. With permission.   Copyright   1974,   American   Medical Association.

2.1–245   POINT-PREVALENCE RATES (%) OF HB Ag AND HB Ab IN PATIENTS AND STAFF OF 15 DIALYSIS CENTERS[a]

Center	Patients				Staff			
	Number tested	% Positive			Number tested	% Positive		
		HB Ag	HB Ab	Either		HB Ag	HB Ab	Either
A	54	13.0	22.2	35.2	70	4.3	20.0	24.3
B	77	11.7	45.4	57.1	26	3.8	61.5	65.4
C	70	12.8	22.8	35.7	36	2.8	44.4	47.2
D	55	20.0	41.8	61.8	43	4.6	32.5	37.2
E	64	26.6	23.4	50.0	22	0.0	9.1	9.1
F	47	14.9	25.5	40.4	43	2.3	39.5	41.9
G	15	26.7	46.7	73.3	22	0.0	27.3	27.3
H	16	25.0	43.7	68.7	18	5.5	38.9	44.4
I	24	0.0	50.0	50.0	16	0.0	50.0	50.0
J	36	5.5	55.5	61.1	12	0.0	50.0	50.0
K	75	29.3	40.0	69.3	10	0.0	20.0	20.0
L	15	0.0	40.0	40.0	7	0.0	28.6	28.6
M	22	4.5	9.1	13.6	22	0.0	0.0	0.0
N	–	–	–	–	95	2.1	28.4	30.5
O	13	7.7	7.7	15.4	9	0.0	44.4	44.4
Total	583	16.1	34.0	50.1	451	2.4	31.3	33.7

[a] In this table subjects with both hepatitis B antigen (HB Ag) and antibody (HB Ab) are shown only as HB Ag carriers.

Source: W. Szmuness et al., *JAMA*, 227, 901, 1974. With permission. Copyright 1974, American Medical Association.

2.1–246   PREVALENCE RATES (%) OF HB Ag OR HB Ab IN RELATION TO ETHNIC BACKGROUND[a]

Subjects	Patients			Staff[b]		
	Whites (N=381)	Nonwhites (N=202)	P	Whites (N=321)	Nonwhites (N=96)	P
With HB Ag	18.9	12.4	<0.05	1.5	5.2	<0.05
With HB Ab	26.5	48.0	<0.001	28.3	42.7	<0.01
With HB Ag or HB Ab	45.4	60.4	<0.05	29.9	47.9	<0.001

[a] In this table subjects with both hepatitis B antigen (HB Ag) and antibody (HB Ab) are shown only as HB Ag carriers.

[b] The ethnic backgrounds for 34 staff members were unknown. HB Ag indicates hepatitis B antigen, HB Ab, hepatitis B antibody.

Source: W. Szmuness et al., *JAMA*, 227, 901, 1974. With permission. Copyright 1974, American Medical Association.

2.1–247   PREVALENCE RATES (%) OF HB Ag OR HB Ab[a] AND HISTORY OF HEPATITIS[b]

Subjects	Patients[c]			Staff[d]		
	Without (N=495)	With (N=79)	P	Without (N=369)	With (N=59)	P
With HB Ag	11.5	50.6	<0.001	2.7	1.7	NS[e]
With HB Ab	37.2	16.5	<0.001	28.2	52.5	<0.001
With HB Ag or HB Ab	48.7	67.1	<0.01	30.9	54.2	<0.005

[a] HB Ag indicates hepatitis B antigen, HB Ab, hepatitis B antibody.

[b] In this table subjects with both hepatitis B antigen (HB Ag) and antibody (HB Ab) are shown only as HB Ag carriers.

[c] Histories for nine patients were unknown.

[d] The histories of 23 staff members were unknown.

[e] NS indicates not significant.

Source: W. Szmuness et al., *JAMA*, 227, 901, 1974. With permission. Copyright 1974, American Medical Association.

2.1–248   CUMULATIVE PREVALENCE RATES (%) OF HB Ag OR HB Ab IN PATIENTS[a]

Duration of treatment, months	Blood units transfused, number			Total	
	0	1–9	≥10	Crude	Adjusted for trans- fusions[b]
<4	35.5	40.0	60.0[c]	38.5	40.2
4–11	60.0	52.2	60.0	55.9	57.4
12–23	45.5	55.5	56.4	54.0	53.1
≥24	25.0[c]	57.9	57.0	55.2	51.9
Total					
Crude	46.9	51.2	57.2		
Adjusted for duration[b]	49.3	51.2	54.2	51.4	

[a] In this table subjects with both hepatitis B antigen (HB Ag) and antibody (HB Ab) are shown only as HB Ag carriers.

[b] By the indirect method.[1]

[c] Denominator <10.

Source: W. Szmuness et al., *JAMA*, 227, 901, 1974. With permission. Copyright 1974, American Medical Association.

REFERENCE

1. A.M. Lilienfeld and J.E. Dowd, *Cancer Epidemiology: Methods of Study*, Johns Hopkins Press, Baltimore, 1967.

2.1–249   HEPATITIS B INFECTION IN HOUSEHOLDS OF DIALYSIS PATIENTS AND CONTROLS[a]

	Index-case		
	Dialysis Patients		
	With HB Ag[b]	Without HB Ag[b]	Blood Donor Without HB Ag[b]
Households tested, number	31	37	107
Family contacts tested, number	46	110	247
With HB Ag or VH, %	13.0	0	0.8[c]
With HB Ab, %	47.8	11.8	7.3
With any of above, %	60.8	11.8	8.1

[a] HB Ag indicates hepatitis B antigen, HB Ab, hepatitis B antibody, and VH, viral hepatitis. In this table subjects with both hepatitis B antigen (HB Ag) and antibody (HB Ab) are shown only as HB Ag carriers.

[b] Or history of viral hepatitis.

[c] By radioimmunoassay only.

Source: W. Szmuness et al., *JAMA*, 227, 901, 1974. With permission. Copyright 1974, American Medical Association.

2.2 STERILIZATION, DISINFECTION, AND CLEANING TECHNIQUES

2.2–1   BUILD-UP OF BACTERIAL CONTAMINATION OVER A 24-hr PERIOD FOLLOWING ROUTINE CLEANING OF A HOSPITAL CORRIDOR FLOOR

Mean colonies per Rodac® plate – based on 60 samples per time period per day

Day	Immediately after cleaning – 8:00 a.m.	2 hr later – 10:00 a.m.	4 hr later – noon	8 hr later – 4 p.m.	16 hr later – midnight	24 hr later – 8:00 a.m.
1	55	121	128	196	261	159
2	67	202	237	316	361	277

Source: D. Vesley and G.S. Michaelsen, *Health Lab. Sci.*, 1, 107, 1964. With permission.

2.2–2   SEASONAL COMPARISON OF BACTERIAL CONTAMINATION BUILD-UP FOLLOWING CLEANING OF A HOSPITAL CORRIDOR FLOOR

Mean colonies per Rodac® – 60 samples per time period per day

Season	Immediately after cleaning	4 hr after cleaning	Mean traffic census per 4-hr period
Winter and early spring February–April; generally cold and wet weather (10 runs)	52	186	552
Late spring and summer May–September; generally warm and dry weather (21 runs)	51	159	500

Source: D. Vesley and G.S. Michaelsen, *Health Lab. Sci.*, 1, 107, 1964. With permission.

2.2–3   COMPARISON OF BACTERIAL CONTAMINATION ON TERRAZZO AND ASPHALT TILE FLOORS IN A HOSPITAL CORRIDOR

Mean colonies per Rodac® plate – 60 samples per time period per day

Description	Mean colonies per Rodac plate before cleaning	Reduction immediately after cleaning	½ hr after cleaning	1 hr after cleaning	2 hr after cleaning	4 hr after cleaning	Mean traffic census per 4-hr period
Terrazzo (8 runs)	188	72.2%	76	90	113	132	452
Asphalt tile (18 runs)	183	69.0%	71	102	132	175	547

Source: D. Vesley and G. S. Michaelsen, *Health Lab. Sci.*, 1, 107, 1964. With permission.

2.2–4   COMPARISON OF BACTERIAL CONTAMINATION ON SEALED AND UNSEALED FLOORS IN A HOSPITAL CORRIDOR

Description	Mean colonies per Rodac plate before cleaning	Reduction immediately after cleaning	Mean colonies per Rodac® plate – 60 samples per time period per day				Mean traffic census per 4-hr period
			½ hr after cleaning	1 hr after cleaning	2 hr after cleaning	4 hr after cleaning	
Sealed floors (12 runs)	180	70.3%	70	97	122	148	496
Unsealed floors (14 runs)	188	70.6%	75	99	129	174	542

Source: D. Vesley and G.S. Michaelsen, *Health Lab. Sci.,* 1, 107, 1964. With permission.

2.2–5   RESULTS OF SIEVE-IN-MOTION SAMPLES ON A TILE FLOOR BEFORE AND AFTER MOPPING

Sample number	Number of colonies	
	Before mopping	After mopping
1	79	15
2	121	11
3	160	32
4	183	19
5	170	23
6	180	14
7	114	9
8	159	25
9	183	18
10	108	12
Average	146	18
Percent reduction		88

Source: J. G. Shaffer, *Health Lab. Sci.,* 3, 73, 1966. With permission.

2.2–6   RESULTS OF WEEKLY AIR SAMPLING ON CARPETED AND UNCARPETED (TILE) AREAS OF FOUR FLOORS OF THE HOSPITAL, SHOWING TOTAL BACTERIA/ft³ OF AIR AND TOTAL *STAPHYLOCOCCUS AUREUS* WITH HEAVY VACUUMING DAILY

		Tile		Carpet	
Date	Floor	Total bacteria per ft³ of air	Total *S. aureus*	Total bacteria per ft³ of air	Total *S. aureus*
Jan. 7–8	2	4	0	6	0
		3	0	9	2
	3	19	0	12	1
		14	0	16	1
	6	10	0	20	1
		40	0	15	3
	7	11	0	8	0
		31	0	18	0
Jan. 14–15	2	4	0	5	0
		4	0	5	0
	3	8	0	7	2
		10	0	17	0
	6	8	5	17	2
		46	0	19	1
	7	39	0	11	0
		46	8	13	0
Jan. 21–22	2	5	0	6	0
		4	0	5	0
	3	12	0	2	0
		5	0	10	0
	6	18	1	10	0
		20	0	8	1
	7	20	1	10	0
		8	3	12	0
Jan. 28–29	2	5	0	6	1
		4	0	5	1
	3	12	0	2	1
		5	1	10	0
	6	18	0	10	0
		20	0	8	0
	7	20	2	10	0
		8	7	12	0
Feb. 4–5	2	3	0	6	0
		2	0	4	0
	3	13	0	22	1
		9	2	7	0
	6	24	1	28	0
		16	0	27	1
	7	6	0	14	1
		19	0	15	0
Feb. 11–12	2	5	0	5	0
		2	0	3	0
	3	10	0	8	0
		8	0	4	0
	6	12	0	12	1
		20	0	18	1
	7	10	0	8	0
		16	0	7	0
Average		14		11	
Total *S. aureus*			31		22

Source: J.G. Shaffer, *Health Lab. Sci.*, 3, 73, 1966. With permission.

2.2−7   RELATION BETWEEN AIRBORNE BACTERIAL COUNTS AND MAINTENANCE SCHEDULE IN CARPETED AREAS, COMPARED TO NONCARPETED (TILE) AREAS

Period	Schedule of vacuuming carpet	Average number bacteria per ft^3 of air			Total number of *Staphylococcus aureus*	
		Tile[a]	Carpet	Difference	Tile	Carpet
Jan. 7−Feb. 11	Heavy vacuuming daily	14	11	−3	31	22
Feb. 18−Mar. 25	Heavy vacuuming on alternate days	11	11	0	14	13
Apr. 1−May 6	Heavy vacuuming every 3rd day	12	15	+3	41	22
June 10−July 8	Heavy vacuuming every 3rd day[b]	13	16	+3	19	24
July 15−Aug. 26	Heavy vacuuming on alternate days	12	14	+2	22	36

[a] The maintenance schedule on the tile floors was constant through the entire period of study.
[b] During the period between May 13 and June 3, the carpet was completely shampooed, beginning on the seventh floor. The vacuuming schedule on the carpet was kept at the 3-day interval to see if the shampooing had any effect.

Source: J.G. Shaffer, *Health Lab. Sci.,* 3, 73, 1966. With permission.

2.2−8   COMPARISON OF THE EFFECTIVENESS OF MOPPING vs. WET VACUUM PICKUP IN THE CONTROL OF BACTERIA ON A HOSPITAL CORRIDOR FLOOR

Description	Mean colonies per Rodac plate before cleaning	Reduction immediately after cleaning	Mean colonies per Rodac® plate − 60 samples per time period per day				Mean traffic census per 4-hr period
			½ hr after cleaning	1 hr after cleaning	2 hr after cleaning	4 hr after cleaning	
Mopping (22 runs)	188	68.9%	76	99	124	160	516
Wet vacuum pick-up (4 runs)	164	78.8%	52	95	139	172	526

Source: D. Vesley and G.S. Michaelsen, *Health Lab. Sci.,* 1, 107, 1964. With permission.

2.2–9 COMPARISON OF VARIOUS PATIENT ROOM FLOOR CLEANING PROCEDURES WITH ARBITRARY GUIDELINES OF A.P.H.A. COMMITTEE ON MICROBIAL CONTAMINATION OF SURFACES

Method	Type of product	Number of participating hospitals	Percent of participating hospitals achieving result		
			Good (<25 colonies per Rodac® after cleaning)	Fair (26–50 colonies per Rodac after cleaning)	Poor (>50 colonies per Rodac after cleaning)
Unstandardized mopping (1965)	Variable	14	7	29	64
Double bucket and mop (1966)	Phenolic	4	0	25	75
Wet-vacuum pick-up – test area only (1966)	Phenolic	8	63	0	38
Wet-vacuum pick-up – entire room – 13-min contact time (1967)	Phenolic	12	67	25	8
Wet-vacuum pick-up – entire room – 3-min contact time (1967)	Phenolic	13	54	31	15
Wet-vacuum pick-up – entire room – 5-min contact time (1968)	Phenolic	12	75	25	0
	Quaternary ammonium	12	83	17	0
	All-purpose cleaner (no disinfectant)	12	50	42	8

Source: D. Vesley et al., *Health Lab. Sci.*, 7, 256, 1970. With permission.

2.2–10 COMPARISON OF 1968 RESULTS (USING SPECIFIED WET-VACUUM PICK-UP METHOD) WITH ARBITRARY COMMITTEE GUIDELINES

By Number of Rooms

Type of product	Number of rooms	Good (<25 colonies per Rodac®)		Fair (26–50 colonies per Rodac)		Poor (>50 colonies per Rodac)	
		Number	%	Number	%	Number	%
Phenolic	88	72	82	8	9	8	9
Quaternary	87	74	85	9	10	4	5
All-purpose cleaner	89	58	65	17	19	14	16

Source: D. Vesley et al., *Health Lab. Sci.*, 7, 256, 1970. With permission.

2.2–11   ARBITRARY GUIDELINES FOR ACHIEVABLE LEVELS OF MICROBIAL CLEANLINESS OF FLOORS IN HOSPITAL PATIENT ROOMS[a]

Colonies per Rodac® plate

Good	Fair	Poor
0–25	26–50	Over 50

[a] Above guidelines are mean colony counts from 15 Rodac plates randomly applied within a floor area of about 8 ft² in a high traffic area immediately after cleaning, with the floor visibly dry. The guidelines are quantitative only, and the presence of specific pathogenic or opportunistic species would have to be interpreted separately.

Source: D. Vesley et al., *Health Lab. Sci.,* 7, 212, 1970. With permission.

2.2–12   MOP COUNTS MADE AFTER 2 DAYS' USE WITH GENERAL CLEANER

Before Routine Mop Decontamination

Mop location	Bacterial count per gram of mop	Rodac® plate count	Preliminary identification of organisms
Maid 4	3,500		*Staphylococcus aureus*
Maid 6	200		*Bacillus* (spore forming)
Maid 7	825		*Bacillus* (spore forming), *Pseudomonas*
Maid 8	17,100		*Bacillus* (spore forming), *Pseudomonas*
Maid 9	1,180		*Bacillus* (spore forming)
Maid 14	356		*Bacillus* (spore forming)
Janitor 5	67,600	500+	*Pseudomonas, S. aureus, Bacillus* (spore forming)

After Routine Mop Contamination

Janitor 5	None	1	*Bacillus* (spore forming)

Source: E. LaFave et al., *Hospitals,* 41, 83, 1967. With permission.

2.2–13 RESULTS OF MICROBIOLOGICAL ASSAYS
OF MOP STRINGS

When mops were assayed	Number of mops assayed	Number of microorganisms removed per string	
		Mean	Range
Before change of laundering procedures	17	4.9×10^3	$6.0 \times 10^1 - 2.9 \times 10^4$
After change of laundering procedures	8	5.6×10^0	$1.0 \times 10^0 - 1.7 \times 10^1$

Source: N.J. Petersen et al., *Health Lab. Sci.,* 10, 23, 1973. With permission.

2.2–14 MOP COUNTS MADE WITH RODAC® PLATES

Mop location	Bacterial count per gram of mop	Rodac plate count	Preliminary identification of organisms
Maid 7	None	1	*Staphylococcus aureus*
Maid 9	4,160	11	*S. aureus, Bacillus* (spore forming)
Maid 13	None	1	*Bacillus* (spore forming)
Janitor 5	67,600	500+	*Pseudomonas, S. aureus, Bacillus* (spore forming)

Source: E. LaFave et al., *Hospitals,* 41, 83, 1967. With permission.

2.2–15 MOP COUNTS MADE ON SAME DAY PHENOLIC
DISINFECTANT WAS USED

After Routine Mop Decontamination

Mop location	Appearance of mops	Bacterial count per gram of mop	Preliminary identification of organisms
Maid 4	Clean, white	None	
Maid 6	Clean, white	None	
Maid 7	Clean, white	None	
Maid 8	Clean, white	None	
Maid 9	Clean, white	None	
Maid 10	Clean, white	None	
Maid 12	Slightly soiled	None	
Janitor 1	Slightly soiled	100	*Bacillus* (spore forming)
Janitor 2	Heavily soiled	8800	*Proteus, Pseudomonas*
Janitor 3	Heavily soiled	3500	*Proteus, Pseudomonas*
Janitor 5	Heavily soiled	None	
Janitor 11	Slightly soiled	1200	*Staphylococcus aureus, Pseudomonas, Bacillus* (spore forming)

Source: E. LaFave et al., *Hospitals,* 41, 83, 1967. With permission.

2.2–16 MOP COUNTS MADE AFTER 5 DAYS' USE WITH GENERAL CLEANER

After Routine Mop Decontamination

Mop location	Bacterial count per gram of mop	Preliminary identification of organisms
Maid 4	None	
Maid 6	None	
Maid 7	None	
Maid 8	100	*Staphylococcus aureus*
Maid 9	100	*Bacillus* (spore forming)
Maid 10	100	*Bacillus* (spore forming)
Maid 12	None	
Janitor 1	300	*Pseudomonas*
Janitor 2	780	*Pseudomonas, Bacillus* (spore forming)
Janitor 3	100	*Bacillus* (spore forming)
Janitor 5	None	
Janitor 11	18,700	*Staphylococcus, Bacillus* (spore forming), *Pseudomonas*

Source: E. LaFave et al., *Hospitals*, 41, 83, 1967. With permission.

2.2–17 PERCENT DISTRIBUTION AFTER CLEANING OF ALL GROUPS OF ORGANISMS. SOLID SURFACES (RODAC® PLATES AND MOPPING ONLY) VS. CARPETS IN RODAC PLATES AND PLUGS

	Solid surface	Surface carpet	Plug carpet
Staphylococcus aureus	2	5	8
S. epidermidis	11	24	42
Enterobacter	4	8	5
Pseudomonas	2	12	6
Yellow rods	0	5	2
A.A.A.[a]	2	10	2
Gram-positive	68	28	32
No growth	11	8	3
	100	100	100

[a] A.A.A.: groups of *Acinetobacter, Alcaligenes,* and *Achromobacter.*

Source: G.J. Bonde, *Health Lab. Sci.,* 10, 308, 1973. With permission.

2.2–18 MICROBIAL FLOOR CONTAMINATION IN HOSPITAL PATIENT ROOMS BEFORE AND AFTER SPECIFIED MOPPING AND WET-VACUUM CLEANING PROCEDURES

	Mopping			Wet-vacuum			
		Mean colonies per Rodac®			Mean colonies per Rodac		
Hospital[a]	Number of rooms	Before	After	Number of rooms	Before	After	Routine predust
1	16	190	57	15	156	13	Yes
2	10	108	67	10	168	8	Yes
3	5	283	231	5	256	83	No
4	–	–	–	10	69	7	No
5	–	–	–	10	238	11	No
6	–	–	–	10	157	162	No
7	–	–	–	10	64	6	No
8	–	–	–	11	348	58	Partial
9	10	127	38	–	–	–	Yes
All hospitals (weighed mean)	41	166	76	81	178	39	

[a] Code numbers in Tables 2.2–18, 2.2–19, and 2.2–38 are independent. Hospitals with similar numbers in the three tables are *not* necessarily the same.

Source: D. Vesley et al., *Health Lab. Sci.,* 7, 256, 1970. With permission.

2.2–19 MICROBIAL FLOOR CONTAMINATION IN HOSPITAL PATIENT ROOMS BEFORE AND AFTER TWO SPECIFIED WET-VACUUM CLEANING PROCEDURES

1967 Data

Hospital[a]	Short contact time (average = 3 min)			Long contact time (average = 13 min)		
		Mean colonies per Rodac® plate			Mean colonies per Rodac plate	
	Number of rooms sampled	Before cleaning	After cleaning	Number of rooms sampled	Before cleaning	After cleaning
1	20	24	19	–	–	–
2	10	184	47	10	224	34
3	12	251	37	8	91	9
4	10	144	9	10	111	7
5	10	239	21	12	184	14
6	10	150	4	10	142	5
7	10	81	2	10	90	6
8	10	151	25	10	87	13
9	10	208	41	10	116	28
10	10	117	5	10	116	10
11	5	150	27	14	191	15
12	10	127	70	10	228	49
13	10	310	312	10	425	85
All hospitals (weighed mean)	137	156	46[b]	124	169	23

[a] Code numbers in Tables 2.2–18, 2.2–19, and 2.2–38 are independent. Hospitals with similar numbers in the three tables are *not* necessarily the same.

[b] The mean of 46 colonies per Rodac plate after the short contact time is disproportionately influenced by the results from Hospital #13. It should be pointed out that a large number of factors could not be controlled in these cooperation studies. Although protocols were written each year in an attempt to produce uniformity it is possible that significant variations in procedure may have occurred because of the large number of investigators, technicians, and hospital personnel who were eventually involved. Thus, it is difficult to make precise data comparisons from year to year or from different hospitals within the same year.

Source: D. Vesley et al., *Health Lab. Sci.,* 7, 256, 1970. With permission.

2.2–20   NUMBERS OF ORGANISMS OBTAINED BY THE PROBE METHOD IN THREE NYLON TILE CARPETED HOSPITAL AREAS

Numbers of Organisms per Square Centimeter

Tests, 1967	Before vacuuming			After vacuuming		
	Elevator	Site 1	Site 17	Elevator	Site 1	Site 17
July 17	260,000	100,000	4,800	100,000	61,000	1700
Aug. 24	58,000	105,000	26,000	41,000	65,000	5300
Sept. 5	119,000	108,000	4,500	93,000	82,000	3200
Oct. 5	71,000	51,000	18,000	66,000	26,000	2800

Source: W.G. Walter and A.H. Stober, *Health Lab. Sci.*, 5, 162, 1968. With permission.

2.2–21   THE AMOUNT OF SURFACE BACTERIA PER 24.6 cm^2 ON A NEEDLE-FELT CARPET BEFORE AND AFTER AN ORDINARY VACUUM CLEANING

Day	Before	After	Percent reduction
A	79	25	69
B	40	11	73
C	91	4	96
D	33	8	76
E	27	2	93
F	24	8	67
Mean value	54	9	79

Source: R. Rylander et al., *Am. J. Public Health*, 64, 163, 1974. With permission.

2.2–22   BACTERIA FOUND (PERCENTAGE OF THE NUMBER INITIALLY FOUND) AFTER CLEANING BY ORDINARY CLEANING EMPLOYEES AND BY LABORATORY TECHNICIANS IN ROOMS WITH PLASTIC TILES, TUFTED CARPETS, AND NEEDLE-FELT CARPETS

Mean Amount from Four to Five Tests with Eight to Ten Determinations in Each

Type of bacteria	Tile floor		Tufted carpet		Needle-felt carpet	
	O[a]	L[b]	O	L	O	L
Surface bacteria	210	0	50	5		7
Total amount			23	4		33

[a] Ordinary cleaning employees.
[b] Laboratory technicians.

Source: R. Rylander et al., *Am. J. Public Health*, 64, 163, 1974. With permission.

2.2–23 COMPARATIVE RESULTS OF RODAC® AND SIEVE-IN-MOTION SAMPLES ON CARPET SOILED FOR 7 DAYS PRIOR TO CULTURE

Sample number	Before vacuuming Number of colonies[a]		After vacuuming Number of colonies	
	Sieve	Rodac	Sieve	Rodac
1	_[b]	65	21	2
2	584	52	37	1
3	540	33	44	2
4	676	30	40	1
5	596	27	51	3
6	592	36	102	3
7	544	108	55	11
8	756	109	74	4
9	516	87	68	1
Average	601	61	55	3
Percent reduction			91.5	95

[a] All plates were incubated 24 hr at 37°C.
[b] Plate dropped before incubation.

Source: J.G. Shaffer, *Health Lab. Sci.,* 3, 73, 1966. With permission.

2.2–24 RESULTS OF SIEVE-IN-MOTION SAMPLES ON CARPET IN USE IN THE HOSPITAL, VACUUMED DAILY FOR 14 DAYS

Sample number	Number of colonies	
	Before vacuuming	After vacuuming
1	120	13
2	111	15
3	147	6
4	107	6
5	167	17
6	114	10
7	136	6
8	52	11
9	97	9
10	90	5
Average	114	10
Percent reduction		91

Source: J.G. Shaffer, *Health Lab. Sci.,* 3, 73, 1966. With permission.

2.2–25 CHARACTERISTICS OF CORRIDOR CARPETS INSTALLED IN BOZEMAN DEACONESS HOSPITAL

Type of carpet	Fiber	Construction	Pile height, inches type pile	Pile weight, ounces per square yard	Backing
Wool	100% natural animal fiber	Tufted	0.25 Looped	38	Polypropylene Polybac®
Veltron®	45 Denier nylon	Electrostatic	0.19 Cut pile	17	Latex-coated jute
Acrilan®	Acrylonitrile units	Woven, velvet	0.25 Looped	41.06	Latex-coated cotton, rayon
Herculon®	Polypropylene olefin	Tufted	0.50 Looped	22	Primary – polypropylene Secondary – jute
Nylon-tile	Polyamide	Tufted	0.25 Looped	16.9	Fiberglass and foam rubber
Nylon (Antron®)	Polyamide	Woven, Wilton	Looped	32	Jute and cotton latex coated

Source: W.G. Walter and A.H. Stober, *Health Lab. Sci.,* 5, 162, 1968. With permission.

2.2–26 EFFECT OF SHAMPOOING ON THE BACTERIAL POPULATION OF HOSPITAL CORRIDOR CARPETS AFTER 1 YEAR'S INSTALLATION

Organisms per Square Centimeter of Carpet

Type of carpet	Before vacuum	After vacuum	After shampoo	% reduction after shampooing
Veltron®a	3,900	2,700	400	90
Acrilan®b	11,300	5,600	4700	42
Wool^c	14,500	13,000	7400	49
Antron®d	5,800	5,100	1300	78
Herculon®e	1,400	200	100	93

[a] 45 Denier nylon.
[b] Acrylonitrile units.
[c] 100% natural animal fiber.
[d] Polyamide.
[e] Polypropylene olefin.

Source: W.G. Walter and A.H. Stober, *Health Lab. Sci.*, 5, 162, 1968. With permission.

2.2–27   EFFECTIVE METHODS FOR CLEANING STAINLESS STEEL

	Cleaning agent[a]	Method of application[b]	Effect on finish
Routine cleaning	Soap, ammonia, or detergent and water	Sponge with cloth, then rinse with clear water and wipe dry.	Satisfactory for use on all finishes.
Smears and fingerprints	Arcal 20, Lac-O-Nu, Lumin® Wash, O'Cedar® Cream Polish, Stainless Shine	Rub with cloth as directed on the package.	Satisfactory for use on all finishes. Provides barrier film to minimize prints.
Stubborn spots and stains, baked-on spatter, and other light discolorations	Allchem Concentrated Cleaner	Apply with damp sponge or cloth.	Satisfactory for use on all finishes.
	Samae, Twinkle®, or Cameo® Copper Cleaner	Rub with damp cloth.	Satisfactory for use on all finishes if rubbing is light.
	Grade FFF Italian pumice, whiting, or talc	Rub with damp cloth.	Use in direction of polish lines on No. 4 (polished) finish. Use light pressure on No. 2 (mill) finishes and on No. 7 and 8 (polished) finishes.
	Liquid NuSteel	Rub with dry cloth using small amount of cleaner.	Use in direction of polish lines on No. 4 (polished) finish. May scratch No. 2 (mill) and No. 7 and 8 (polished) finishes.
	Paste NuSteel or DuBois® Temp	Rub with dry cloth using small amount of cleaner.	Use in direction of polish lines on No. 4 (polished) finish. May scratch No. 2 (mill) and No. 7 and 8 (polished) finishes.
	Cooper's Stainless Steel Cleaner, Revere Stainless Steel Cleaner	Apply with damp sponge or cloth.	Use in direction of polish lines on No. 4 (polished) finish. May scratch No. 2 (mill) and No. 7 and 8 (polished) finishes.
	Household cleansers, such as Old Dutch®, Lighthouse®, Sunbrite, Wyandotte®, Bab-O, Gold Dust®, Sapolio, Bon Ami® Ajax®, or Comet®	Rub with damp cloth. May contain chlorine bleaches. Rinse thoroughly after use.	Use in direction of polish lines on No. 4 (polished) finish. May scratch No. 2 (mill) and No. 7 and 8 (polished) finishes.
	Grade F Italian pumice, Steel Bright, Lumin Cleaner, Zud®, Restoro, Sta-Clean®, or Highlite	Rub with damp cloth.	Use in direction of polish lines on No. 4 (polished) finish. May scratch No. 2 (mill) and No. 7 and 8 (polished) finishes.
	Penny-Brite or Copper-Brite®	Rub with dry cloth using small amount of cleaner.	Use in direction of polish lines on No. 4 (polished) finish. May scratch No. 2 (mill) and No. 7 and 8 (polished) finishes.
Heat tint or heavy discoloration	Penny-Brite or Copper Brite	Rub with dry cloth.	Use in direction of polish lines on No. 4 (polished) finish. May scratch No. 2 (mill) and No. 7 and 8 (polished) finishes.
	Paste NuSteel, DuBois Temp, or Tarnite	Rub with dry cloth or stainless steel wool.	Use in direction of polish lines on No. 4 (polished) finish. May scratch No. 2 (mill) and No. 7 and 8 (polished) finishes.

2.2—27 EFFECTIVE METHODS FOR CLEANING STAINLESS STEEL (continued)

	Cleaning agent[a]	Method of application[b]	Effect on finish
Heat tint or heavy discoloration (continued)	Revere Stainless Steel Cleaner	Apply with damp sponge or cloth.	Use in direction of polish lines on No. 4 (polished) finish. May scratch No. 2 (mill) and No. 7 and 8 (polished) finishes.
	Allen Polish, Steel Bright, Wyandotte, Bab-O, or Zud	Rub with damp cloth.	Use in direction of polish lines on No. 4 (polished) finish. May scratch No. 2 (mill) and No. 7 and 8 (polished) finishes.
Burnt-on foods and grease, fatty acids, milkstone (where swabbing or rubbing is not practical)	Easy-Off®; De-Grease-It®; 4 to 6% hot solution of such agents as trisodium phosphate or sodium tripolyphosphate, or 5–15% caustic soda solution	Apply generous coating. Allow to stand for 10–15 min. Rinse. Repeated application may be necessary.	Excellent removal; satisfactory for use on all finishes.
Tenacious deposits, rusty discolorations, industrial atmospheric stains	Oakite® No. 33, Dilac®, Texo® 12, Texo N.Y., Flash-Klenz, Caddy Claner, Turco® Scale 4368, or Permag® 57	Swab and soak with clean cloth. Let stand 15 min. or more according to directions on package, then rinse and dry.	Satisfactory for use on all finishes.
Hard-water spots and scale	Vinegar	Swab or wipe with cloth. Rinse with water and dry.	Satisfactory for use on all finishes.
	5% oxalic aid, 5% sulfamic acid, 5–10% phosphoric acid, Dilac, Oakite No. 33, Texo 12, or Texo N.Y.	Swab or soak with cloth. Let stand 10–15 min. Always follow with neutralizer rinse and dry.	Satisfactory for use on all finishes. Effective on tenacious deposits or where scale has built up.
Grease and oil	Organic solvents such as carbon tetrachloride, trichloroethylene, acetone, kerosene, gasoline, benzene, alcohol, or chloroethane	Rub with cloth. Organic solvents may be flammable and/or toxic. *Observe all precautions against fire.*, Do not smoke while vapors are present, and *be sure area is well ventilated.*	Satisfactory for use on all finishes.

a   Use of proprietary names is intended only to indicate a type of cleaner and *does not* constitute an endorsement; omission of any proprietary cleanser is not intended to imply its inadequacy. All products should be used in strict accordance with instructions on package.

b   In all applications stainless steel wool or sponge or fibrous brushes or pads are recommended. Avoid use of ordinary steel wool or steel brushes for scouring stainless steel.

Source: E.S. Kopecki, *Hospitals*, 41, 84, 1967. With permission.

2.2–28   THE EFFECT OF ULTRAVIOLET BARRIERS ON THE ESCAPE OF ORGANISMS FROM A PLENUM-VENTILATED CUBICLE

Total micrococci in 12 ft³ of air

Time of sampling	Doors	Inside cubicle		Corridor outside cubicle	
		UV off	UV on	UV off	UV on
Before release of organisms	Closed	21	10	4	13
2 min after release		532	497	17	15
4 min after release		461	534	17	14
6 min after release		231	284	39	13
8 min after release		212	270	30	11
Before release of organisms	Open	13	10	14	58
2 min after release		370	293	484	153
4 min after release		114	76	275	73
6 min after release		50	46	103	57
8 min after release		41	31	84	34

Source: G.A.J. Ayliffe et al., *J. Hyg.,* 69, 511, 1971. With permission of Cambridge University Press.

2.2–29   THE EFFECT OF ULTRAVIOLET LIGHT BARRIERS IN AN AIRLOCK

Slit-sampling Studies in Plenum-ventilated Units

Mean bacterial counts in 50 ft³/air

Site of sampling	UV on			UV off		
	Number of plates	Total organisms	Total *Staphylococcus aureus*	Number of plates	Total organisms	Total *S. aureus*
Cubicle	16	434.6 ± 121.7	6.3	13	309 ± 103.9	2.8
Airlock	16	120 ± 19.9	0.4	13	250.6 ± 35.3	1.5

Source: G.A.J. Ayliffe et al., *J. Hyg.,* 69, 511, 1971. With permission of Cambridge University Press.

2.2–30   THE EFFECT OF ULTRAVIOLET LIGHT BARRIERS IN AN AIRLOCK

Contact Plates from Floors of Plenum-ventilated Units
Mean bacterial count per plate

Site of sampling	UV on			UV off		
	Number of plates	Total organisms	Total *Staphylococcus aureus*	Number of plates	Total organisms	Total *S. aureus*
Cubicle	71	106.8 ± 18.8	3.6	68	269.9 ± 28.1	3.5
Airlock	81	12.8 ± 2.2	0.02	75	190.8 ± 19.1	3.6

Source: G.A.J. Ayliffe et al., *J. Hyg.,* 69, 511, 1971. With permission of Cambridge University Press.

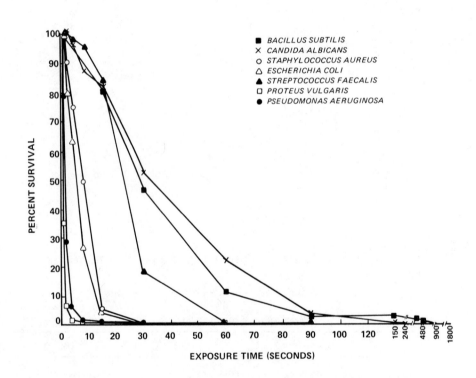

2.2–31   Germicidal efficiencies of the ultraviolet irradiated pass-through chambers of the patient isolator as determined for a select group of microbial species. The exposure distance of monocontaminated surfaces from two ceiling-mounted G-8T5[a] germicidal lamps was 12 in.; effective light transmission of the protective guards enclosing the lamps was 60%.

[a] Lamp Division, General Electric Co.

Source: S. Shadomy et al., *Arch. Environ. Health,* 11, 183, 1965. With permission. Copyright 1965, American Medical Association.

2.2–32   EFFICIENCIES OF SELECTED GERMICIDAL AGENTS IN SANITIZING MONOCONTAMINATED POLYVINYL CHLORIDE PLASTIC SURFACES

Effective removal rates

Agent	Agent concentration, %	Escherichia coli	Dev[a]	Pseudomonas aeruginosa	Dev[a]	Proteus vulgaris	Dev[a]
Saline	0.9	91.53	—	95.40	—	97.09	—
Wescodyne G®[b]	0.5	99.31	7.78	98.51	3.11	97.75	0.66
Tergisyl®[c]	1.0	98.84	7.31	97.93	2.53	99.22	2.13
Benzalkonium chloride[d]	1:750	99.18	7.65	99.42	4.02	98.81	1.72
Amphyl®[c]	0.5	98.08	6.55	99.18	3.78	99.06	1.97
SBT[e]	0.5	96.13	4.60	99.56	4.16	97.70	0.6i
SBT-24[e]	0.5	96.02	4.49	99.15	3.75	96.26	-0.83
Ves-phene®[f]	1.0	97.30	5.77	99.64	4.24	96.08	-1.01
Overall average percent removal of individual microbial contaminants by the various agents	—	97.84	6.31	99.06	3.66	97.84	0.75

Effective removal rates

Agent	Agent concentration, %	Staphylococcus aureus	Dev[a]	Streptococcus faecalis	Dev[a]	Group C β hemolytic streptococci	Dev[a]
Saline	0.9	93.40	—	87.45	—	97.74	—
Wescodyne G	0.5	99.47	6.07	88.14	0.69	96.17	1.57
Tergisyl	1.0	98.68	5.28	98.93	11.48	99.51	1.77
Benzalkonium chloride	1:750	99.41	6.01	97.96	10.51	99.93	2.19
Amphyl	0.5	96.15	2.75	96.50	9.05	99.38	1.64
SBT	0.5	93.33	-0.07	94.02	6.57	96.11	-1.63
SBT-24	0.5	94.61	1.21	87.46	0.01	97.48	-0.26
Ves-phene	1.0	98.35	5.05	99.17	11.72	98.53	0.79
Overall average percent removal of individual microbial contaminants by the various agents	—	97.14	3.74	94.60	7.15	98.16	0.42

2.2–32 EFFICIENCIES OF SELECTED GERMICIDAL AGENTS IN SANITIZING MONOCONTAMINATED POLYVINYL CHLORIDE PLASTIC SURFACES (continued)

Agent	Agent concentration %	Effective removal rates		Overall effective removal rate (average)[g]	Dev[a]
		Bacillus subtilis	Dev[a]		
Saline	0.9	96.30	–	94.13	–
Wescodyne G	0.5	95.16	-1.14	96.36	2.23
Tergisyl	1.0	99.60	3.30	98.96	4.83
Benzalkonium chloride	1:750	98.66	2.36	99.05	4.92
Amphyl	0.5	98.95	2.65	98.19	4.06
SBT	0.5	97.52	1.22	96.34	2.21
SBT-24	0.5	96.81	0.50	95.40	1.27
Ves-phene	1.0	99.59	3.29	98.38	4.25
Overall average percent removal of individual microbial contaminants by the various agents	—	98.04	1.74	97.53	3.40

Note: Evaluation of germicides for use in concurrent sanitization of the occupied patient isolator. The efficiencies of the different candidate germicidal agents in effecting sanitization of monocontaminated plastic (polyvinyl chloride) strips were determined for a select group of microbial species. The efficiencies were evaluated in terms of the number of organisms removed from the contaminated plastic surfaces as compared with the removal rates obtained for the same organisms with sterile saline applied under identical conditions. The individual effective removal rate values represent the percentage differences between the number of viable cells recovered from monocontaminated plastic strips following treatment with either test agent or saline and the number recovered from the corresponding untreated control strips. The various effective removal rate values represent arithmetic means derived from a minimum of three repetitive experiments for each series of test agent and bacterial contaminant combinations. The original percentages of recovery from which the various mean removal values are derived were selected on the basis of having a combined value for Sk of ± 1.0 or less.

a Dev: the difference between the effective removal rate obtained for the individual germicide-contaminant combination and that obtained for the corresponding saline control-contaminant combination as treated under mechanically identical conditions. The values expressed as Dev are taken as representing the actual difference between the physical removal of surface contaminants produced by technique of assay and the true germicidal destruction or elimination as effected by the test agent under evaluation.

b West Chemical Products, Inc.

c Lehn and Fink Products Corp.

d Lever Bros. Co., Industrial Division.

e Certified Laboratories, Inc.

f Vestal Laboratories.

g Overall effective removal rate: the average of the effective removal rates obtained for the complete battery of surface contaminants for each of the germicidal agents.

Source: S. Shadomy, et al., *Arch. Environ. Health*, 11, 183, 1965. With permission. Copyright 1965, American Medical Association.

2.2–33   SUMMARY OF INDICATED USES FOR VARIOUS CHEMICAL GERMICIDES

Chemical germicide	Use concentration	Indicated uses	Comments
Alcohol (isopropyl or ethyl)	70–90%	Clean, hard surfaces (thermometers); clean polyethylene tubing; lensed instruments; anesthesia equipment; skin antiseptic	Volatile, no residual, dries skin, not sporicidal
Chlorine	4–5%	Bathrooms; lavatories; laundry bleach	Flash action, inactivated readily by organic matter, corrosive, irritating
Iodine Tincture	0.5–8% iodine, 47–83% alcohol	Skin antiseptic	Stains and irritates skin
Iodophor	75–450 ppm	Smooth, hard surfaces: tubing; lensed instruments: floors; furnishings; plastic mattress covers	Newer products reduce staining, some sporicidal activity at high concentrations
Phenolics Germicide	1–3%	Smooth, hard surfaces; plastic; rubber tubing; lensed and hinged instruments; etc.	Somewhat corrosive and irritating, not sporicidal
Germicidal detergent	0.5–3%	Floors; walls; furnishings	Stable, not readily inactivated by organic matter, irritating to skin, not sporicidal
Hexachlorophene (bisphenol)	1%	Skin antiseptic; hand scrub	Selective for Gram-positive bacteria, consistent use recommended, may be bacteriostatic only
Quaternary ammonium compounds	1:1000–1:500	Smooth, hard surfaces; instruments; floors; walls; furnishings	Noncorrosive, nonirritating, some formulations not effective against tubercle bacilli or Gram-negative bacteria, not sporicidal
Glutaraldehyde	2% aqueous	Smooth, hard surfaces; tubing; lensed instruments; anesthesia equipment; forceps; inhalation therapy equipment	Sporicidal in 3 to 10 hr unstable (2 weeks maximum), must be rinsed thoroughly
Formaldehyde	3–8% aqueous	Emergency spills of infectious materials	Sporicidal, irritating, toxic, corrosive
Formaldehyde + alcohol (Bard-Parker solution)	8% formaldehyde, 60% ethanol, 10% methanol	Smooth, hard surfaces; tubing; lensed and hinged instruments; thermometers	Sporicidal, volatile, toxic
Chlorhexidine	0.5% in water or alcohol	Bacteriostatic agent in ophthalmic solutions; skin antiseptic, particularly for urology, gynecology	Relatively nontoxic, requires higher concentration against Gram-negatives, popular in British Commonwealth but not U.S.

Source: R.G. Bond et al., Eds., *Environmental Health and Safety in Health-Care Facilities,* Macmillan, New York, © 1973, 57. With permission.

2.2–34   BACTERIAL FLOOR COUNTS AFTER MOPPING WARD FLOOR

| | Mean total impression plate counts on covered floor 1 hr after cleaning with | | | |
| | Soap and water | | Sudol 1/100 | |
Time of sampling floor	Number of plates	Mean total count	Number of plates	Mean total count
Before cleaning	30	337	9	325
After cleaning first area	18	6	6	4
After cleaning one third of ward	24	32	6	5
After cleaning two thirds of ward	30	104	6	4

Source: G.A.J. Ayliffe et al., *J. Hyg.,* 65, 515, 1967. With permission of Cambridge University Press.

2.2–35   BACTERIAL COUNTS FROM MOP WATER DURING AND AFTER MOPPING A WARD FLOOR

| | Mean viable counts on treatment of floor with | | | |
| | Soap and water | | Sudol 1/100 | |
	Number of samples	Mean total counts per milliliter	Number of samples	Mean total counts per milliliter
Before cleaning	5	10	1	20
After cleaning one third of ward	5	650	1	10
After cleaning two thirds of ward	5	15,000	1	30
After cleaning complete ward	4	34,000	1	20

Source: G.A.J. Ayliffe et al., *J. Hyg.,* 65, 515, 1967. With permission of Cambridge University Press.

2.2–36   COMPARISON OF EFFECTIVENESS OF THREE GERMICIDAL PRODUCTS IN THE CONTROL OF BACTERIA ON A HOSPITAL CORRIDOR FLOOR

Description	Mean colonies per Rodac plate before cleaning	Reduction immediately after cleaning	Mean colonies per Rodac® plate based on 60 samples per time period per day			
			½ hr after cleaning	1 hr after cleaning	2 hr after cleaning	4 hr after cleaning
Product A – (2 runs) A quaternary ammonium nonionic detergent germicide.	131	66.3%	58	92	126	174
Product B – (2 runs) A synthetic multiphenolic detergent germicide	173	70.8%	72	86	118	156
Product C – (2 runs) A synthetic detergent with a small percent of phenolic germicide	177	51.8%	90	116	149	205

Source: D. Vesley and G.S. Michaelsen, *Health Lab. Sci.,* 1, 107, 1964. With permission.

2.2–37   COMPARISON OF THE EFFECTIVENESS OF HOT WATER, DETERGENTS, AND GERMICIDES IN THE CONTROL OF BACTERIA ON A HOSPITAL CORRIDOR FLOOR

Description	Mean colonies per Rodac plate before cleaning	Reduction immediately after cleaning	Mean colonies per Rodac® plate – 60 samples per time period per day				Mean traffic census per 4-hr period
			½ hr after cleaning	1 hr after cleaning	2 hr after cleaning	4 hr after cleaning	
Three germicides[a] Product A – 2 runs Product B – 10 runs Product C – 2 runs	181	71.1%	69	91	120	165	540
Non germicidal detergent (based on 2 runs)	180	74.4%	63	82	115	171	Data not available
Hot tap water (based on 10 runs)	190	68.7%	79	102	137	156	477

[a] Product A: a quaternary ammonium nonionic detergent germicide. Product B: a synthetic multiphenolic detergent germicide. Product C: a synthetic detergent with a small percent of phenolic germicide.

Source: D. Vesley and G.S. Michaelsen, *Health Lab. Sci.,* 1, 107, 1964. With permission.

2.2–38   MICROBIAL FLOOR CONTAMINATION IN HOSPITAL PATIENT ROOMS BEFORE AND AFTER A STANDARDIZED WET-VACUUM CLEANING PROCEDURE COMPARING THREE DIFFERENT TYPES OF PRODUCTS

1968 Data

	A (phenolic)			B (all-purpose cleaner)			C (quaternary)		
		Mean colonies Per Rodac®			Mean colonies per Rodac			Mean colonies per Rodac	
Hospital[a]	Number of rooms	Before	After	Number of rooms	Before	After	Number of rooms	Before	After
1	8	66	11	8	45	8	8	114	14
2	8	81	9	8	122	25	8	148	18
3	8	119	19	8	100	29	7	45	4
4	7	34	8	7	17	35	8	22	11
5	6	71	3	6	121	3	6	68	4
6	8	70	27	8	147	38	8	79	10
7	7	145	23	7	162	18	7	141	18
8	8	49	4	8	59	12	4	79	43
9	8	118	30	8	111	48	8	95	17
10	7	88	13	8	109	24	8	156	6
11	4	57	20	3	59	111	5	66	14
12	9	179	48	10	211	38	10	95	40
All hospitals (weighed mean)	88	93	19	89	110	29	87	94	17

[a] Code numbers in Tables 2.2–18, 2.2–19, and 2.2–38 are independent. Hospitals with similar numbers in the three tables are *not* necessarily the same.

Source: D. Vesley et al., *Health Lab. Sci.,* 7, 256, 1970. With permission.

2.2–39   MICROBIAL FLOOR CONTAMINATION IN HOSPITAL PATIENT ROOMS BEFORE AND AFTER CLEANING BY VARIOUS METHODS

1965–1968 Data

Method	Type of product	Mean colonies per Rodac® plate	
		Before cleaning	After cleaning
Unspecified mopping procedure (1965 data)	Variable	230	99
Specified mop and double bucket – entire room (1966 data)	Phenolic	166	76
Specified wet-vacuum spray and pick-up – test area only (1966 data)	Phenolic	178	39
Specified wet-vacuum spray and pick-up – entire room – ~3-min contact time (1967 data)	Phenolic	156	46[a]
Specified wet-vacuum spray and pick-up – entire room – ~13-min contact time (1967 data)	Phenolic	169	23
Specified wet-vacuum spray and pick-up – entire room – 5-min contact time (1968 data)	Phenolic	93	19
	Quaternary ammonium	94	17
	General purpose cleaner (no disinfectant)	110	29

[a] The mean of 46 colonies per Rodac plate after the short contact time is disproportionately influenced by the results from one hospital. It should be pointed out that a large number of factors could not be controlled in these cooperation studies. Although protocols were written each year in an attempt to produce uniformity it is possible that significant variations in procedure may have occurred because of the large number of investigators, technicians, and hospital personnel who were eventually involved. Thus, it is difficult to make precise data comparisons from year to year or from different hospitals within the same year.

Source: D. Vesley et al., *Health Lab. Sci.*, 7, 256, 1970. With permission.

2.2–40   BACTERIA IN THE CARPET AFTER DISINFECTION AS PERCENTAGE OF AMOUNT IN UNTREATED CONTROL CARPET

Contaminant	Treatment	Needle-felt	Tufted	Wool
Staphylococcus aureus	Chloramine 3%	0		
in blood	VA[a] 0.4%	16		
Escherichia coli	Chloramine 3%	0		
in urine	Ethanol 70% × 2	0		
	VA 0.6%	7	1	19
	0.6 × 2	2	<1	
	1%	7	<1	2
	2%	1	<1	2
	Acetic acid			
	0.1%		7	80
	0.5%		8	40
	Chlorhexidine 3.3%	<1		

[a] An orthophenylphenol preparation.

Source: R. Rylander et al., *Am. J. Public Health*, 64, 163, 1974. With permission.

2.2–41   *ESCHERICHIA COLI* IN NaCl AND URINE 15 min AFTER DISINFECTION WITH 0.08% VA[a] AT APPLICATION OF CARPET YARN TO THE SUSPENSION EXPRESSED AS A PERCENTAGE OF THE ORIGINAL AMOUNT

	Additive	
	NaCl	Urine
Control	<1	21
Carpet yarn	38	72
Carpet yarn—filtered before disinfection	63	

[a] An orthophenylphenol preparation.

Source: R. Rylander et al., *Am. J. Public Health*, 64, 163, 1974. With permission.

2.2–42   BACTERIAL REDUCTION ON HUBBARD TANK SURFACE WITH AND WITHOUT DISINFECTION

| | | Mean contact plate count[a] | | | | |
Treatment	Location	After emptying tank	After scrub rinse	Reduction, %	After chlorine disinfection	Total reduction, %
With chlorine disinfection	Side 1	382	11	97	11	97
	Side 2	364	18	95	9	98
	Bottom	370	37	90	12	97
	Combined	372	22	94	11	97
Without chlorine disinfection	Side 1	196	3	98	–	98
	Side 2	244	5	98	–	98
	Bottom	296	15	95	–	95
	Combined	245	8	97	–	97

[a] Each count represents an average from 100 contact plates.

Source: A.G. Turner et al., *Arch. Environ. Health*, 28, 101, 1974. With permission. Copyright 1974, American Medical Association.

2.2–43   SELECTED BACTERIAL SPECIES ISOLATED FROM HUBBARD TANK BATH WATER OF 11 PATIENTS

| Bacteria[a] | Number of negative samples | Positive samples, counts per milliliter of water | |
		Minimum	Maximum
Coliforms	4	3	87,333
Staphylococci	7	3	7,080
Fungi	18	3	9,850
Pseudomonas	27	8	681

[a] Characterized by selective media only.

Source: A.G. Turner et al., *Arch. Environ. Health*, 28, 101, 1974. With permission. Copyright 1974, American Medical Association.

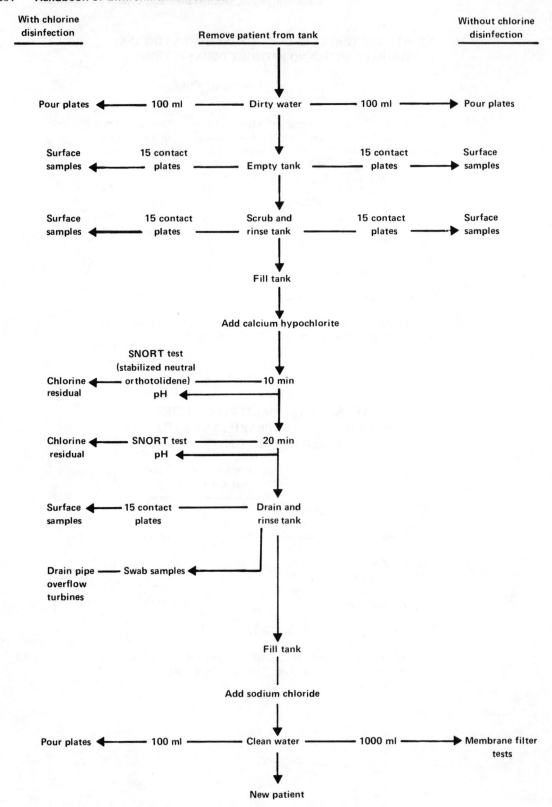

2.2–44   Sampling protocol for Hubbard Tank. A detergent-germicide was always used in scrubbing tank.

Source: A.G. Turner et al., *Arch. Environ. Health*, 28, 101, 1974. With permission. Copyright 1974, American Medical Association.

2.2--45   ISOLATIONS OF GRAM-POSITIVE AND GRAM-NEGATIVE BACTERIAL SPECIES FROM PUS, BLOOD, AND CEREBROSPINAL FLUID

	Units with skin disinfection			Units without skin disinfection		
	Jan. 1970	Aug. 1970	$x^2_{(1)}$ P	Jan. 1970	Aug. 1970	$x^2_{(1)}$ P
Total number of cultures	53	29	–	40	28	–
Number of cultures from which Gram-positive bacteria were isolated	33	8	7.68 $0.005 < P < 0.01$	20	5	6.00 $0.01 < P < 0.02$
Number of cultures from which Gram-negative bacteria were isolated	9	13	6.05 $0.01 < P < 0.025$	3	7	2.74 $0.05 < P < 0.10$

Source: M.F. Michel and C.C. Priem, *J. Hyg.*, 69, 453, 1971. With permission of Cambridge University Press.

2.2—46   DISEASE CATEGORIES AND ISOLATION INSTRUCTIONS

Disease	Instructions	Period of isolation
Strict Isolation		
Diphtheria		At least 7 days and until two nose and throat cultures are negative
Staphylococcal pneumonia, exudative	Wear gloves during care.	Until lesions are healed
Streptococcal infections (group A, β hemolytic)	Wear gloves during care.	Until 24 hr of adequate penicillin therapy
Meningococcal infections	Staff extensively and intimately exposed to patient may receive penicillin prophylaxis.	Until 2 days after start of amtimicrobial therapy
Smallpox		Until diagnosis disproven
Tuberculosis (suspect or proven active when open[a])		Until two or three consecutive sputum or gastric aspirates are negative on staining
Wound and Skin Isolation		
Staphylococcus, Pseudomonas, or other gross wound infection	Wear gloves during care.	Until clinical recovery
Severe dermatitis with broken skin	Wear gloves during care.	Until clinical recovery
Gas gangrene	Linen autoclaved before laundering, minimal 45 min at 15 lb of pressure; instruments must be thoroughly scrubbed before autoclaving.	Until lesions clear
Syphilis (with skin or mucous membrane lesions)	Care of secretions, wear gloves, autoclave syringes.	Until lesions clear
Fungus infections, ringworm of the scalp only[b]		Until lesions clear
Enteric Isolation		
Leptospirosis	Urine precautions only	Until clinical recovery
Amebiasis	Stool precautions	Until clinical recovery
Hepatitis, serum and infectious	Discard needles and thermometers, autoclave syringes, stool precautions for infectious hepatitis only.	Total duration of hospital stay
Shigellosis (bacillary disentery)	Stool precautions	Until three consecutive stool cultures are negative

2.2–46   DISEASE CATEGORIES AND ISOLATION INSTRUCTIONS (continued)

Disease	Instructions	Period of isolation
	Enteric Isolation (continued)	
Salmonellosis (typhoid and paratyphoid)	Stool precautions	Until three consecutive stool cultures are negative
Poliomyelitis	Stool precautions	At least 4 weeks after onset
Aseptic meningitis	Stool precautions	At least 4 weeks after onset
Viral myocarditis or pericarditis	Stool precautions	At least 4 weeks after onset
	Protective Isolation	

For infants: premature, newborn, and in 1st year of life
For acutely ill patients with low resistance to bacterial infections such as patients with agranulocytosis, acute renal failure, and burns. (Use of topical agents for treatment of burns may make isolation unnecessary in the future.)
Close adherence to handwashing technique

[a]Open tuberculosis is defined as suspect or proven cavitary pulmonary disease or a draining lesion anywhere. Primary miliary meningeal, genitourinary, or bone tuberculosis is not considered to be open unless a draining lesion is present. In genitourinary tuberculosis, urine should be voided into commode. If it is collected in a vessel, this should be considered contaminated.
[b]Except for ringworm of the scalp, spread from man to man is not well documented.

Source: C.M. Kunin, *Hospitals*, 42, 96, 1968. With permission.

2.2–47   CONTENTS OF DOOR SIGNS USED ON ISOLATION DOORS

Color of sign	Category of isolation	Instructions
Yellow	Strict	Gowns and masks: to be worn by all persons entering room and discarded before leaving the room Hands: to be washed thoroughly on entering and leaving room Articles: to be discarded, or washed and disinfected, or wrapped for autoclaving before taken from room
Pink	Wound and skin	Gowns and gloves: to be worn only by persons having direct contact with patient Masks: to be worn by all persons entering room Hands: to be washed thoroughly on entering and leaving room Articles: to be discarded, or washed and disinfected, or wrapped for autoclaving before taken from room
Brown	Enteric	Gowns and gloves: to be worn only by persons having direct contact with patient Masks: not necessary Hands: to be washed thoroughly on entering and leaving room Articles: to be discarded, or washed and disinfected, or wrapped for autoclaving before taken from room. For patients with hepatitis, needles are to be discarded, syringes must be autoclaved.
Green	Protective	Gowns and masks: to be worn by all persons entering room Hands: to be washed thoroughly on entering room

Source: C.M. Kunin, *Hospitals*, 42, 96, 1968. With permission.

2.2–48 SUPPORT SERVICES FOR LAMINAR FLOW ROOMS

Service	Additional staff	Additional hours per patient	Training	Special supplies	Space	Turnover	Problem areas
Nursing	2 RNs and 1 LPN per 4 rooms, or 1 RN per 2 rooms; 1 RN per critically ill patient	–	5 days to 2 months (depends on sterile techniques background)	–	Twice as much room space per patient	Some immediately because of complicated protocols	Maintaining supply levels; staffing for critically ill patients
Central supply	2 per 4 rooms (with disposables); 4 per 4 rooms (no disposables); 1 1/3 per 2 rooms (with disposables); 20 hr per week per 2 beds for wrapping	–	2 days if personnel have experience	Disposables; gas sterilizer; aeration chamber	20 × 20 ft storage room area for wrapping	Low	Storage; projecting needs; maintaining stock
Pharmacy	None	1–5 hr per patient week	None	Special antibiotics; i.v. tubing	No additional	Low	None
Dietary	2 technicians, ½ dietitian (4 rooms, sterile diet); 1 technician, ¼ dietitian (2 rooms, cooked food diet)	–	3–4 weeks	Bags for sterilizing; individual portions of canned goods	Sterile diet kitchen or room for preparation and storage	Low	Storage, palatability of food
Housekeeping	Only for cleaning after patient discharge	–	4 hr instruction, plus supervised practice	Cleaning solutions, sterile mops, etc.	Storage closet for supplies	Low	Waste generation
Maintenance	None	1 hr per week (4 rooms)	1 week	Replacement parts	Parts storage	Low	Crisis maintenance must be inhouse; maintaining parts inventory

2.2–48   SUPPORT SERVICES FOR LAMINAR FLOW ROOMS (continued)

Service	Additional staff	Additional hours per patient	Training	Special supplies	Space	Turnover	Problem areas
Bacteriology	1 technician per 4 rooms; biologist or bacteriologist	–	Minimal if personnel have experience	–	–	Low	Careful monitoring of standing water required
Social and psychiatric services	Usually none	2–4 hr per week with patients and staff	None	–	–	Low	Providing recreational activities for patients; high referral rate of patients due to close staff observation

Source: E.C. Drazen and A.S. Levine, *Hospitals*, 48, 89, 1974. With permission.

2.2–49   EFFECT OF VARIOUS LAUNDRY PROCESSES ON VIABLE MICROORGANISMS

Process	Temperature and treatment time	Effect
Sudsing cycle	150–170° F/30 min	Equivalent to disinfection cycle
Rinsing cycle	165–170° F/10–12 min	Additional lethal effect
Bleach	~100 ppm chlorine/10–15 min	Lethal to vegetative organisms
Sour	Reduces pH from ~10 to ~5 in short time	Lethal to some Gram-negative organisms
Hot-air dryers	~160° F for 20–30 min (dry heat)	Lethal to some Gram-negative organisms
Hot ironing	~330°F for a few seconds	Additional lethal effect
Special cycle for synthetics or highly colored linens	~120° F/30 min	Cannot be depended on to destroy vegetative bacteria

Source: R.G. Bond et al., Eds., *Environmental Health and Safety in Health-Care Facilities*, Macmillan, New York, © 1973, 339. With permission.

Section 3

Environmental Hygiene and
Radiological Health

---

3. ENVIRONMENTAL HYGIENE AND RADIOLOGICAL HEALTH

3.1   *VENTILATION AND AIR CONDITIONING*

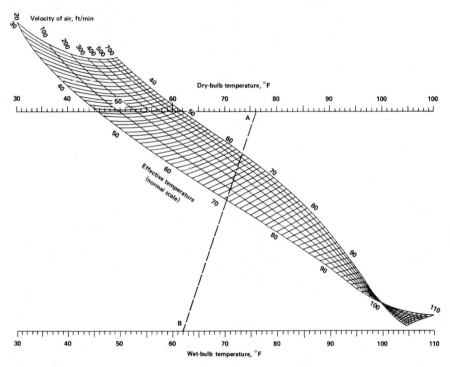

3.1–1   Effective temperature chart for normally clothed sedentary individuals.

Note: To use the chart, draw line A–B through measured dry-bulb and wet-bulb temperature. Read effective temperature or velocity at desired intersections with line A–B. Example: Given 67°F dry bulb and 62°F wet bulb, read 69 *ET* at 100 fpm velocity, or 340 fpm required for 66 *ET*.

Source: American Society of Heating, Refrigerating, and Air Conditioning Engineers, *ASHRAE Guide and Data Book: Handbook of Fundamentals,* ASHRAE, New York, 1967. With permission.

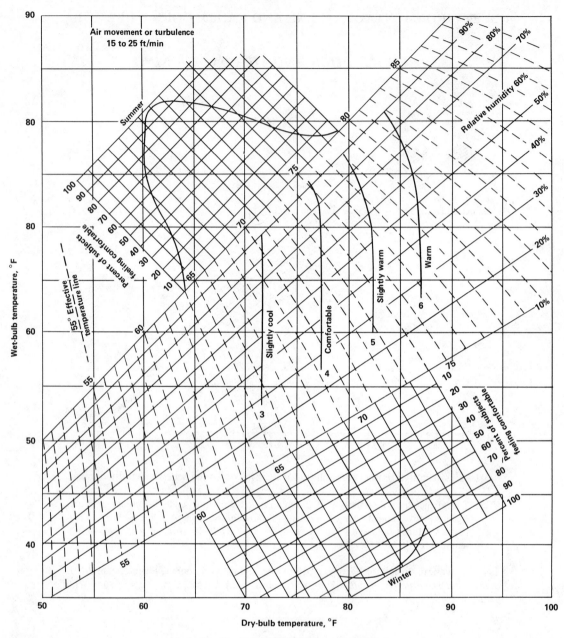

3.1—2   Revised ASHRAE comfort chart.

Source: American Society of Heating, Refrigerating, and Air Conditioning Engineers, *ASHRAE Guide and Data Book: Handbook of Fundamentals,* ASHRAE, New York, 1967. With permission

3.1—3 EVALUATION OF HEAT-STRESS INDEX

Index of heat stress	Physiologic and hygienic implications of 8-hr exposures to various heat stresses
−20 −10	Mild cold strain. This condition frequently exists in areas where men recover from exposure to heat.
0	No thermal strain.
+10 20 30	Mild to moderate heat strain. Where a job involves higher intellectual functions, dexterity, or alertness, subtle to substantial decrements in performance may be expected. In performance of heavy physical work, little decrement expected unless ability of individuals to perform such work under no thermal stress is marginal.
40 50 60	Severe heat strain, involving a threat to health unless men are physically fit. Break-in period required for men not previously acclimatized. Some decrement in performance of physical work is to be expected. Medical selection of personnel desirable because these conditions are unsuitable for those with cardiovascular or respiratory impairment or with chronic dermatitis. These working conditions are also unsuitable for activities requiring sustained mental effort.
70 80 90	Very severe heat strain. Only a small percentage of the population may be expected to qualify for this work. Personnel should be selected (1) by medical examination and (2) by trial on the job (after acclimatization). Special measures are needed to assure adequate water and salt intake. Amelioration of working conditions by any feasible means is highly desirable, and may be expected to decrease the health hazard while increasing efficiency on the job. Slight indisposition which in most jobs would be insufficient to affect performance may render workers unfit for this exposure.
100	The maximum strain tolerated daily by fit, acclimatized young men.

Source: American Society of Heating, Refrigerating, and Air Conditioning Engineers, *ASHRAE Guide and Data Book: Handbook of Fundamentals,* ASHRAE, New York, 1967, 120. With permission.

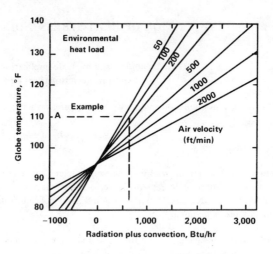

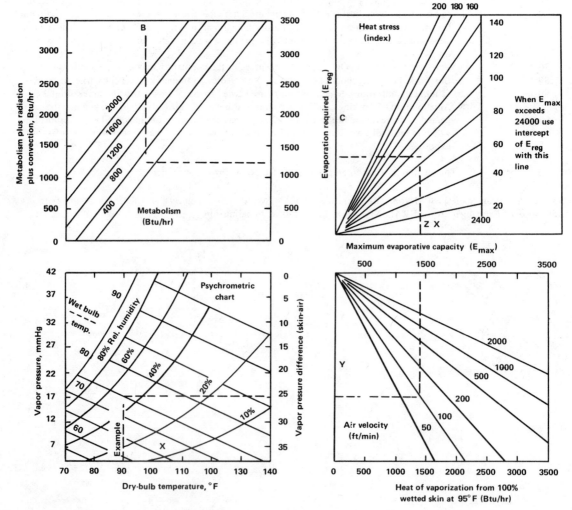

3.1–4   Charts for determining heat stress.

Note: Data are for an 8-hr day exposure, whereas a 24-hr exposure will be the norm for institutional patients.

Source: American Society of Heating, Refrigerating, and Air Conditioning Engineers, *ASHRAE Guide and Data Book: Handbook of Fundamentals,* ASHRAE, New York, 1967. With permission.

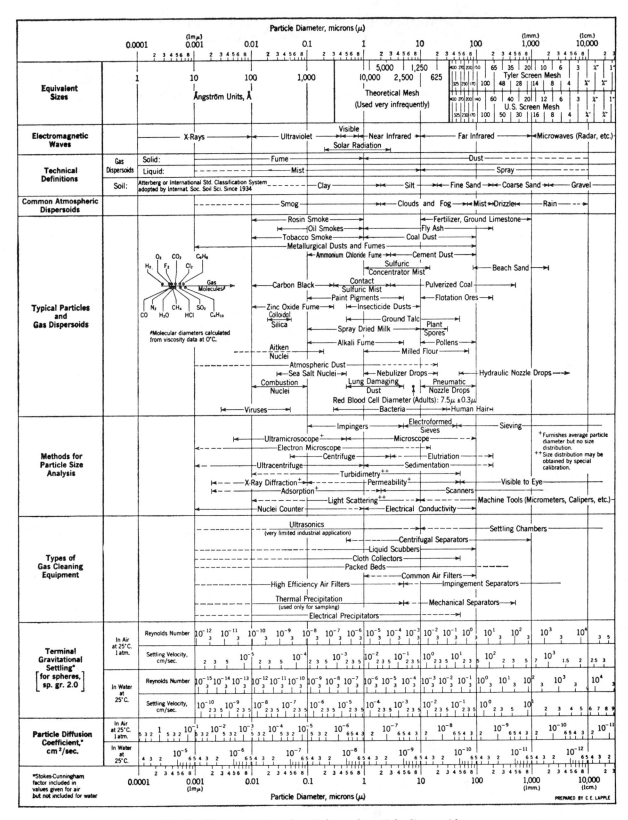

3.1—5   Characteristics of particles and particle dispersoids.

Source: C.E. Lapple, *Stanford Res. Inst. J.,* 5, 95, 1961. With permission.

3.1−6   MINIMUM ODOR-FREE AIR REQUIREMENTS TO REMOVE OBJECTIONABLE BODY ODORS UNDER LABORATORY CONDITIONS

Type of occupants	Air space per person in cubic feet	Odor-free air supply, cfm per person	Type of occupants	Air space per person in cubic feet	Odor-free air supply, cfm per person
Heating season with or without recirculation. Air not conditioned. Sedentary adults of average socioeconomic status			Heating season. Air humidified by means of centrifugal humidifier. Water atomization rate 8 to 10 gph. Total air circulation 30 cfm per person. Sedentary adults		
	100	25			
	200	16			
	300	12			
	500	7		200	12
Laborers	200	23			
Grade-school children of average socioeconomic status			Summer season. Air cooled and dehumidified by means of a spray dehumidifier. Spray water changed daily. Total air circulation 30 cfm per person. Sedentary adults		
	100	29			
	200	21			
	300	17			
	500	11			
Grade-school children of lower socio-economic status	200	38		200	4
Children attending private grade schools	100	22			

Source: American Society of Heating Refrigerating, and Air Conditioning Engineers, New York, 1967. With permission.

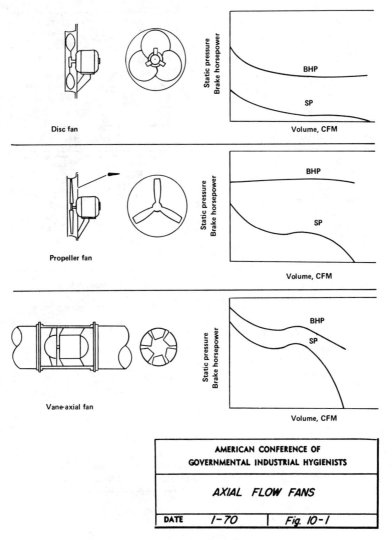

3.1−7   Axial-flow fans.

Source: American Conference of Governmental Industrial Hygienists, Committee on Industrial Ventilation, *Industrial Ventilation — A Manual of Recommended Practice,* 11th ed., ACGIH, Lansing, Mich., 1970. With permission.

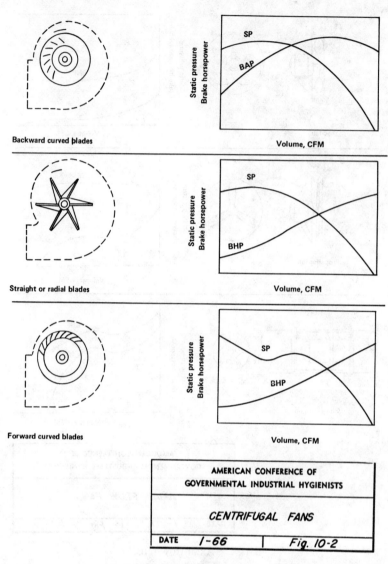

3.1—8   Centrifugal fans.

Source: American Conference of Governmental Industrial Hygienists, Committee on Industrial Ventilation, *Industrial Ventilation — A Manual of Recommended Practice,* 11th ed., ACGIH, Lansing, Mich., 1970. With permission.

3.1−9   AIR MIXING WITHIN THE OPERATING ROOM WITH THE THREE VENTILATING SYSTEMS

Values of the Performance Index[a] Using Nitrous Oxide as a Tracer Gas

System	Turbulent air velocity 1 ft above operating table, feet per minute	Tracer found over table; tracer liberated at			Tracer liberated around table at 3-ft level; tracer found at		
		5-ft-6-in. level	3-ft level	1-ft-6-in. level	5-ft-6-in. level	3-ft level	1-ft-6-in. level
A	9	3.6	1.1	0.4	0.6	0.9	2.3
B	25	1.6	1.1	0.4	1.3	1.4	About 2.0
C	9	Not done	1.0	Not done	0.6	0.7	About 2.8
[P]	Small	Large	Near unity	Very small	Very small	Small	Large
[M]	Large	1.0	1.0	1.0	1.0	1.0	1.0

Notes: These measurements were made during the third period of the investigation when good temperature control had been established and the improved diffusers (perforated metal shells) had been fitted to the downward displacement system (System A). The row of the table labeled [P] shows the values that would be expected in an effective downward displacement system. That labeled [M] shows that those would result from complete turbulent mixing.

System A. Downward displacement: the ventilating air was introduced through six diffusers A1−A6 spaced, as far as possible, evenly over the ceiling.

System B. Moderate velocity turbulence: the air entered the room through three grilles B1−B3 fitted with directional deflectors. These deflectors were adjusted so as to direct all three air streams towards and above the operating table. The velocity of air movement 1 ft above the table averaged around 25 ft/min when the rate of air supply was 1400 ft³/min.

System C. Low velocity turbulence: the air was introduced vertically downwards at low velocity through three large grilles C1−C3 in the ceiling along one side.

[a] The performance index is the ratio of the quantity of tracer recovered at the sampling point to that quantity that would have been recovered, with the same volume of ventilating air, if air turbulence in the room had been high enough to ensure perfect mixing of air in the room at all times. A low value, therefore, corresponds to less gas reaching the sampling point.[1]

Source: O.M. Lidwell et al., *J. Hyg.,* 65, 193, 1967. With permission of Cambridge University Press.

REFERENCE

1.   O.M. Lidwell, *J. Hyg.,* 58, 297, 1960.

3.1–10   EFFECT OF VENTILATION RATE ON
BACTERIAL AIR CONTAMINATION

Number of sessions	Ventilation rate, ft^3/min	Mean total count per ft^3	Rate of dispersal (colony forming units per minute)	Median *Staphylococcus aureus* count per 100 ft^3
8[a]	550–900	12.3	13.3×10^3	1.1
19[a]	1050–1400	5.5	8.5×10^3	1.0
23[a]	1420–1700	6.6	12.4×10^3	0.9
273[b]	1375 (mean rate)	5.7	9.6×10^3	–

Rate of dispersal was calculated from the formula

rate of dispersal (colony forming units per minute) = colonies isolated per ft^3 × (ventilation rate + floor area of room)

This assumes complete mixing of the air in the room and a mean settling rate for the airborne particles of 1 ft/min (see Table 3.1–14). The average number of persons present in the operating room during sampling was six so that the mean rate of dispersal (all species of organisms) arising from their activities was about 1600 colony forming units per minute per person.

[a] The first three rows of the table are derived from observations made during the first period of the investigation when, owing to mechanical difficulties with the plant, the ventilation rate varied substantially.

[b] The fourth row gives the mean results obtained during the second and third periods of observation, when the ventilation rate was relatively constant. The results for all three systems have been combined.

Source: O.M. Lidwell et al., *J. Hyg.,* 65, 193, 1967. With permission of Cambridge University Press.

3.1–11   BACTERIAL CONTAMINATION OF THE AIR

Ventilation system	Second period, 157 operating sessions, about 36,000 ft^3 of air sampled				Third period, 116 operating sessions, about 27,000 ft^3 of air sampled			
	A	B	C	All together	A	B	C	All together
Average ventilation, ft^3/min	1380	1370	1390	1380	1290	1390	1420	1370
Total count, colonies/ft^3								
At operating table	5.7	4.6	5.2	5.2	7.0	6.8	5.2	6.4
At side of room	6.7	6.0	5.1	5.9	6.2	5.9	4.4	5.5
Staphylococcus aureus, colonies/100 ft^3								
At operating table	2.6	1.7	2.8	2.4	1.1	0.6	0.4	0.7
At side of room	3.7	2.2	2.4	2.8	1.1	0.7	0.7	0.8

Note: During the second period temperature control of the ventilating air was poor. During the third period this had been remedied and improved diffusers (perforated metal shells) fitted to the downward displacement system (System A).

Source: O.M. Lidwell et al., *J. Hyg.,* 65, 193, 1967. With permission of Cambridge University Press.

3.1–12   APPARENT RATE OF DISPERSAL OF
STAPHYLOCOCCUS AUREUS INTO THE OPERATING ROOM
(COLONY FORMING UNITS PER CARRIER PER HOUR)

	Ventilation system			All
Carrier disperser	A	B	C	together
Heavy carrier "X" at operating table	2600	5300	2700	2900
at side of room	8900	3100	4400	5600
Other staff carriers at operating table	210	220	270	240
at side of room	360	270	290	300
Patients	530	110	640	440

Note: The heavy carrier-disperser appeared to be the source of 75% of the *S. aureus* colonies isolated during the second period of the investigation and of 41% during the third period or of 70% during the two periods taken together.

Rate of dispersal was calculated from the formula

$$\text{rate of dispersal} = \frac{\text{number of colonies}}{\text{number of hours carrier present}}$$

$$\times \frac{\text{ventilation rate + floor area of room}}{\text{sampling rate}}$$

For the average ventilation rate of 1375 ft³/min, a floor area of 320 ft² and a sampling rate of 4 ft³/min:

rate of dispersal (colony forming units per carrier per hour)

$$= 420 \times \frac{\text{number of colonies isolated}}{\text{number of hours carrier present}}$$

This formula assumes complete mixing into the air of the room and a mean settling rate for the airborne particles of 1 ft/min.

Source: O.M. Lidwell et al., *J. Hyg.,* 65, 193, 1967. With permission of Cambridge University Press.

3.1–13   PERCENTAGE DISTRIBUTION OF SOURCES OF
STAPHYLOCOCCUS AUREUS RECOVERED FROM THE AIR

	Source				
Sampling site	Patient	Staff at operating table	Staff at side of room	Uncertain or other	Unknown
At table	6	16	51	20	9
At side	5	12	55	20	8

Note: Staff at the operating table comprised the scrubbed staff together with the anesthetist. The source was uncertain when the same strain was carried by several individuals falling into more than one category.

Source: O.M. Lidwell et al., *J. Hyg.,* 65, 193, 1967. With permission of Cambridge University Press.

3.1–14   NUMBERS OF ORGANISMS (ALL SPECIES) SETTLING PER SQUARE FOOT OF EXPOSED SURFACE PER MINUTE

		Ventilation system		
		A (downward displacement)	B (moderate turbulence)	C (low turbulence)
At table }	during operation	4.4	6.2	3.1
At side }		3.6	4.5	2.7
At table }	after operation	20.2	25.6	16.6
At side }		12.4	21.7	8.8

Note: The numerical similarity between the rate of settling (per ft^2/min) and the numbers of organisms recovered simultaneously from 1 ft^3 of air (see Table 3.1–11) indicate that the average settling rate of the bacteria-carrying airborne particles approximated to 1 ft/min.

Source: O.M. Lidwell et al., *J. Hyg.,* 65, 193, 1967. With permission of Cambridge University Press.

3.1–15   COMPARISON OF FILTER-EFFICIENCY TEST METHODS[a]

	Percent efficiency obtained by test method					
Class of filter	DOP	NBS Atmos.	AFI Atmos.	NBS Artificial	AFI Wt.	Bacterial
Viscous, oiled fibrous impingement filter (furnace filter)	0–2	5–12	3–10	50–60	65–75	10–60
Viscous, oiled, metallic impingement filter	0–2	5–12	3–8	55–60	70–75	10–60
Dry, fibrous medium efficiency; commercial electrostatic precipitator	45–55	80–85	60–70	98–99	NA[a]	90–95
Dry, fibrous high efficiency; commercial electrostatic precipitator	65–75	90–95	78–88	99+	NA	95–99
Dry, fibrous "hospital" type	95–98	99+	99+	NA	NA	99+
High interception; high-efficiency particulate air filter (HEPA)	99.97	NA	NA	NA	NA	99.99+

[a] NA = not applicable; method not sufficiently sensitive to yield meaningful results.

Adapted from Mine Safety Appliances Company, *Aerosol Filter Efficiency Ratings and What They Mean,* MSA Tech. Rep. No. F-101A, The Company, Pittsburgh, 1962.

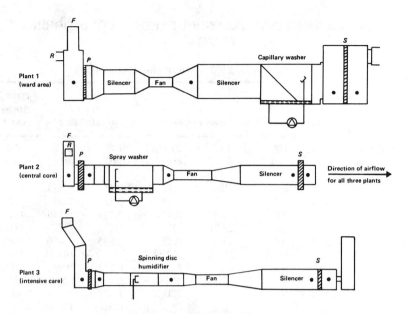

3.1–16   Diagram of the three air conditioning plants. ●: sampling point; *P*: primary filter; *S*: secondary filter; *F*: fresh-air duct; *R*: recirculatory duct.

TYPES OF FILTER IN THE THREE AIR CONDITIONING PLANTS STUDIED

Air conditioning plant	Primary filter	Secondary filter
Plant 1 (ward area)	Vokes® Supervee	Under full fresh air – Vokes Univee Grade C Under recirculation – Vokes Univee Grade A
Plant 2 (central core)	Vokes Miniroll Standard	Vokes Miniroll Standard
Plant 3 (intensive care)	Vokes Supervee	Vokes Univee Grade C

Source: W. Whyte, *J. Hyg.*, 66, 567, 1968. With permission of Cambridge University Press.

3.1–17   DUST RETAINING EFFICIENCY OF FILTERS (EXPRESSED AS THE PERCENTAGE OF DUST RETAINED BY THE FILTERS)

	B.S. 2831 specifications		
Type of filter	Test dust No. 1	Test dust No. 2	Test dust No. 3
Vokes® Supervee	5–10	93.5	50.6
Vokes Univee Grade A	75–82	99.35	99.0
Vokes Univee Grade C	18–37	91.8	26.0
Vokes Miniroll Standard	10–30	87.5	68.0
Vokes K 600 Kompak	20–40	97.4	46.6

Source: W. Whyte, *J. Hyg.*, 66, 567, 1968. With permission of Cambridge University Press.

REFERENCE

K.D. Mulcaster and E.A. Stokes, *J. Inst. Heat. Vent. Eng.*, 34, 197, 1966.

3.1–18   AIR SAMPLING THROUGHOUT THE AIR CONDITIONING PLANTS

		Air conditioning plant					
		Plant 1 (ward area)			Plant 2 (central core)		Plant 3 (intensive care)
Amount of fresh air, %		100	33 (C)[a]	33 (A)[b]	100	75	100
Number of tests		60	60	60	40	30	40
Intake	B	0.42	2.16	1.93	0.485	0.873	0.985
	B + A + M	1.71	3.14	2.08	2.51	3.21	4.21
After primary filter	B	0.071	0.557	0.424	0.064	0.146	0.152
		(16.9)	(25.8)	(22.0)	(13.1)	(16.7)	(15.4)
	B + A + M	0.563	1.000	0.885	0.845	1.450	2.800
		(32.9)	(31.8)	(42.5)	(33.7)	(45.2)	(66.5)
After humidifier	B	N.D.[c]	N.D.	N.D.	0.061	0.091	0.096
					(12.5)	(10.4)	(11.1)
	B + A + M	N.D.	N.D.	N.D.	1.050	0.854	1.665
					(41.8)	(26.6)	(39.5)
Before secondary filter	B	0.035	0.318	0.183	0.061	0.121	0.105
		(8.3)	(14.7)	(9.5)	(12.5)	(13.9)	(10.3)
	B + A + M	0.320	0.700	0.264	0.970	0.994	1.615
		(18.7)	(22.3)	(12.7)	(38.7)	(31.0)	(38.4)
After secondary filter	B	0.035	0.198	0.013	0.046	0.079	0.091
		(8.2)	(9.2)	(0.7)	(9.5)	(9.0)	(9.2)
	B + A + M	0.237	0.494	0.021	0.790	0.972	1.290
		(13.9)	(15.7)	(1.0)	(31.5)	(30.3)	(30.6)

Note: Mean bacteria counts (B) and mean microorganism counts (B + A + M) are expressed as numbers per cubic foot. Figures in parentheses indicate percentage penetration. One test was 15 min or 15 cubic feet of air.

[a] Final filtration through Grade C filter.
[b] Final filtration through Grade A flter.
[c] N.D.: not done.

Source: W. Whyte, *J. Hyg.,* 66, 567, 1968. With permission of Cambridge University Press.

3.1–19   COMPARISON OF THE REMOVAL EFFICIENCIES OF THE SECONDARY FILTERS AGAINST BACTERIA AND ALOXITE[®] 50

Type of filter	Percentage efficiency against bacteria	Percentage efficiency against Aloxite 50
Univee Grade "A"	99.3	99.35
Univee Grade "C"	90.8, 91.8, 90.8	91.8
Miniroll Standard	90.5, 91.0	87.5

Source: W. Whyte, *J. Hyg.,* 66, 567, 1968. With permission of Cambridge University Press.

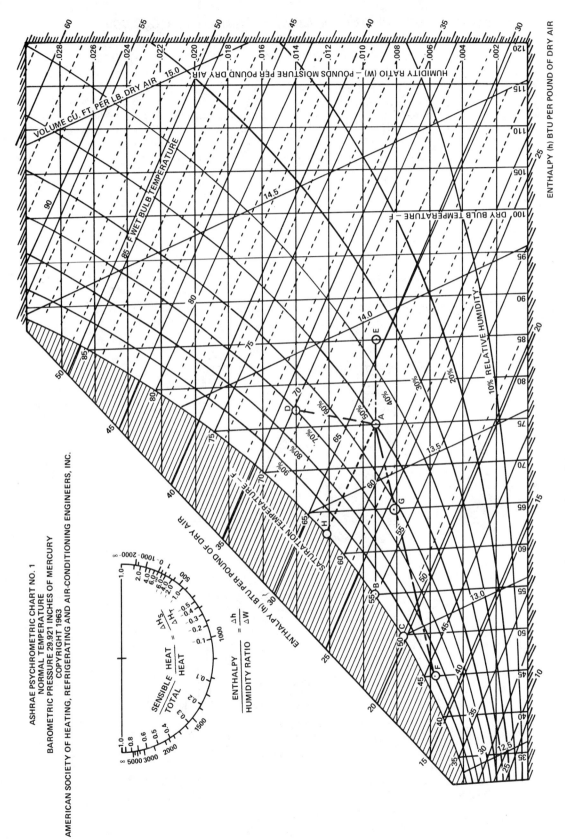

3.1–20   Psychrometric chart. (Illustrative examples added to ASHRAE chart.)

3.1−21   CONCENTRATION OF *BACILLUS SUBTILIS* VAR. *GLOBIGII* EMITTED FROM TWO HUMIDIFIERS COMPARED WITH THE AMOUNT IN THE RESERVOIR

One Test is a 15-ft^3 Sample of Air

Humidifier type	Number of tests	Bacterial count per milliliter of water in reservoir	Number of bacteria per cubic foot of air
Spray type	6	3.0×10^3	0.189
	4	5.2×10^3	0.213
	10	9.0×10^3	0.62
	6	8.6×10^4	1.335
	12	2.4×10^5	3.32
Capillary washer	6	2.5×10^3	0.100
	6	3.0×10^3	0.133
	8	5.8×10^4	0.732
	6	3.5×10^4	0.80

Source: W. Whyte, *J. Hyg.*, 66, 567, 1968. With permission of Cambridge University Press.

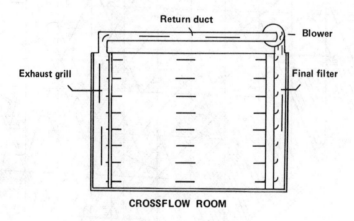

CROSSFLOW ROOM

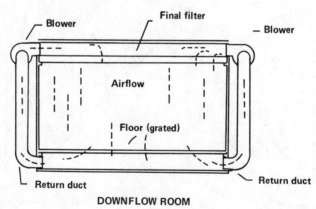

DOWNFLOW ROOM

3.1−22   Diagrammatic drawing of laminar crossflow and downflow rooms.

Source: D.G. Fox, A Study of the Application of Laminar Flow Ventilation to Operating Rooms, PHS Monogr. No. 78, Public Health Service, U.S. Department of Health, Education, and Welfare, Washington, D.C., 1969.

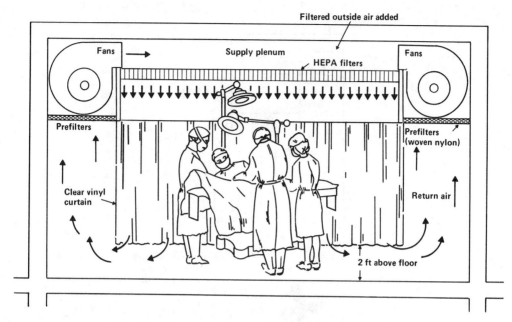

3.1–23   Schematic of vertical laminar-flow shroud area.

Note: HEPA = high-efficiency particulate air.

Source: W. Viesmann, *Heat. Piping Air Cond.,* 40, 61, 1968. With permission.

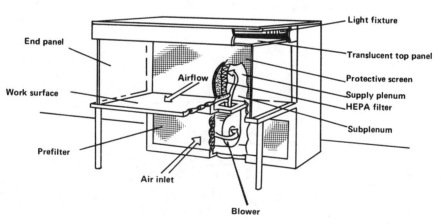

3.1–24   Horizontal-flow bench.

Note: HEPA = high-efficiency particulate air.

Source: W.J. Whitfield and K.J. Lindell, *Contam. Control,* 8, 10, 1969. With permission.

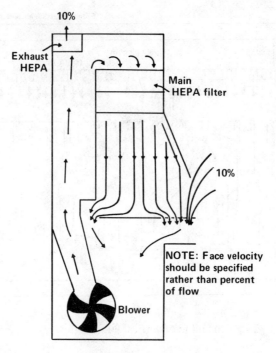

10%

Exhaust
HEPA

Main
HEPA filter

10%

NOTE: Face velocity
should be specified
rather than percent
of flow

Blower

3.1–25   A biohazard hood.

Note: HEPA = high-efficiency particulate air.

Source: L.L. Coriell and G.J. McGarrity, *Appl. Microbiol.*,
16, 1895, 1968. With permission.

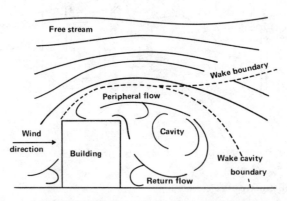

Free stream

Wake boundary

Peripheral flow

Cavity

Wind
direction

Building

Wake cavity
boundary

Return flow

3.1–26   Airflow patterns over a building.

Source: American Society of Heating, Refrigerating, and
Air Conditioning Engineers, *ASHRAE Guide and Data
Book,* ASHRAE, New York, 1968. With permission.

3.1–27 HILL-BURTON ACT
REQUIREMENTS FOR AIR CONDITIONING

General Hospitals

1. Temperatures and humidities
   a. The systems shall be designed to provide the temperatures and humidities shown below.

Area designation	Temperature, °F	RH, %
Operating	70–76[a]	50–60
Delivery	70–76[a]	50–60
Recovery	75	50–60
Nursery (observation)	75	50
Nursery (full-term)	75	50
Nursery (premature)	75–80[a]	50–60[a]
Intensive care	70–80[a]	30–60

[a] Variable range required.

   b. For all other occupied areas, a minimum temperature of 75°F shall be provided at winter design conditions.
2. Ventilation system details. All air-supply and air-exhaust systems shall be mechanically operated. All fans serving exhaust systems shall be located at the discharge end of the system. The ventilation rates shown on Table 3.1–28 shall be considered as minimum acceptable rates and shall not be construed as precluding the use of higher ventilation rates if they are required to meet design conditions.
   a. Outdoor ventilation air intakes, other than for individual room units, shall be located as far away as practicable but not less than 25 ft from the exhaust from any ventilating system or combustion equipment. The bottom of outdoor intakes serving central air systems shall be located as high as possible but not less than 8 ft above the ground level or, if installed through the roof, 3 ft above roof level.
   b. The ventilation systems shall be designed and balanced to provide the general pressure relationship to adjacent areas shown in Table 3.1–28.
   c. All air supplied to sensitive areas such as operating and delivery rooms and nurseries shall be delivered at or near the ceiling of the area served, and all air exhausted from the area shall be removed near floor level. At least two exhaust outlets shall be used in all operating and delivery rooms. Exhaust outlets shall be located not less than 3 in. above the floor.
   d. Room supply air inlets, recirculation, and exhaust air outlets installed in nonsensitive areas shall be located not less than 3 in. above the floor.
   e. Corridors shall not be used to supply air to or exhaust air from any room, except that exhaust air from corridors may be used to ventilate bathrooms, toilet rooms, or janitor's closets opening directly on corridors.
   f. Filters. The ventilation systems serving sensitive areas such as operating rooms, delivery rooms, nurseries, isolation rooms, and laboratory sterile rooms, and recirculated central air systems serving other hospital areas, shall be equipped with a minimum of two filter beds. Filter bed #1 shall be located upstream of the conditioning equipment and shall have a minimum efficiency of 30%. Filter bed #2 shall be located downstream of the conditioning equipment and shall have a minimum efficiency of 90%.

   Central systems using 100% outdoor air and serving other than sensitive areas, except as noted in Section 8–23D2n, shall be provided with filters rated at 80% efficiency.

   The above filter efficiencies shall be warranted by the manufacturer and shall be based on the National Bureau of Standards Dust Spot Test Method with Atmospheric Dust.

   The exhausts from all laboratory hoods in which infectious or radioactive materials are processed shall be equipped with filters having a 99% efficiency based on the DOP (dioctylphthalate) test method.

   Filter frames shall be durable and carefully dimensioned, and shall provide an air-tight fit with the enclosing ductwork. All joints between filter segments and the enclosing ductwork shall be gasketed or sealed to provide a positive seal against air leakage.
   g. A manometer shall be installed across each filter bed serving central air systems.
   h. Ducts shall be constructed of iron, steel, aluminum, or other approved metal or materials such as clay or asbestos cement.
   i. Ducts which penetrate construction intended for X-ray or other ray protection shall not impair the effectiveness of the protection.
   j. Duct linings shall meet the Erosion Test Method described in UL Pub. No. 181. Duct linings, coverings, vapor barriers, and the adhesives used for applying them shall have a flame spread classification of not more than 25 and a smoke-developed rating of not more than 50.
   k. Acoustical lining materials shall not be used in the interior of duct systems serving sensitive areas such as operating and delivery rooms, nurseries, and isolation rooms.
   l. Ducts which pass through fire walls shall be provided with approved automatic fire doors on both sides of the wall except that $\frac{3}{8}$-in. steel plates may be used in lieu of fire doors for openings not exceeding 1 ft 6 in. in diameter. An approved fire damper shall be provided on each opening through each fire partition and on each opening through the walls of a vertical shaft. Ducts which pass

3.1–27   HILL-BURTON ACT
REQUIREMENTS FOR AIR CONDITIONING (continued)

General Hospitals (continued)

through a required smoke barrier shall be provided with dampers which are actuated by products of combustion other than heat. Access for maintenance shall be provided at all dampers.

m. Cold-air ducts shall be insulated wherever necessary to maintain the efficiency of the system or to minimize condensation problems.

n. Laboratories shall be provided with outdoor air at a rate of two air changes per hour. If this ventilation rate does not provide the air required to ventilate fume hoods and safety cabinets, additional air shall be provided. A filter with 90% efficiency shall be installed in the air supply system at its entrance to the media transfer room.

o. Laboratory hoods for general use shall have a minimum average face velocity of 75 ft per minute. Hoods in which infectious or highly radioactive materials are processed shall have a face velocity of 100 ft per minute and each shall have an independent exhaust system with the fan installed at the discharge point of the system. Hoods used for processing infectious materials shall be equipped with a means for disinfection.

p. Duct systems serving hoods shall be constructed of corrosion-resistant material. Duct systems serving hoods in which highly radioactive materials and strong oxidizing agents are used shall be constructed of stainless steel for a minimum distance of 10 ft from the hood and shall be equipped with washdown facilities.

q. The air from dining areas may be used to ventilate the food preparation areas only after it has passed through a filter with 80% efficiency.

r. Exhaust hoods in food preparation centers shall have a minimum exhaust rate of 100 ft³ per minute per square foot of hood face area. All hoods over cooking ranges shall be equipped with fire extinguishing systems and heat-actuated fan controls. Cleanout openings shall be provided every 20 ft 0 in. in horizontal exhaust duct systems serving hoods.

s. The ventilation system for anesthesia storage rooms shall conform to the requirements of NFPA Standard No. 56.

t. Boiler rooms shall be provided with sufficient outdoor air to maintain combustion rates of equipment and reasonable temperatures in the rooms and in adjoining areas.

Nursing Homes

1. Temperatures. A minimum temperature of 75°F shall be provided for all occupied areas at winter design conditions.

2. Ventilation system details. All air-supply and air-exhaust systems shall be mechanically operated. All fans serving exhaust systems shall be located at or near the point of discharge from the building. The ventilation rates shown on Table 3.1–29 shall be considered as minimum acceptable rates and shall not be construed as precluding the use of higher ventilation rates if they are required to meet design conditions.

a. Outdoor ventilation air intakes, other than for individual room units, shall be located as far away as practicable but not less than 25 ft 0 in. from the exhausts from any ventilating system or combustion equipment. The bottom of outdoor intakes serving central air systems shall be located as high as possible but not less than 8 ft 0 in. above the ground level or, if installed through the roof, 3 ft 0 in. above roof level.

b. The ventilation systems shall be designed and balanced to provide the general pressure relationship to adjacent areas shown in Table 3.1–29.

c. Room supply air inlets, recirculation, and exhaust air outlets shall be located not less than 3 in. above the floor.

d. Corridors shall not be used to supply air to or exhaust air from any room, except that exhaust air from corridors may be used to ventilate rooms such as bathrooms, toilet rooms, or janitor's closets which open directly on corridors.

e. Filters. Central systems designed for recirculation of air shall be equipped with a minimum of two filter beds. Filter bed #1 shall be located upstream of the conditioning equipment and shall have a minimum efficiency of 30%. Filter bed #2 shall be located downstream of the conditioning equipment and shall have a minimum efficiency of 90%.

Central systems using 100% outdoor air shall be provided with filters rated at 80% efficiency.

The above filter efficiencies shall be warranted by the manufacturer and shall be based on the National Bureau of Standards Dust Spot Test Method with Atmospheric Dust.

Filter frames shall be durable and carefully dimensioned, and shall provide an air-tight fit with the enclosing ductwork. All joints between filter segments and the enclosing ductwork shall be gasketed or sealed to provide a positive seal against air leakage.

f. A manometer shall be installed across each filter bed serving central air systems.

g. Ducts shall be constructed of iron, steel, aluminum, or other approved metal or materials such as clay or asbestos cement.

3.1–27 HILL-BURTON ACT
REQUIREMENTS FOR AIR CONDITIONING (continued)

Nursing Homes (continued)

h. Duct linings shall meet the Erosion Test Method described in UL Pub. No. 181. Duct linings, coverings, vapor barriers, and the adhesives used for applying them shall have a flame spread classification of not more than 25 and a smoke-developed rating not more than 50.

i. Ducts which pass through fire walls shall be provided with approved automatic fire doors on both sides of the wall except that $\frac{3}{8}$-in. steel plates may be used in lieu of fire doors for openings not exceeding 18 in. in diameter. An approved fire damper shall be provided on each opening through each fire partition and on each opening through the walls of a vertical shaft. Ducts which pass through a required smoke barrier shall be provided with dampers which are actuated by products of combustion other than heat. Access for maintenance shall be provided at all dampers.

j. Cold air ducts shall be insulated wherever necessary to maintain the efficiency of the system or to minimize condensation problems.

k. The air from dining areas may be used to ventilate the food preparation areas only after it has passed through a filter with 80% efficiency.

l. Exhaust hoods in food preparation centers shall have a minimum exhaust rate of 100 ft^3 per minute per square foot of hood face area. All hoods over cooking ranges shall be equipped with fire extinguishing systems and heat-actuated fan controls. Cleanout openings shall be provided every 20 ft 0 in. in horizontal exhaust duct systems serving hoods.

m. Boiler rooms shall be provided with sufficient outdoor air to maintain combustion rates of equipment and reasonable temperatures in the rooms and in adjoining areas.

Source: Department of Health, Education, and Welfare, General Standards of Construction and Equipment for Hospital and Medical Facilities, PHS Pub. No. 930-A-7, U.S. Public Health Service, Washington, D.C., 1969.

3.1–28 PRESSURE RELATIONSHIPS AND VENTILATION OF CERTAIN HOSPITAL AREAS

Area designation	Pressure relationship to adjacent areas	All supply air from outdoors	Minimum air changes of outdoor air per hour	Minimum total air changes per hour	All air exhausted directly to outdoors	Recirculated within room
Operating room	+	—	5	12	—	No
Emergency operating room	+	—	5	12	—	No
Delivery room	+	—	5	12	—	No
Nursery	+	—	5	12	—	No
Recovery	0	—	2	6	Yes	No
Intensive care	+	—	2	6	—	No
Patient room	0	—	2	2	—	—
Patient area corridor	0	—	2	4	—	—
Isolation room	0	—	2	6	Yes	No
Isolation anteroom	0	—	2	6	Yes	No
Treatment room	0	—	2	6	—	No
X-ray, fluoroscopy room	—	—	2	6	Yes	No
X-ray, treatment room	0	—	2	6	—	—
Physical therapy and hydrotherapy	—	—	2	6	—	No
Soiled workroom	—	—	2	4	—	No
Clean workroom	+	—	2	4	—	—
Autopsy and darkroom	—	—	2	12	Yes	No
Toilet room	—	—	—	10	Yes	No
Bedpan room	—	—	—	10	Yes	No
Bathroom	—	—	—	10	Yes	No
Janitor's closet	—	—	—	10	Yes	No
Sterilizer equipment room	—	—	—	10	Yes	No
Linen and trash chute rooms	—	—	—	10	Yes	No
Laboratory, general[a]	—	—	2	6	—	—
Laboratory, media transfer[b]	+	—	2	4	—	No
Food preparation centers[c]	0	—	2	10	Yes	No
Dishwashing room	—	—	—	10	Yes	No
Dietary day storage	0	—	—	2	—	No
Laundry, general	0	—	2	10	Yes	No
Soiled linen sorting and storage	—	—	—	10	Yes	No

Note: + = positive, – = negative, 0 = equal, — = optional.

[a]See sec. 8-23D2n and sec. 8-23D2o for additional requirements.
[b]See sec. 8-23D2n for additional requirements.
[c]See sec. 8-23D2q for exceptions.

3.1–28   PRESSURE RELATIONSHIPS AND VENTILATION OF CERTAIN HOSPITAL AREAS (continued)

Area designation	Pressure relationship to adjacent areas	All supply air from outdoors	Minimum air changes of outdoor air per hour	Minimum total air changes per hour	All air exhausted directly to outdoors	Recirculated within room
Clean linen storage	+	—	2	2	—	—
Anesthesia storage[d]	0	—	—	8	Yes	No
Central medical and surgical supply						
Soiled or decontamination room	—	—	2	4	—	No
Clean workroom	+	—	2	4	—	—
Unsterile supply storage	0	—	2	2	—	—

[d]   See sec. 8-23D2s for additional requirements.

Source: Department of Health, Education, and Welfare, General Standards of Construction and Equipment for Hospital and Medical Facilities, PHS Pub. No. 930-A-7, Public Health Service, U.S. Department of Health, Education, and Welfare, Washington, D.C., 1969.

3.1–29 PRESSURE RELATIONSHIPS AND VENTILATION OF CERTAIN NURSING HOME AREAS

Area designation	Pressure relationship to adjacent areas	All supply air from outdoors	Minimum air changes of outdoor air per hour	Minimum total air changes per hour	All air exhausted directly to outdoors	Recirculated within room
Patient room	0	–	2	2	–	–
Patient area corridor	0	–	2	4	–	–
Special purpose room	0	–	2	6	Yes	No
Physical therapy and hydrotherapy	–	–	2	6	–	–
Soiled workroom	–	–	2	4	–	No
Clean workroom	+	–	2	4	–	–
Toilet room	–	–	–	10	Yes	No
Bedpan room	–	–	–	10	Yes	No
Bathroom	–	–	–	10	Yes	No
Janitor's closet	–	–	–	10	Yes	No
Sterilizer equipment room	–	–	–	10	Yes	No
Linen and trash chute rooms	–	–	–	10	Yes	No
Food preparation center	0	–	2	10	Yes	No
Dishwashing room	–	–	–	10	Yes	No
Dietary day storage	0	–	–	2	–	No
Laundry, general	0	–	2	10	Yes	No
Soiled linen sorting and storage	–	–	–	10	Yes	No
Clean linen storage	+	–	2	2	–	–

Note: + = positive, – = negative, 0 = equal, – = optional.

Source: Department of Health, Education, and Welfare, General Standards of Construction and Equipment for Hospital and Medical Facilities, PHS Pub. No. 930-A-7, Public Health Service, U.S. Department of Health, Education, and Welfare, Washington, D.C., 1969.

3.2   TOXIC AGENTS

3.2—1   TYPES OF AIRBORNE CONTAMINANTS

Dusts
Toxic mineral dusts
  Pneumoconiosis
     Asbestosis — repair of asbestos insulation
    Silicosis
Toxic metallic dusts
  Lead — soldering, grinding, polishing brazing
  Cadmium — soldering, cutting, welding
  Galvanized iron — brazing, welding
  Mercury — see Vapors
Nuisance dusts
  Mineral dusts — road dust, cement
  Organic dusts — grain, wood, pollen

Fumes
Toxic metallic dusts — see above

Vapors
Organic solvents
  Hydrocarbons
    Aromatic — benzene, toluene, xylene
    Aliphatic — petroleum ethers, hexane, naphtha, mineral spirits, gasoline, Stoddard's solvent
    Halogenated — monochloromethane, dichloromethane, chloroform, carbon tetrachloride, methyl chloroform (1,1,1-trichloroethane)

Alcohols
Ethers
Esters   } Found in laboratories, maintenance shops, housekeeping departments
Aldehydes
Oleofins
Ketones
Mercury — broken mercury thermometers, laboratory uses, dental care facilities

Gases
Chlorine — disinfectant in recreational and therapeutic swimming pools; disinfectant housecleaning activities
Ammonia — housekeeping activities, refrigerant
Nitrogen dioxide — produced in maintenance shops in cleaning metals with nitric acid and in electric-arc welding
Ozone — used for odor control, use of ozone discouraged
Asphyxiation — maintenance operation
          — carbon monoxide — incomplete combustion

Mists
Aerosol sprays — insecticides, rodenticides, caustic cleaning compounds, sanitizing agents

Adapted from G. S. Michaelsen, Toxic aspects, in *Environmental Health and Safety in Health-Care Facilities,* R. G. Bond et al., Eds., Macmillan, New York, © 1973.

3.2–2   CONTROL OF AIRBORNE CONTAMINANTS

1. Local exhaust ventilation
2. General dilution ventilation
3. Substitution
   a. Less toxic agent substituted for toxic substance
   b. Use of less hazardous procedure
4. Protective devices
   a. Dust respirators
   b. Gas masks
   c. Chemical cartridge respirators
   d. Airline respirators
   e. Hose masks
   f. Self-contained breathing apparatus
5. Good housekeeping

Adapted from G. S. Michaelsen, Toxic aspects, in *Environmental Health and Safety in Health-Care Facilities,* R. G. Bond et al., Eds., Macmillan, New York, © 1973.

3.2–3   MERCURY CONCENTRATIONS IN DOCTORS AND DENTISTS OFFICES AND HOSPITALS IN THE DALLAS AREA

Location	Date	Mercury concentration, ng/m^3	Remarks
Doctor's room 1	12/13/71	4950	Hg thermometer broken in the past
Doctor's room 2	12/13/71	5680	Hg thermometer broken in the past
Doctor's room 3	12/13/71	4550	Hg thermometer broken in the past
Dentist 1, room 1	12/23/71	5550	Mixing area for Ag amalgam
Dentist 1, room 2	12/23/71	5030	
Dentist 1, room 3	12/23/71	4770	
Dentist 2, room 1	12/28/71	1295	Inactive for previous 4 days
Dentist 2, room 2	12/28/71	1135	
Dentist 2, room 3	12/28/71	1160	
Hospital laboratory	12/13/71	307	
Hospital ward	12/13/71	336	

Source: R.S. Foote, *Science,* 177, 513, 1972. With permission. Copyright 1972 by The American Association for the Advancement of Science.

3.2–4 MERCURY VAPOR CONCENTRATIONS IN VARIOUS LOCATIONS AT THE UNIVERSITY OF TENNESSEE COLLEGE OF DENTISTRY AND IN PRIVATE DENTAL OFFICES

Work area	Location	Number of samples	Range, μg Hg/m^3	Mean, μg Hg/m^3
Dental office				
Operative dentistry student technique laboratory	Workbench	12	15−64	31
	Air conditioner outlet	12	22−83	64
Operative dentistry clinic	Operating unit	12	5−32	18
	Mercury dispensing area	12	15−38	27
	Air conditioner outlet	12	16−53	42
Oral surgery clinic	Operating unit	12	0−18	14
Private dental offices				
Operatories	Operating unit	10	32−75	56
	Air conditioner outlet	10	42−64	46
Waiting rooms	Center of room	10	33−70	47

Source: J.P. McGinnis et al., *J. Am. Dent. Assoc.,* 88, 785, 1974. Copyright by the American Dental Association. Reprinted by permission.

3.2–5 URINE MERCURY LEVELS FOR DENTAL STUDENTS, FACULTY, NONACADEMIC STAFF, AND PRIVATE DENTISTS

Groups	Number	Range, μg% Hg	Average, μg% Hg	Range, μg/24 hr	Average, μg/24 hr
Basic science dental students	4	0.4−1.3	0.72	3.9−21.4	9.95
Preclinical (operative technique) dental students	4	0.2−2.2	1.02	3.0−22.8	11.22
Clinical dental students	16	0.0−4.0	1.36	0.0−52.8	12.02
Clinical dental instructors	7	0.2−4.3	1.27	3.7−35.6	13.44
University dental assistants	7	0.2−3.0	1.04	2.4−18.0	9.07
Supply (mercury) dispensers	6	0.1−4.0	1.86	2.4−26.0	13.03
Private dentists	10	0.1−6.5	1.97	1.7−78.0	20.40

Source: J.P. McGinnis et al., *J. Am. Dent. Assoc.,* 88, 785, 1974. Copyright by The American Dental Association. Reprinted by permission.

3.2–6 POUNDS OF MERCURY PER YEAR VS. EXPOSURE[a]

Mercury used, pounds/year	Exposures in mg/m³							
	0.005–0.009		0.01–0.045		0.05–0.095		0.1 and greater	
	D[b]	A[c]	D	A	D	A	D	A
1–3	6	2	4	1				
3–5	2	2	8	4	2	1		
5–10	5	5	13	10	1	1	1	
10–20	1	1	8	6	3	2	5	4
Total	14	10	33	21	6	4	6	4

[a] TLV (8-hr daily exposure) = 0.1 mg/m³.
[b] D = dentists.
[c] A = assistants.

Source: P.A. Gronka et al., *J. Am. Dent. Assoc.*, 81, 923, 1970. Copyright by the American Dental Association. Reprinted by permission.

3.2–7 EXPOSURES VS. EXCRETION

Exposure range, mg/m³	Mercury in urine in mg/l									
	0.01–0.45		0.05–0.095		0.1		0.2		0.3	
	D[a]	A[b]	D	A	D	A	D	A	D	A
0.005–0.009	3		3	2						
0.010–0.045	3	2								
0.050–0.095	1		1	1	1					
0.1 and greater			1	1			2	1	1	

[a] D = dentist.
[b] A = assistant.

Source: P.A. Gronka et al., *J. Am. Dent. Assoc.*, 81, 923, 1970. Copyright by the American Dental Association. Reprinted by permission.

3.2–8   POUNDS OF MERCURY USED PER YEAR VS. EXCRETION

Pounds/year	Mercury in urine in mg/l									
	0.01–0.45		0.05–0.095		0.1		0.2		0.3	
	D[a]	A[b]	D	A	D	A	D	A	D	A
3	2	2	1							
3–5	1	1								
5–10	3		2	3	1		2			1
10–15			1						1	
15–20			1	1						

[a] D = dentist.
[b] A = assistant.

Source: P.A. Gronka et al., *J. Am. Dent. Assoc.,* 81, 923, 1970. Copyright by the American Dental Association. Reprinted by permission.

3.2–9   TLVs[®] THRESHOLD LIMIT VALUES FOR CHEMICAL SUBSTANCES IN WORKROOM AIR ADOPTED BY ACGIH FOR 1975[a]

1975 TLV AIRBORNE CONTAMINANTS COMMITTEE

Hector P. Blejer, M.D., D.I.H.

Paul E. Caplan, P.E., M.P.H.

Hervey B. Elkins, Ph.D.

W.G. Fredrick, Sc.D.

Paul Gross, M.D.

John W. Knauber, M.P.H.

Jesse Lieberman, P.E.

Trent R. Lewis, Ph.D.

Keith R. Long, Ph.D.

Frederick T. McDermott, P.E.

E. Mastromatteo, M.D. (Can.)

Col. Walter W. Melvin, Jr., M.D.

Meier Schneider, P.E., C.I.H.

Col. Marshall Steinberg, Ph.D.

Gordon J. Stopps, M.B.

John F. Summersett

Ralph C. Wands, M.S.

CONSULTANTS

James F. Morgan

Theodore R. Torkelson, Sc.D.

Mitchell R. Zavon, M.D.

Herbert E. Stokinger, Ph.D., Chairman

William D. Wagner, Recording Secretary

P.O. Box 1937

Cincinnati OH 45201

---

[a]This publication is reprinted in its entirety by permission of the American Conference of Governmental Industrial Hygienists, Cincinnati, Ohio.

PREFACE
CHEMICAL CONTAMINANTS

Threshold limit values refer to airborne concentrations of substances and represent conditions under which it is believed that nearly all workers may be repeatedly exposed day after day without adverse effect. Because of wide variation in individual susceptibility, however, a small percentage of workers may experience discomfort from some substances at concentrations at or below the threshold limit; a smaller percentage may be affected more seriously by aggravation of a pre-existing condition or by development of an occupational illness.

Simple tests are now available (J. Occup. Med. 15: 564, 1973; Ann. N.Y. Acad. Sci., *151, Art. 2:* 968, 1968) that may be used to detect those individuals hypersusceptible to a variety of industrial chemicals (respiratory irritants, hemolytic chemicals, organic isocyanates, carbon disulfide). These tests may be used to screen out by appropriate job placement the hyperreactive worker and thus in effect improve the "coverage" of the TLVs.

Threshold limit values refer to time-weighted concentrations for a 7- or 8-hour workday and 40-hour workweek. They should be used as guides in the control of health hazards and should not be used as fine lines between safe and dangerous concentrations. (Exceptions are the substances listed in Appendices E and F, and those substances designated with a "C" or Ceiling value, Appendix D).

Time-weighted averages permit excursions above the limit provided they are compensated by equivalent excursions below the limit during the workday. In some instances it may be permissible to calculate the average concentration for a workweek rather than for a workday. The degree of permissible excursion is related to the magnitude of the threshold limit value of a particular substance as given in Appendix D. The relationship between threshold limit and permissible excursion is a rule of thumb and in certain cases may not apply. The amount by which threshold limits may be exceeded for short periods without injury to health depends upon a number of factors such as the nature of the contaminant, whether very high concentrations — even

for short periods — produce acute poisoning, whether the effects are cumulative, the frequency with which high concentrations occur, and the duration of such periods. All factors must be taken into consideration in arriving at a decision as to whether a hazardous condition exists.

Threshold limits are based on the best available information from industrial experience, from experimental human and animal studies, and, when possible, from a combination of the three. The basis on which the values are established may differ from substance to substance; protection against impairment of health may be a guiding factor for some, whereas reasonable freedom from irritation, narcosis, nuisance or other forms of stress may form the basis for others.

The amount and nature of the information available for establishing a TLV varies from substance to substance; consequently, the precision of the estimated TLV is also subject to variation and the latest *Documentation* should be consulted in order to assess the extent of the data available for a given substance.

The committee holds to the opinion that limits based on physical irritation should be considered no less binding than those based on physical impairment. There is increasing evidence that physical irritation may initiate, promote or accelerate physical impairment through interaction with other chemical or biologic agents.

In spite of the fact that serious injury is not believed likely as a result of exposure to the threshold limit concentrations, the best practice is to maintain concentrations of all atmospheric contaminants as low as is practical.

These limits are intended for use in the practice of industrial hygiene and should be interpreted and applied only by a person trained in this discipline. They are not intended for use, or for modification for use, (1) as a relative index of hazard or toxicity, (2) in the evaluation or control of community air pollution nuisances, (3) in estimating the toxic potential of continuous, uninterrupted exposures, (4) as proof or disproof of an existing disease or physical condition, or (5) for adoption by countries whose working conditions differ from those in the United States of America and where substances and processes differ.

Ceiling vs Time-Weighted Average Limits. Although the time-weighted average concentration provides the most satisfactory, practical way of monitoring airborne agents for compliance with the limits, there are certain substances for which it is inappropriate. In the latter group are substances which are predominantly fast acting and whose threshold limit is more appropriately based on this particular response. Substances with this type of response are best controlled by a ceiling "C" limit that should not be exceeded. It is implicit in these definitions that the manner of sampling to determine noncompliance with the limits for each group must differ; a single brief sample, that is applicable to a "C" limit, is not appropriate to the time-weighted limit; here, a sufficient number of samples are needed to permit a time-weighted average concentration throughout a complete cycle of operations or throughout the work shift.

Whereas the ceiling limit places a definite boundary which concentrations should not be permitted to exceed, the time-weighted average limit requires an explicit limit to the excursions that are permissible above the listed values. The magnitude of these excursions may be pegged to the magnitude of the threshold limit by an appropriate factor shown in Appendix D. It should be noted that the same factors are used by the Committee in making a judgment whether to include or exclude a substance for a "C" listing.

"Skin" Notation. Listed substances followed by the designation "Skin" refer to the potential contribution to the overall exposure by the cutaneous route including mucous membranes and eye, either by airborne, or more particularly, by direct contact with the substance. Vehicles can alter skin absorption. This attention-calling designation is intended to suggest appropriate measures for the prevention of cutaneous absorption so that the threshold limit is not invalidated.

Mixtures. Special consideration should be given also to the application of the TLVs in assessing the health hazards which may be associated with exposure to mixtures of two or more substances. A brief discussion of basic considerations involved in developing threshold limit values for mixtures, and methods for their development, amplified by specific examples are given in Appendix C.

Nuisance Particulates. In contrast to fibrogenic dusts which cause scar tissue to be formed in lungs when inhaled in excessive amounts, so-called "nuisance" dusts have a long history of little adverse effect on lungs and do not produce significant organic disease or toxic effect when exposures are kept under reasonable control. The nuisance dusts have also been called (biologically) "inert" dusts, but the latter term is inappropriate to the extent that there is no dust which does not evoke some cellular response in the lung when inhaled in sufficient amount. However, the lung-tissue reaction caused by inhalation of nuisance dusts has the following characteristics: 1) The architecture of the air spaces remains intact. 2) Collagen (scar tissue) is not formed to a significant extent. 3) The tissue reaction is potentially reversible.

Excessive concentrations of nuisance dusts in the workroom air may seriously reduce visibility, may cause unpleasant deposits in the eyes, ears and nasal passages (Portland Cement dust), or cause injury to the skin or mucous membranes by chemical or mechanical action per se or by the rigorous skin cleansing procedures necessary for their removal.

A threshold limit of 10 mg/m³, or 30 mppcf, of total dust < 1% quartz is recommended for substances in these categories and for which no specific threshold limits have been assigned. This limit, for a normal workday, does not apply to brief exposures at higher concentrations. Neither does it apply to those substances which may cause physiologic impairment at lower concentrations but for which a threshold limit has not yet been adopted. Some nuisance particulates are given in Appendix E.

Simple Asphyxiants — "Inert" Gases or Vapors. A number of gases and vapors, when present in high concentrations in air, act primarily as simple asphyxiants without other significant physiologic effects. A TLV may not be recommended for each simple asphyxiant because the limiting factor is the available oxygen. The minimal oxygen content should be 18 percent by volume under normal atmospheric pressure (equivalent to a partial pressure, pO_2 of 135 mm Hg). Atmospheres deficient in O_2 do not provide adequate warning and most simple asphyxiants are odorless. Several simple asphyxiants present an explosion hazard. Account should be taken of this factor in limiting the concentration of the asphyxiant. Specific examples are listed in Appendix F.

Short-Term Limits (STLs). Because many industrial exposures are not continuous, 8-hour daily exposures, but are short-term, or intermittent, to which the TLVs do not necessarily apply, STLs for 5, 15, or 30 minutes for 142 substances have been put into the regulations of the Pennsylvania Department of Health (Chapter 4, Art. 432, Revised Jan. 25, 1968). These STLs represent the maximal average atmospheric concentration of a contaminant to which a worker may be exposed for the stipulated time. The concentration represents an upper limit of exposure and assumes that there is sufficient recovery between exposures before another is initiated. The daily average exposure including that provided by the STL shall be such that the TLV shall not be exceeded.

Similar STLs for a more restricted number of substances have been recommended by the American National Standards Institute. This standard-setting body refers to these short-term limits as "peaks."

Physical Factors. It is recognized that such physical factors as heat, ultraviolet and ionizing radiation, humidity, abnormal pressure (altitude) and the like may place added stress on the body so that the effects from exposure at a threshold limit may be altered. Most of these stresses act adversely to increase the toxic response of a substance. Although most threshold limits have built-in safety factors to guard against adverse effects to moderate deviations from normal environments, the safety factors of most substances are not of such a magnitude as to take care of gross deviations. For example, continuous work at temperatures above 90°F, or overtime extending the workweek more than 25%, might be considered gross deviations. In such instances judgment must be exercised in the proper adjustments of the Threshold Limit Values. Brief & Scala (AIHAJ. 26, 467, 1975) have proposed formulae for calculating the TLV Reduction Factor for novel work schedules, i.e. 10-hr workday.

Biologic Limit Values (BLVs). Other means exist and may be necessary for monitoring worker exposure other than reliance on the Threshold Limit Values for industrial air, namely, the Biologic Limit Values. These values represent limiting amounts of substances (or their effects) to which the worker may be exposed without hazard to health or well-being as determined in his tissues and fluids or in his exhaled breath. The biologic measurements on which the BLVs are based can furnish two kinds of information useful in the control of worker exposure: (1) measure of the individual worker's over-all exposure; (2) measure of the worker's individual and characteristic response. Measurements of response furnish a superior estimate of the physiologic status of the worker, and may be made of (a) changes in amount of some critical biochemical constituent, (b) changes in activity of a critical enzyme, (c) changes in some physiologic function. Measurement of exposure may be made by (1) determining in blood, urine, hair, nails, in body tissues and fluids, the amount of substance to which the worker was exposed; (2) determination of the amount of the metabolite(s) of the substance in tissues and fluids; (3) determination of the amount of the substance in the exhaled breath. The biologic limits may be used as an adjunct to the TLVs for air, or in place of them. The BTLs, and their associated procedures for determining compliance with them, should thus be regarded as an effective means of providing health surveillance of the worker.

Unlisted substances. There are a number of reasons why a substance does not appear in the Threshold Limit list; either insufficient information is available or it has not been brought to the attention of the Threshold Limits Committee from which a limit can be developed, or it is a substance that could be included in the Appendices E and F pertaining to Nuisance Particulates and Simple Asphyxiants. Substances appearing in these appendices serve as examples only; the appendices are not intended to be inclusive.

"Notice of Intent." At the beginning of each year, proposed actions of the Committee for the forthcoming year are issued in the form of a "Notice of Intended Changes." This Notice provides not only an opportunity for comment, but solicits suggestions of substances to be added to the list. The

suggestions should be accompanied by sub-stantiating evidence. The list of Intended Changes follows the Adopted Values in the TLV booklet.

Legal Status. By publication in the Federal Register (Vol. 36, No. 105, May 29, 1971) the Threshold Limit Values for 1968 are now official federal standards for industrial air.

Reprint Permission. This publication may be reprinted provided that written permission is obtained from the Secretary-Treasurer of the Conference and that it be published in its entirety.

ADOPTED VALUES
See Documentation for basis of TLVs (1974)

Substance	ppm[a]	mg/m³[b]
Abate	—	10
Acetaldehyde	100	180
Acetic acid	10	25
C Acetic anhydride	5	20
Acetone	1,000	2,400
Acetonitrile	40	70
Acetylene	F	—
Acetylene dichloride, see 1, 2-Dichloroethylene	—	—
Acetylene tetrabromide	1	14
Acrolein	0.1	0.25
Acrylamide — Skin	—	0.3
Acrylonitrile — Skin	20	45
Aldrin — Skin	—	0.25
Allyl alcohol — Skin	2	5
Allyl chloride	1	3
Allyl glycidyl ether (AGE) — Skin	5	22
Allyl propyl disulfide	2	12
Alundum (Al₂O₃)	—	E
4-Aminodiphenyl—Skin	—	A1b
2-Aminoethanol, see Ethanolamine	—	—
2-Aminopyridine	0.5	2
Ammonia	25	18
Ammonium chloride, fume	—	10
Ammonium sulfamate (Ammate)	—	10
n-Amyl acetate	100	525
sec-Amyl acetate	125	650
Aniline — Skin	5	19
Anisidine (o-, p-isomers) — Skin	0.1	0.5
** Antimony & compounds (as Sb)	—	(0.5)
ANTU (alpha naphthyl thiourea)	—	0.3
Argon	F	—

Substance	ppm[a]	mg/m³[b]
** Arsenic & compounds (as As)	—	(0.5)
Arsine	0.05	0.2
Asbestos (all forms)		A1a
Asphalt (petroleum) fumes	—	5
Azinphos methyl — Skin	—	0.2
Baygon (Propoxur)	—	0.5
Barium (soluble compounds)	—	0.5
**C Benzene—Skin	(25)	(80)
Benzidine production — Skin	—	A1b
p-Benzoquinone, see Quinone	—	—
Benzoyl peroxide	—	5
Benzyl chloride	1	5
Beryllium	—	0.002
Biphenyl	0.2	1
Bismuth telluride	—	10
Bismuth telluride (Se-doped)	—	5
Boron oxide	—	10
Boron tribromide	1	10
C Boron trifluoride	1	3
Bromine	0.1	0.7
Bromine pentafluoride	0.1	0.7
Bromoform — Skin	0.5	5
Butadiene (1, 3-butadiene)	1,000	2,200
** Butane	(500)	(1200)
Butanethiol, see Butyl mercaptan	—	—
2-Butanone	200	590
2-Butoxy ethanol (Butyl Cellosolve) — Skin	50	240
Butyl acetate (n-butyl acetate)	150	710
sec-Butyl acetate	200	950
tert-Butyl acetate	200	950

Capital letters refer to Appendices.
Footnotes (a thru h) see Page 234.
**See Notice of Intended Changes.

ADOPTED VALUES (continued)

Substance	ppm[a]	mg/m³ [b]	Substance	ppm[a]	mg/m³ [b]
** n-Butyl alcohol—Skin...	(100)	(300)	Chlorobromomethane...	200	1,050
sec-Butyl alcohol.......	150	450	2-Chloro-1, 3-butadiene		
tert-Butyl alcohol......	100	300	see Chloroprene......	—	—
C Butylamine — Skin.....	5	15	Chlorodifluoromethane..	1,000	3,500
C tert-Butyl chromate (as			Chlorodiphenyl (42%		
CrO₃) — Skin.......	—	0.1	Chlorine) — Skin.....	—	1
n-Butyl glycidyl ether			Chlorodiphenyl (54%		
(BGE).............	50	270	Chlorine) — Skin.....	—	0.5
** Butyl lactate..........	(1)	(5)	1-Chloro, 2, 3- epoxy-		
Butyl mercaptan.......	0.5	1.5	propane, see		
p-tert-Butyltoluene.....	10	60	Epichlorhydrin.......	—	—
** Cadmium (Metal dust			2-Chloroethanol, see		
and soluble salts, as Cd)	—	(0.2)	Ethylene chlorohydrin.	—	—
C Cadmium oxide fume (as			Chloroethylene, see		
Cd)................	—	0.05	Vinyl chloride........	—	—
Calcium carbonate......	—	E	Chloroform		
Calcium arsenate, as As.	—	1	(Trichloromethane)...	25	120
Calcium oxide..........	—	5	bis-Chloromethyl ether..	0.001	A1a
Camphor (Synthetic)...	2	12	1-Chloro-1-nitropropane.	20	100
Caprolactam			Chloropicrin...........	0.1	0.7
Dust................	—	1	Chloroprene (2-chloro-1,		
Vapor..............	5	20	3-butadiene) — Skin..	25	90
Carbaryl (Sevin®)......	—	5	* Chlorpyrifos (Dursban®)		
Carbon black..........	—	3.5	— Skin	—	0.2
Carbon dioxide.........	5,000†	9,000	* o-Chlorostyrene........	50	285
Carbon disulfide — Skin.	20	60	o-Chlorotoluene........	50	250
Carbon monoxide.......	50	55	*2-Chloro-6-(trichlor-		
* Carbon tetrabromide....	0.1	1.4	omethyl) pyridine (N-		
Carbon tetrachloride —			Serve®)	—	10
Skin...............	10	65	Chromates, certain insol-		
Cellulose (paper fiber)...	—	E	uble forms..........	—	A1a
* Cesium hydroxide......	—	2	Chromic acid and		
Chlordane — Skin......	—	0.5	chromates (as CrO₃)..	—	0.1
Chlorinated camphene —			Chromium, sol. chromic,		
Skin...............	—	0.5	chromous salts as Cr..	—	0.5
Chlorinated diphenyl			* Clopidol (Coyden®)	—	10
oxide................	—	0.5	Coal tar pitch volatiles		
Chlorine..............	1	3	(see Particulate Poly-		
Chlorine dioxide........	0.1	0.3	cyclic Organic Matter		
C Chlorine trifluoride.....	0.1	0.4	(PPOM)	—	—
C Chloroacetaldehyde.....	1	3			
α-Chloroacetophenone			** Cobalt, metal fume &		
(phenacylchloride)....	0.05	0.3	dust	—	(0.1)
Chlorobenzene			* Copper fume..........	—	0.2
(monochlorobenzene)..	75	350	Dusts and Mists....	—	1
o-Chlorobenzylidene			Corundum (Al₂O₃)......	—	E
malononitrile (OCBM)			Cotton Dust (raw).....	—	0.2[m]
— Skin.............	0.05	0.4	Crag® herbicide.......	—	10

Capital letters refer to Appendices.
Footnotes (a thru h) see Page 234.
† See 1974 Revised Documentation.
*1975 Addition.
**See Notice of Intended Changes.
m) see p. 236.

ADOPTED VALUES (continued)

Substance	ppm[a]	mg/m³ [b]
Cresol (all isomers) — Skin	5	22
Crotonaldehyde	2	6
*Crufomate (Ruelene®)	—	50
Cumene — Skin	50	245
Cyanide (as CN) — Skin	—	5
Cyanogen	10	20
Cyclohexane	300	1,050
Cyclohexanol	50	200
Cyclohexanone	50	200
Cyclohexene	300	1,015
Cyclohexylamine — Skin	10	40
Cyclopentadiene	75	200
2,4-D	—	10
DDT	—	1
DDVP, see Dichlorvos	—	—
Decaborane — Skin	0.05	0.3
Demeton® — Skin	0.01	0.1
Diacetone alcohol (4-hydroxy-4-methyl-2-pentanone)	50	240
1,2-Diaminoethane, see Ethylenediamine	—	—
Diazinon — Skin	—	0.1
Diazomethane	0.2	0.4
Diborane	0.1	0.1
1,2-Dibromoethane (ethylene dibromide) — Skin	20	145
Dibrom®	—	3
2-N Dibutylaminoethanol — Skin	2	14
Dibutyl phosphate	1	5
Dibutylphthalate	—	5
C Dichloracetylene	0.1	0.4
C o-Dichlorobenzene	50	300
p-Dichlorobenzene	75	450
Dichlorobenzidine — Skin	—	A1b
Dichlorodifluoromethane	1,000	4,950
1,3-Dichloro-5,5-dimethyl hydantoin	—	0.2
1,1-Dichloroethane	200	820
1,2-Dichloroethane	50	200
1,2-Dichloroethylene	200	790
Dichloroethyl ether — Skin	5	30
Dichloromethane, see Methylene chloride	—	—
Dichloromonofluoromethane	1,000	4,200
C 1,1-Dichloro-1-nitroethane	10	60
1,2-Dichloropropane, see Propylenedichloride	—	—
Dichlorotetrafluoroethane	1,000	7,000
Dichlorvos (DDVP) — Skin	0.1	1
*Dicyclopentadienyl-iron	—	10
Dieldrin — Skin	—	0.25
Diethylamine	25	75
Diethylamino ethanol — Skin	10	50
Diethylene triamine — Skin	1	4
Diethylether, see Ethyl ether	—	—
Diethylphthalate	—	5
Difluorodibromomethane	100	860
C Diglycidyl ether (DGE)	0.5	2.8
Dihydroxybenzene, see Hydroquinone	—	—
Diisobutyl ketone	25	150
Diisopropylamine — Skin	5	20
Dimethoxymethane, see Methylal	—	—
Dimethyl acetamide — Skin	10	35
Dimethylamine	10	18
Dimethylaminobenzene, see Xylidene	—	—
Dimethylaniline (N-dimethylaniline) — Skin	5	25
Dimethylbenzene, see Xylene	—	—
Dimethyl 1,2-dibromo-2-dichloroethyl phosphate, see DiBrom	—	—
Dimethylformamide — Skin	10	30
2,6-Dimethylheptanone, see Diisobutyl ketone	—	—

Capital letters refer to Appendices.
Footnotes (a thru h) see Page 234.
*1975 Addition.

ADOPTED VALUES (continued)

Substance	ppm[a]	mg/m³[b]
1, 1-Dimethylhydrazine — Skin	0.5	1
Dimethylphthalate	—	5
** Dimethyl sulfate—Skin.	(0.01)	(A2)
Dinitrobenzene (all isomers) — Skin	0.15	1
Dinitro-o-cresol — Skin.	—	0.2
* 3,5-Dinitro-o-toluam-ide (Zoalene®)	—	5.0
Dinitrotoluene — Skin	—	1.5
Dioxane, technical grade —Skin	50	180
Diphenyl, see Biphenyl	—	—
Diphenyl amine	—	10
Diphenylmethane diisocyanate, see Methylene bisphenyl isocyanate (MDI)	—	—
Dipropylene glycol methyl ether — Skin	100	600
Diquat	—	0.5
Di-sec, octyl phthalate (Di-2-ethylhexyl-phthalate)	—	5
Disyston—Skin	—	0.1
* 2,6-Ditert-butyl-p-cresol	—	10
Emery	—	E
Endosulfan (Thiodan®) — Skin	—	0.1
Endrin — Skin	—	0.1
Epichlorhydrin — Skin	5	19
EPN — Skin	—	0.5
1, 2-Epoxypropane, see Propylene oxide	—	—
2, 3-Epoxy-1-propanol, see Glycidol	—	—
Ethane	F	—
Ethanethiol, see Ethylmercaptan	—	—
Ethanolamine	3	6
2-Ethoxyethanol—Skin.	100	370
2-Ethoxyethylacetate (Cellosolve acetate) — Skin	100	540
Ethyl acetate	400	1,400
Ethyl acrylate — Skin	25	100
Ethyl alcohol (ethanol)	1,000	1,900
Ethylamine	10	18

Substance	ppm[a]	mg/m³[b]
Ethyl sec-amyl ketone (5-methyl-3-heptanone)	25	130
Ethyl benzene	100	435
Ethyl bromide	200	890
Ethyl butyl ketone (3-Heptanone)	50	230
Ethyl chloride	1,000	2,600
Ethyl ether	400	1,200
Ethyl formate	100	300
Ethyl mercaptan	0.5	1
Ethyl silicate	100	850
Ethylene	F	—
** Ethylene chlorohydrin — Skin	(5)	(16)
Ethylenediamine	10	25
Ethylene dibromide, see 1, 2-Dibromoethane	—	—
Ethylene dichloride, see 1, 2-Dichloroethane	—	—
Ethylene glycol, particulate	—	10
Ethylene glycol, vapor	100	260
C Ethylene glycol dinitrate and/or Nitroglycerin—Skin	0.2[d]	—
Ethylene glycol mono-methyl ether acetate (Methyl cellosolve ace-tate) — Skin	25	120
Ethylene oxide	50	90
Ethylenimine — Skin	0.5	1
Ethylidine chloride, see 1, 1-Dichloroethane	—	—
C Ethylidene norbornene	5	25
N-Ethylmorpholine — Skin	20	94
Ferbam	—	10
Ferrovanadium dust	—	1
Fluoride (as F)	—	2.5
Fluorine	1	2
Fluorotrichloromethane	1,000	5,600
C Formaldehyde	2	3
* Formamide	20	30
Formic acid	5	9
Furfural — Skin	5	20
Furfuryl alcohol	5	20
Gasoline	—	B²
Germanium tetrahydride	0.2	0.6
Glass, fibrous[e] or dust	—	E

Capital letters refer to Appendices.
Footnotes (a thru h) see Page 234.
*1975 Addition.
**See Notice of Intended Changes.

ADOPTED VALUES (continued)

Substance	ppm[a]	mg/m³[b]	Substance	ppm[a]	mg/m³[b]
Glycerin mist..........	—	E	Isopropyl acetate.......	250	950
Glycidol (2, 3-Epoxy-1-propanol)...........	50	150	Isopropyl alcohol—Skin.	400	980
			Isopropylamine........	5	12
Glycol monoethyl ether, see 2-Ethoxyethanol..	—	—	Isopropyl ether........	250	1,050
Graphite, (Synthetic)...	—	E	Isopropyl glycidyl ether (IGE)..............	50	240
Guthion,® see Azinphos-methyl..............	—	—	Kaolin...............	—	E
Gypsum.............	—	E	Ketene...............	0.5	0.9
Hafnium.............	—	0.5	Lead, inorg., fumes and dusts, as Pb.........	—	0.15
Helium...............	F	—	Lead arsenate, as Pb....	—	0.15
Heptachlor — Skin.....	—	0.5	Limestone.............	—	E
** Heptane (n-heptane)....	(500)	(2,000)	Lindane—Skin	—	0.5
Hexachlorocyclopenta-diene...............	0.01	0.11	Lithium hydride........	—	0.025
Hexachloroethane—Skin...............	1	10	L.P.G. (Liquified petroleum gas).............	1,000	1,800
Hexachloronaphthalene — Skin.............	—	0.2	Magnesite.............	—	E
Hexafluoroacetone......	0.1	0.7	Magnesium oxide fume..	—	10
** Hexane (n-hexane)......	(500)	(1,800)	Malathion — Skin......	—	10
** 2-Hexanone (Methylbutyl ketone) — Skin...	(100)	(410)	Maleic anhydride.......	0.25	1
Hexone (Methyl isobutyl ketone) — Skin.......	100	410	C Manganese and compounds, as Mn.......	—	5
sec-Hexyl acetate.......	50	300	Manganese cyclopentadienyl tricarbonyl (as Mn)—Skin.......	—	0.1
**Hydrazine — Skin......	(1)	(1.3)			
Hydrogen.............	F	—	Marble...............	—	E
Hydrogen bromide......	3	10	Mercury (Alkyl compounds) — Skin, as Hg	0.001	0.01
C Hydrogen chloride......	5	7	Mercury (All forms except alkyl) as Hg....	—	0.05
Hydrogen cyanide—Skin...............	10	11	Mesityl oxide..........	25	100
Hydrogen fluoride......	3	2	Methane.............	F	—
Hydrogen peroxide.....	1	1.4	Methanethiol, see Methyl mercaptan...........	—	—
Hydrogen selenide......	0.05	0.2	Methoxychlor..........	—	10
Hydrogen sulfide.......	10	15	2-Methoxyethanol—Skin (Methyl cellosolve)...	25	80
Hydroquinone.........	—	2	Methyl acetate.........	200	610
Indene...............	10	45	Methyl acetylene (propyne)...............	1,000	1,650
Indium and compounds, as In...............	—	0.1	Methyl acetylene-propadiene mixture (MAPP)...........	1,000	1,800
C Iodine...............	0.1	1			
* Iron oxide fume........	B⁴	5	Methyl acrylate — Skin.	10	35
Iron pentacarbonyl.....	0.01	0.08	Methyl acrylonitrile — Skin...............	1	3
Iron salts, soluble, as Fe.	—	1			
Isoamyl acetate........	100	525	Methylal (dimethoxymethane)...........	1,000	3,100
Isoamyl alcohol........	100	360			
Isobutyl acetate........	150	700			
** Isobutyl alcohol........	(100)	(300)			
** Isophorone...........	(10)	(50)			

Capital letters refer to Appendices.
*1975 Addition.
**See Notice of Intended Changes.

ADOPTED VALUES (continued)

Substance	ppm[a]	mg/m³[b]	Substance	ppm[a]	mg/m³[b]
Methyl alcohol (methanol)—Skin	200	260	C Methylene bisphenyl isocyanate (MDI)	0.02	0.2
Methylamine	10	12	** Methylene chloride (dichloromethane)	(100)	(360)
Methyl amyl alcohol, see Methyl isobutyl carbinol	—	—	4,4′-Methylene bis (2-chloraniline)—Skin	0.02	A2
Methyl 2-cyanoacrylate	2	8	C Methylene bis (4-cyclohexylisocyanate)	0.01	0.11
Methyl isoamyl ketone	100	475	Molybdenum, as Mo		
Methyl n-amyl ketone (2-Heptanone)	100	465	Soluble compounds	—	5
Methyl bromide — Skin	15	60	Insoluble compounds	—	10
Methyl butyl ketone, see 2-Hexanone	—	—	Monomethyl aniline — Skin	2	9
Methyl cellosolve — Skin see 2-Methoxyethanol	—	—	C Monomethyl hydrazine — Skin	0.2	0.35
Methyl cellosolve acetate — Skin, see Ethylene glycol monomethyl ether acetate	—	—	Morpholine — Skin	20	70
			Naphthalene	10	50
			β-Naphthylamine	—	A1b
Methyl chloride	100	210	Neon	F	—
Methyl chloroform	350	1,900	** Nickel carbonyl	(0.001 A1a)	(0.007)
** Methylcyclohexane	(500)	(2000)	Nickel, metal and insoluble compounds (as Ni)	—	1
Methylcyclohexanol	50	235	Nicotine — Skin	—	0.5
o-Methylcyclohexanone—Skin	50	230	Nitric acid	2	5
			Nitric oxide	25	30
Methylcyclopentadienyl manganese tricarbonyl (as Mn) — Skin	0.1	0.2	p-Nitroaniline — Skin	1	6
			Nitrobenzene — Skin	1	5
Methyl demeton — Skin	—	0.5	p-Nitrochlorobenzene — Skin	—	1
Methyl ethyl ketone (MEK), see 2-Butanone	—	—	4-Nitrodiphenyl	—	A1b
			Nitroethane	100	310
C Methyl ethyl ketone peroxide	0.2	1.5	C Nitrogen dioxide	5	9
Methyl formate	100	250	Nitrogen trifluoride	10	29
Methyl iodide — Skin	5	28	Nitroglycerin[d] — Skin	0.2	2
Methyl isobutyl carbinol — Skin	25	100	Nitromethane	100	250
			1-Nitropropane	25	90
			2-Nitropropane	25	90
Methyl isobutyl ketone, see Hexone	—	—	N-Nitrosodimethylamine (dimethylnitrosoamine) — Skin	—	A2
Methyl isocyanate — Skin	0.02	0.05	Nitrotoluene — Skin	5	30
Methyl mercaptan	0.5	1	Nitrotrichloromethane, see Chloropicrin	—	—
Methyl methacrylate	100	410	Nitrous oxide	F	—
Methyl parathion—Skin	—	0.2	Octachloronaphthalene — Skin	—	0.1
Methyl propyl ketone, see 2-Pentanone	—	—	** Octane	(400)	(1,900)
C Methyl silicate	5	30	Oil mist, particulate	—	5[f]
Cα-Methyl styrene	100	480	Oil mist, vapor	[g]B²	—

Capital letters refer to Appendices.
Footnotes (a thru h) see Page 234.
**See Notice of Intended Changes.

ADOPTED VALUES (continued)

Substance	ppm[a]	mg/m³[b]	Substance	ppm[a]	mg/m³[b]
Osmium tetroxide, as Os	0.0002	0.002	** Phthalic anhydride.....	(2)	(12)
Oxalic acid............	—	1	*Picloram (Tordon®)....	—	10
Oxygen difluoride.......	0.05	0.1	Picric acid — Skin......	—	0.1
Ozone................	0.1	0.2	Pival® (2-Pivalyl-1, 3-		
* Paraffin wax fume......	—	2	indandione)..........	—	0.1
Paraquat — Skin.......	—	0.5	Plaster of Paris.........	—	E
Parathion — Skin.......	—	0.1	Platinum (Soluble Salts)		
Particulate polycyclic or-			as Pt................	—	0.002
ganic matter (PPOM)			Polychlorobiphenyls, see		
as benzene solubles...	—	0.2	Chlorodiphenyls......	—	—
Pentaborane...........	0.005	0.01	Polytetrafluoroethylene		
Pentachloronaphthalene			decomposition prod-		
— Skin..............	—	0.5	ucts................	—	B¹
Pentachlorophenol			C Potassium hydroxide....	—	2
— Skin.............	—	0.5	Propane..............	F	—
Pentaerythritol.........	—	E	β Propiolactone........	—	A2
** Pentane..............	(500)	(1,500)	Propargyl alcohol—Skin.	1	2
2-Pentanone...........	200	700	n-Propyl acetate.......	200	840
Perchloroethylene–Skin .	100	670	Propyl alcohol—Skin...	200	500
Perchloromethyl			n-Propyl nitrate........	25	110
mercaptan..........	0.1	0.8	Propylene dichloride (1,		
Perchloryl fluoride......	3	14	2-Dichloropropane)...	75	350
Petroleum Distillates			Propylene glycol mono-		
(naphtha)...........	[g]B³	—	methyl ether.........	100	360
Phenol — Skin.........	5	19	Propylene imine — Skin.	2	5
Phenothiazine — Skin...	—	5	Propylene oxide.......	100	240
p-Phenylene diamine —			Propyne, see Methyl-		
Skin................	—	0.1	acetylene...........	—	—
Phenyl ether (vapor)....	1	7	Pyrethrum............	—	5
Phenyl ether-Diphenyl			Pyridine..............	5	15
mixture (vapor)......	1	7	Quinone.............	0.1	0.4
Phenylethylene, see			RDX — Skin..........	—	1.5
Styrene.............	—	—	Rhodium, Metal fume		
Phenyl glycidyl ether			and dusts (as Rh)....	—	0.1
(PGE)..............	10	60	Soluble salts.......	—	0.001
Phenylhydrazine — Skin.	5	22	Ronnel...............	—	10
C Phenylphosphine.......	0.05	0.25	Rosin Core Solder		
Phorate (Thimet®)—Skin	—	0.05	pyrolysis products (as		
Phosdrin (Mevinphos®)			formaldehyde).......	—	0.1
— Skin..............	0.01	0.1	Rotenone (commercial)..	—	5
** Phosgene (carbonyl chlo-			Rouge...............	—	E
ride)................	(0.10)	(0.4)	Selenium compounds (as		
Phosphine.............	0.3	0.4	Se).................	—	0.2
Phosphoric acid........	—	1	Selenium hexafluoride, as		
Phosphorus (yellow)....	—	0.1	Se..................	0.05	0.4
Phosphorus pentachlo-			Sevin® (see Carbaryl)...	—	—
ride................	—	1	Silane (see Silicon tetra-		
Phosphorus pentasulfide.	—	1	hydride).............	—	—
Phosphorus trichloride..	0.5	3	Silicon...............	—	E

Capital letters refer to Appendices.
Footnotes (a thru h) see Page 234.
*1975 Addition.
**See Notice of Intended Changes.

ADOPTED VALUES (continued)

Substance	ppm[a)]	mg/m[3b)]	Substance	ppm[a)]	mg/m[3b)]
Silicon carbide.........	—	E	Tetraethyl lead (as Pb)		
Silicon tetrahydride			— Skin..............	—	0.100[h)]
(Silane).............	0.5	0.7	Tetrahydrofuran.......	200	590
Silver, metal and soluble			Tetramethyl lead (as Pb)		
compounds, as Ag....	—	0.01	— Skin..............	—	0.150[h)]
Sodium fluoroacetate			Tetramethyl succinoni-		
(1080) — Skin........	—	0.05	trile — Skin..........	0.5	3
C Sodium hydroxide......	—	2.0	Tetranitromethane......	1	8
Starch...............	—	E	Tetryl (2, 4, 6-		
Stibine...............	0.1	0.5	trinitrophenyl-		
** Stoddard solvent.......	(200)	(1150)	methylnitramine) —		
Strychnine............	—	0.15	Skin...............	—	1.5
Styrene, monomer			Thallium (soluble com-		
(Phenylethylene)	100	420	pounds)—Skin (as Tl).	—	0.1
*C Subtilisins (Proteolytic			Thiram®..............	—	5
enzymes as 100% pure			Tin (inorganic com-		
crystalline enzyme)...	—	0.00006[o)]	pounds, except SnH₄		
Sucrose...............	—	E	and SnO₂) as Sn......	—	2
Sulfur dioxide..........	5	13	Tin (organic compounds)		
Sulfur hexafluoride.....	1,000	6,000	— Skin (as Sn).......	—	0.1
Sulfuric acid...........	—	1	Tin oxide.............	—	E
Sulfur monochloride....	1	6	Titanium dioxide.......	—	E
Sulfur pentafluoride.....	0.025	0.25	Toluene (toluol) — Skin.	100	375
Sulfur tetrafluoride.....	0.1	0.4			
Sulfuryl fluoride........	5	20	C Toluene-2, 4-		
Systox, see Demeton®...	—	—	diisocyanate (TDI)...	0.02	0.14
2, 4, 5-T..............	—	10	o-Toluidine............	5	22
Tantalum.............	—	5	Toxaphene, see Chlori-		
TEDP — Skin.........	—	0.2	nated camphene......	—	—
Teflon® decomposition			Tributyl phosphate.....	—	5
products............	—	B¹	1, 1, 1-Trichloroethane,		
Tellurium.............	—	0.1	see Methyl chloroform.	—	—
Tellurium hexafluoride,			1, 1, 2-Trichloroethane		
as Te...............	0.02	0.2	— Skin..............	10	45
TEPP — Skin..........	0.004	0.05	Trichloroethylene.......	100	535
C Terphenyls............	1	9	Trichloromethane, see		
1, 1, 1, 2-Tetrachloro-2,			Chloroform..........	—	—
2-difluoroethane......	500	4,170	Trichloronaphthalene —		
1, 1, 2, 2-Tetrachloro-1,			Skin...............	—	5
2-difluoroethane......	500	4,170	1, 2, 3-Trichloropropane.	50	300
1, 1, 2, 2-Tetrachloro-			1, 1, 2-Trichloro 1, 2, 2-		
ethane — Skin........	5	35	trifluoroethane.......	1,000	7,600
Tetrachloroethylene, see			Triethylamine..........	25	100
Perchloroethylene....	—	—			
Tetrachloromethane, see			* Tricyclohexyltin hy-		
Carbon tetrachloride..	—	—	droxide (Plictran®)...	—	5
Tetrachloronaphthalene			Trifluoromonobromo-		
— Skin..............	—	2	methane.............	1,000	6,100
			Trimethyl benzene......	25	120

Capital letters refer to Appendices.
Footnotes (a thru h) see Page 234.
*1975 Addition.
**See Notice of Intended Changes.
o) See p. 236.

ADOPTED VALUES (continued)

Substance	ppm[a]	mg/m³[b]
2, 4, 6-Trinitrophenol, see Picric acid.	—	—
2, 4, 6 — Trinitrophenyl-methylnitramine, see Tetryl.	—	—
Trinitrotoluene — Skin. .	0.2	1.5
Triorthocresyl phosphate.	—	0.1
Triphenyl phosphate. . . .	—	3
Tungsten & compounds, as W		
Soluble.	—	1
Insoluble.	—	5
Turpentine.	100	560
Uranium (natural) soluble & insoluble compounds, as U.	—	0.2
Vanadium (V₂O₅), as V		
Dust.	—	0.5
C Fume.	—	0.05
Vinyl acetate.	10	30

Substance	ppm[a]	mg/m³[b]
Vinyl benzene, see Styrene.	—	—
Vinyl bromide.	250	1,100
**Vinyl chloride.	(200)	(510)
Vinyl cyanide, see Acrylonitrile.	—	—
Vinylidene chloride.	10	40
Vinyl toluene.	100	480
Warfarin.	—	0.1
Wood dust (nonallergenic).	—	5
Xylene (o-,m-,p-isomers) — Skin.	100	435
Xylidine — Skin.	5	25
Yttrium.	—	1
Zinc chloride fume.	—	1
Zinc oxide fume.	—	5
Zinc stearate.	—	E
Zirconium compounds (as Zr).	—	5

a) Parts of vapor or gas per million parts of contaminated air by volume at 25 °C and 760 mm. Hg. pressure.

b) Approximate milligrams of substance per cubic meter of air.

d) An atmospheric concentration of not more than 0.02 ppm, or personal protection may be necessary to avoid headache.

e) <7 μm in diameter.

f) As sampled by method that does not collect vapor.

g) According to analytically determined composition.

h) For control of general room air, biologic monitoring is essential for personnel control.

Radioactivity: For permissible concentrations of radioisotopes in air, see U.S. Department of Commerce, National Bureau of Standards Handbook 69, "Maximum Permissible Body Burdens and Maximum Permissible Concentrations of Radionuclides in Air and in Water for Occupational Exposure," June 5, 1969. Also, see U.S. Department of Commerce National Bureau of Standards, Handbook 59, "Permissible Dose from External Sources of Ionizing Radiation," September 24, 1954, and addendum of April 15, 1958. A report, Basic Radiation Protection Criteria, published by the National Committee on Radiation Protection, revises and modernizes the concept of the NCRP standards of 1954, 1957 and 1958; obtainable as NCRP Rept. No. 39, P.O. Box 4867, Washington, D.C. 20008.

MINERAL DUSTS

Substance

SILICA, SiO_2

Crystalline

Quartz

TLV in mppcf[i]:

$$\frac{300^{j)}}{\% \text{ quartz} + 10}$$

TLV for respirable dust in mg/m^3:

$$\frac{10 \text{ mg/m}^{3\,k)}}{\% \text{ Respirable quartz} + 2}$$

TLV for "total dust," respirable and nonrespirable:

$$\frac{30 \text{ mg/m}^3}{\% \text{ quartz} + 3}$$

Cristobalite . . . . . Use one-half the value calculated from the count or mass formulae for quartz.

Tridymite   Use one-half the value calculated from formulae for quartz.

Silica, fused   Use quartz formulae.

Tripoli   Use respirable[p] mass quartz formula

**Amorphous . . . . 20 mppcf[i]

SILICATES (< *1%* quartz)

Asbestos, all forms†	5 fibers/cc > $5\mu m$ in length[n]; A1a
Graphite (natural)	15 mppcf
Mica	20 mppcf
Mineral wool fiber	10 mg/m³
Perlite	30 mppcf
Portland Cement	30 mppcf
Soapstone	20 mppcf
Talc (nonasbestiform)	20 mppcf

Talc (fibrous) use Asbestos limit.

Tremolite, see Asbestos.

COAL DUST

(bituminous). 2 mg/m³ (respirable dust fraction < 5% quartz).

If > 5% quartz, use respirable mass formula.

**See Notice of Intended Changes.

n) See p. 236

NUISANCE PARTICULATES
(see Appendix E)

30 m p p c f or 10 mg/m$^{3\,l)}$
of total dust $<$ 1% quartz

Conversion factors:

mppcf $\times$ 35.3 = million particles per
cubic meter
= particles per c.c.

i) Millions of particles per cubic foot of air, based on impinger samples counted by light-field technics.

j) The percentage of quartz in the formula is the amount determined from airborne samples, except in those instances in which other methods have been shown to be applicable.

k) Both concentration and percent quartz for the application of this limit are to be determined from the fraction passing a size-selector with the following characteristics:

Aerodynamic Diameter (μm) (unit density sphere)	% passing selector
$\gtrless$ 2	90
2.5	75
3.5	50
5.0	25
10	0

l) containing $<$1% quartz; if quartz content $>$ 1%, use formulae for quartz.

m) Lint-free dust as measured by the vertical-elutriator, cotton-dust sampler described in the Transactions of the National Conference on Cotton Dust, J. R. Lynch, pg. 33, May 2, 1970.

n) As determined by the membrane filter method at 400–450X magnification (4 mm objective) phase contrast illumination.

o) Based on "high volume" sampling.

p) "Respirable" dust as defined by the British Medical Research Council Criteria (1) and as sampled by a device producing equivalent results (2).

(1) Hatch, T. E. and Gross, P., Pulmonary Deposition and Retention of Inhaled Aerosols, p. 149. Academic Press, New York, New York, 1964.

(2) Interim Guide for Respirable Mass Sampling, AIHA Aerosol Technology Committee, AHIA J. *31:* 2, 1970, p. 133.

†A more stringent TLV for crocidolite may be required.

NOTICE OF INTENDED CHANGES
(for 1975)

These substances, with their corresponding values, comprise those for which either a limit has been proposed for the first time, or for which a change in the "Adopted" listing has been proposed. In both cases, the proposed limits should be considered trial limits that will remain in the listing for a period of at least two years. If, after two years no evidence comes to light that questions the appropriateness of the values herein, the values will be reconsidered for the "Adopted" list. Documentation is available for each of these substances.

Substance	ppm[a)]	mg/m³[b)]
+ Antimony trioxide, handling & use, as Sb	—	0.5
+ Antimony trioxide production	A2	0.05
+ Arsenic troxide, handling & use	—	0.25
+ Arsenic trioxide production	A1a	0.05
+ Benzene — Skin	10A2	30
+ Borates, tetra, sodium salts		
Anhydrous	—	1
Decahydrate	—	5
Pentahydrate	—	1
Butane	600	1,450
C n-Butyl alcohol	50	150
n-Butyl lactate	5	25
Cadmium dusts & salts, as Cd	—	0.05
+C Cadmium oxide production	A1a	0.05
Calcium cyanamide	—	0.5
Calcium hydroxide	—	2
Captan	—	5
+ Captafol (Difolatan®) — Skin	—	0.1
Carbofuran	—	0.1
+ Catechol (Pyrocatechol)	5	20
+ Chromite ore processing (as CrO₃)	A1a	0.1
+ Cyanamide	—	2
+ Dicrotophos (Bidrin®) — Skin	—	0.25
Dicyclopentadiene	5	30
+C Dimethyl sulfate	1	A2
+ Dioxathion (Delnav®)	—	0.2
Disulfuram	—	2
+ Diuron	—	10
Dyfonate	—	0.1
Ethion (Nialate®)—Skin	—	0.4

Substance	ppm[a)]	mg/m³[b)]
C Ethylene chlorohydrin — Skin	1	3
Fensulfothion (Dasanit)	—	0.1
Formamide	20	30
C Glutaraldehyde	2	8
C Glutaraldehyde (Alkaline activated)	—	0.25
Heptane	400	1,600
Hexane	100	360
2-Hexanone (Methyl butyl ketone)	25	100
C Hexylene glycol	25	125
+ Hydrazine	0.1	0.1
Hydrogenated terphenyls	0.5	5.0
Iodoform	0.2	3.0
Isobutyl alcohol	50	150
C Isophorone	5	25
+ Isophorone diisocyanate— Skin	0.01	0.06
+ Methomyl (Lannate®) — Skin	—	2.5
Methylcyclohexane	400	1,600
Methylene chloride (Dichloromethane)	200	720
+ Monocrotophos (Azodrin®)	—	0.25
Nickel, soluble salts(as Ni)	—	0.1
+ Nickel carbonyl	0.05	0.35
Nonane	200	1,050
Octane	300	1,450
Pentane	600	1,800
C Phosgene	0.05	0.2
+ Phthalic anhydride	1	6
+ m-Phthalodinitrile	—	5
+C 1, 2-Propylene glycol dinitrate — Skin	0.05	0.35
Resorcinol	10	45
+ Rubber Solvent	400	—

Capital letters refer to Appendices.
+1975 Revision or Addition.
Footnotes (a thru h) see Page 234.

NOTICE OF INTENDED CHANGES (continued)

Substance	ppm[a]	mg/m³[b]	Substance	ppm[a]	mg/m³[b]
C Sodium azide...........	0.1	0.3	Vinyl chloride...........Pending		Alc
Stoddard Solvent........	100	175	+ Vinyl cyclohexene dioxide.	10	60
Succindialdehyde (see Glutaraldehyde).......	—	—	+ VM & P Naphtha........	200	—
4, 4'-Thiobis (6-tert butyl-m-cresol)............	—	10	Welding fumes (Total Particulate)...........	—	5 & B[4]
C 1, 2, 4-Trichlorobenzene..	5	40	C m-Xylene α, α' diamine...	—	0.1
Triphenylamine.........	—	5	+ Zinc chromate, as CrO₃..	—	0.1

MINERAL DUSTS

Substance	TLV
+ Silica, amorphous (including natural Diatomaceous Earth)	3 mg/m³ Total dust (all sampled sizes) 1 mg/m³ Respirable dust (<5μm)

Capital letters refer to Appendices.
Footnotes (a thru h) see Page 234.
+ 1975 Revision or Addition.

APPENDIX A
Occupational Carcinogens

The Committee lists below those substances in industrial use that have proven carcinogenic in man, or have induced cancer in animals under appropriate experimental conditions. Present listing of those substances carcinogenic for man takes three forms, those for which a TLV has been assigned (1a), those for which environmental conditions have not been sufficiently defined to assign a TLV (1b), and 1c, those whose reassignment of a TLV is awaiting more definitive data, and hence should be treated as a 1b carcinogen.

1a. *Human Carcinogens.* Substances, or substances associated with occupational processes, recognized to have carcinogenic or cocarcinogenic potential, with an assigned TLV.

TLV

Arsenic trioxide production	As₂O₃, 0.05 mg/m³ as As SO₂, C 5.0 ppm Sb₂O₃, 0.05 mg/m³ as Sb
Asbestos, all forms*	5 fibers/cc, > 5μ in length

Cadmium oxide production	0.05 mg/m³ as Cd
bis (Chloromethyl) ether	1.0 ppb
Chromite ore processing	0.1 mg/m³ as CrO₃
Particulate Polycyclic Organic Matter	0.2 mg/m³, as benzene solubles

1b. *Human Carcinogens.* Substances, or substances associated with industrial processes, recognized to have carcinogenic potential without an assigned TLV:

4-Aminodiphenyl (p-Xenylamine)
Benzidine production — Skin
beta-Naphthylamine
4-Nitrodiphenyl

1c. *Human Carcinogens.* Substances with recognized carcinogenic potential awaiting reassignment of TLV pending further data acquisition:

Vinyl chloride

For the substances in 1b, no exposure or contact by any route, respiratory, skin or oral, as detected by the most sensitive methods, shall be permitted.

*Cigarette smoking may substantially enhance the incidence of bronchogenic carcinoma from this and others of these listed substances or processes.

APPENDIX A (continued)

"No exposure or contact" means hermitizing the process or operation by the best practicable engineering methods. The worker should be properly equipped to insure virtually no contact with the carcinogen.

A2. *Occupational Substances Suspect of Oncogenic Potential for Workers.* Chemical substances, or substances associated with occupational processes, which are suspected of inducing malignant neoplasms, based on either a) limited epidemiologic evidence, exclusive of clinical reports of single cases, or b) demonstration of benign or malignant growths in one or more animal species by appropriate methods.

Present evidence indicates that the assigned TLVs are below the threshold of response for inducing cancer in workers under ordinary conditions of employment.

For those substances without an assigned TLV, exposure by all routes should be carefully controlled to levels consistent with the animal and human experience data (see Documentation).

Substance or Operation	TLV
Antimony trioxide production[a]	0.05 mg/m^3
Benzene[a]	10 ppm
Benzo(a)pyrene[b]—Skin	
Beryllium[a][b]	2.0 μg/m^3
3, 3' Dichlorobenzidine[b] —Skin	
Dimethyl sulfate[b]	C 1.0 ppm
Hydrazine[b] — Skin	0.1 ppm
4, 4'-Methylene bis (2-chloroaniline)[b] — Skin	0.02 ppm
Nickel production (Ni$_3$S$_2$)[a][b]	
Nitrosamines (Dimethyl nitrosamine)[b]	
Propane sultone[b]	
beta-Propiolactone[b]	
Zinc Chromate[a]	0.1 mg/m^3 as CrO$_3$

Letters in parentheses refer to kinds of informational bases qualifying the substance or operation to be included in A2. (See Documentation)
Footnotes (a thru h) see Page 234.

APPENDIX B

B[1] *Polytetrafluoroethylene* decomposition products.* Thermal decomposition of the fluorocarbon chain in air leads to the formation of oxidized products containing carbon, fluorine and oxygen. Because these products decompose in part by hydrolysis in alkaline solution, they can be quantitatively determined in air as fluoride to provide an index of exposure. No TLV is recommended pending determination of the toxicity of the products, but air concentrations should be minimal.

B[2] *Gasoline.* The composition of gasoline varies greatly and thus a single TLV for all types of these materials is no longer applicable. In general, the aromatic hydrocarbon content will determine what TLV applies. Consequently the content of benzene, other aromatics and additives should be determined to arrive at the appropriate TLV (Elkins, et al. A.I.H.A.J. *24*:99, 1963); Runion, ibid. *36*, 338, 1975)

B[3] *Petroleum Distillates.* For petroleum distillates for which no specific TLV's are listed, approximate values can be obtained by use of the following equation:

*Trade Names: Algoflon, Fluon, Halon, Teflon, Tetran.

APPENDIX B (continued)

$$TLV = \cfrac{100}{\cfrac{\% \ Al}{3.6(200 - B.P.\,°C.) + 20} + \cfrac{\% \ Ar}{1.3(200 - B.P.\,°C.) + 10}} \ ppm$$

where Al = aliphatic component
      Ar = aromatic component
      B.P. = mean boiling point in degrees centigrade (normally the 50% distillation temperature).

The equation cannot be used if the benzene content of the fraction exceeds 1%, nor if the mean boiling point is above 200°C.

It may also lead to error if there are large amounts of hexane or cyclohexane in the distillate.

If the molecular weight (average) is not known for the mixture, it can be approximated by the following equation:

M.W. = %Al + 0.88%Ar + 0.5(B.P.°C. − 100)

B[4] *Welding Fumes — Total Particulate (NOC)**
TLV, 5 mg/m³

Welding fumes cannot be classified simply. The composition and quantity of both are dependent on the alloy being welded and the process and electrodes used. Reliable analysis of fumes cannot be made without considering the nature of the welding process and system being examined; reactive metals and alloys such as aluminum and titanium are arc-welded in a protective, inert atmosphere such as argon. These arcs create relatively little fume, but an intense radiation which can produce ozone. Similar processes are used to arc-weld steels, also creating a relatively low level of fumes. Ferrous alloys also are arc-welded in oxidizing environments which generate considerable fume, and can produce carbon monoxide instead of ozone. Such fumes generally are composed of discreet particles of amorphous slags containing iron, manganese, silicon and other metallic constituents depending on the alloy system involved. Chromium and nickel compounds are found in fumes when stainless steels are arc-welded. Some coated and flux-cored electrodes are formulated with fluorides and the fumes associated with them can contain significantly more fluorides than oxides. Because of the above factors, arc-welding fumes frequently must be tested for individual constituents which are likely to be present to determine whether specific TLV's are exceeded. Conclusions based on total fume concentration are generally adequate if no toxic elements are present in welding rod, metal, or metal coating and conditions are not conducive to the formation of toxic gases.

Most welding, even with primitive ventilation, does not produce exposures inside the welding helmet above 5 mg/m³. That which does, should be controlled.

*Not otherwise classified.

APPENDIX C

C.1 THRESHOLD LIMIT VALUES FOR MIXTURES

When two or more hazardous substances are present, their combined effect, rather than that of either individually, should be given primary consideration. In the absence of information to the contrary, the effects of the different hazards should be considered as additive. That is, if the sum of the following fractions,

$$\frac{C_1}{T_1} + \frac{C_2}{T_2} + \cdots \frac{C_n}{T_n}$$

exceeds unity, then the threshold limit of the mixture should be considered as being exceeded. C_1 indicates the observed atmospheric concentration, and T_1 the corresponding threshold limit (See Example 1A.a. and 1A.c.).

Exceptions to the above rule may be made when there is good reason to believe that the chief effects of the different harmful substances are not in fact additive, but *independent* as when purely local effects on different organs of the body are produced by the various components of the mixture. In such cases the threshold limit ordinarily is exceeded only when at least one member of the series $\left(\dfrac{C_1}{T_1} + \text{ or } + \dfrac{C_2}{T_2} \text{ etc.}\right)$ itself has a value exceeding unity (See Example 1A. c.).

Antagonistic action or potentiation may occur with some combinations of atmospheric contaminants. Such cases at present must be determined individually. Potentiating or antagonistic agents are not necessarily harmful by themselves. Potentiating effects of exposure to such agents by routes other than that of inhalation is also possible, e.g. imbibed alcohol and inhaled narcotic (trichloroethylene). Potentiation is characteristically exhibited at high concentrations, less probably at low.

When a given operation or process characteristically emits a number of harmful dusts, fumes, vapors or gases, it will frequently be only feasible to attempt to evaluate the hazard by measurement of a single substance. In such cases, the threshold limit used for this substance should be reduced by a suitable factor, the magnitude of which will depend on the number, toxicity and relative quantity of the other contaminants ordinarily present.

Examples of processes which are typically associated with two or more harmful atmospheric contaminants are welding, automobile repair, blasting, painting, lacquering, certain foundry operations, diesel exhausts, etc.

C.1A Examples of THRESHOLD LIMIT VALUES FOR MIXTURES

The following formulae apply only when the components in a mixture have similar toxicologic effects; they should not be used for mixtures with widely differing reactivities, e.g. hydrogen cyanide & sulfur dioxide. In such case the formula for Independent Effects (1A.c.) should be used.

1A.a. General case, where air is analyzed for each component:

    a. *Additive effects. (Note: It is essential that the atmosphere be analyzed both qualitatively and quantitatively for each component present, in order to evaluate compliance or noncompliance with this calculated TLV.)*

$$\frac{C_1}{T_1} + \frac{C_2}{T_2} + \frac{C_3}{T_3} + \cdots = 1$$

Example No. 1A.a: Air contains 5 ppm of carbontetrachloride (TLV = 10 ppm) 20 ppm of 1, 2-dichloroethane (TLV = 50 ppm) and 10 ppm of 1, 2-dibromoethane (TLV = 20 ppm)

Atmospheric concentration of mixture = 5 + 20 + 10 = 35 ppm of mixture

APPENDIX C (continued)

$$\frac{5}{10} + \frac{20}{50} + \frac{10}{20} = \frac{25 + 20 + 25}{50} = 1.4$$

Threshold Limit is exceeded. Furthermore, the TLV of this mixture may be calculated by reducing the total fraction to 1.0; i.e.

$$\text{TLV of mixture} = \frac{35}{1.4} = 25 \text{ ppm}$$

1A.b. Special case when the source of contaminant is a liquid mixture and the atmospheric composition is *assumed* to be similar to that of the original material; e.g. on a time-weighted average exposure basis, all of the liquid (solvent) mixture eventually evaporates.

Additive effects (approximate solution)

1. The percent composition (by weight) of the liquid mixture is known, the TLVs of the constituents must be listed in mg/m³.

(*Note: In order to evaluate compliance with this TLV, field sampling instruments should be calibrated, in the laboratory, for response to this specific quantitative and qualitative air-vapor mixture, and also to fractional concentrations of this mixture; e.g., 1/2 the TLV; 1/10 the TLV; 2 × the TLV; 10 × the TLV; etc.*)

TLV of mixture =

$$\frac{1}{\dfrac{f_a}{TLV_a} + \dfrac{f_b}{TLV_b} + \dfrac{f_c}{TLV_c} + \cdots \dfrac{f_n}{TLV_n}}$$

Example No. 1: Liquid solvent contains (by weight) 50% heptane (TLV = 1600 mg/m³) 30% methylene chloride (TLV = 720 mg/m³) 20% perchloroethylene (TLV = 670 mg/m³)

$$\text{TLV of mixture} = \frac{1}{\dfrac{0.5}{1600} + \dfrac{0.3}{720} + \dfrac{0.2}{670}}$$

$$= \frac{1}{0.00031 + 0.00042 + 0.0003} = \frac{1}{0.00103}$$

= 1000 mg/m³ (approximately)

To convert to ppm, consult TLV list:

50% heptane = 200 ppm
30% methylene chloride = 33 ppm
20% perchloroethylene = 20 ppm

TLV of mixture: 200 + 60 + 20 = 280 ppm

1A.c. *Independent effects.*

Air contains 0.15 mg/m³ of lead (TLV, 0.15) and 0.7 mg/m³ of sulfuric acid (TLV, 1).

$$\frac{0.15}{0.15} = 1; \qquad \frac{0.7}{1} = 0.7$$

Threshold limit is not exceeded.

1B. TLV for Mixtures of Mineral Dusts.

For mixtures of biologically active mineral dusts the general formula for mixtures may be used.

For a mixture containing 80% non-asbestiform talc and 20% quartz, the TLV for 100% of the mixture is given by:

$$\text{TLV} = \frac{1}{\dfrac{0.8}{20} + \dfrac{0.2}{10}} = 16 \text{ mppcf}$$

Essentially the same result will be obtained if the limit of the more (most) toxic component is used provided the effects are additive. In the above example the limit for 20% quartz is 10 mppcf.

For another mixture of 25% quartz, 25% amorphous silica and 50% talc:

$$\text{TLV} = \frac{1}{\dfrac{0.25}{10} + \dfrac{0.25}{20} + \dfrac{0.5}{20}} = 16 \text{ mppcf}$$

The limit for 25% quartz approximates 9 mppcf.

APPENDIX D
PERMISSIBLE EXCURSIONS FOR TIME-WEIGHTED AVERAGE (TWA) LIMITS

The Excursion TLV Factor in the Table automatically defines the magnitude of the permissible excursion above the limit for those substances not given a "C" designation; i.e., the TWA limits. Examples in the Table show that nitrobenzene, the TLV for which is 1 ppm, should never be allowed to exceed 3 ppm. Similarly, carbon tetrachloride, TLV = 10 ppm, should never be allowed to exceed 20 ppm. By contrast, those substances with a "C" designation are not subject to the excursion factor and must be kept at or below the TLV ceiling.

These limiting excursions are to be considered to provide a "rule-of-thumb" guidance for listed substances generally, and may not provide the most appropriate excursion for a particular substance e.g., the permissible excursion for CO is 400 ppm for 15 minutes.

For appropriate excursions for 142 substances consult Pa. Rules & Regs., Chap. 4, Art. 432, and "Acceptable Concentrations," ANSI.

Substance	TLV	Excursion Factor	Max. Conc. Permitted for short time
	ppm		ppm
Nitro-benzene	1	3	3
Carbon tetra-chloride	10	2	20
o-Dichloro-benzene	50	1.5	75
Acetone	1000	1.25	1250
Boron trifluoride	C 1	—	1
Butylamine	C 5	—	5

EXCURSION FACTORS

For all substances not bearing C notation

		Excursion	
TLV >0–1 (ppm or mg/m³),	Factor	= 3
TLV >1–10 "	"	= 2
TLV >10–100 "	"	= 1.5
TLV >100–1000 "	"	= 1.25

The number of times the excursion above the TLV is permitted is governed by conformity with the Time-Weighted Average TLV.

INTERPRETATION OF MEASURED PEAK CONCENTRATIONS

With increasing use of rapid, direct-reading analytical instruments for airborne contaminants in the work area, the question of interpretation of essentially "instantaneous" peaks arises. Although no general statement can be made covering all occupational substances, the following guidelines should prove helpful, assuming peak excursions conform to time-weighted average TLV as stated above.

The toxicologic importance of momentary peak concentrations depends on whether the substance is fast or slow acting. If slow acting, as for quartz, lead, or carbon monoxide, momentary peaks are of no toxicologic concern provided, of course, they are not astronomic. On the other hand, fast acting substances that rapidly produce disabling narcosis, e.g., H_2S, or intolerable irritation or asphyxiation, NH_3, SO_2, CO_2, or initiate sensitization — the organic isocyanates, even "instantaneous" peaks, appreciably above the permissible excursion, should not be permitted, unless information exists to the contrary. Other more specific excursions will be developed in the future.

Handbook of Environmental Control

BASIS FOR ASSIGNING LIMITING "C" VALUES

By definition in the Preface, a listed value bearing a "C" designation refers to a "ceiling" value that should not be exceeded; all values should fluctuate below the listed value. This, in effect, makes the "C" designation a maximal allowable concentration (MAC).

In general, the bases for assigning or not assigning a "C" value rest on whether excursions of concentration above a proposed limit *for periods up to 15 minutes* may result in a) intolerable irritation, b) chronic, or irreversible tissue change, or c) narcosis of sufficient degree to increase accident proneness, impair self-rescue or materially reduce work efficiency.

APPENDIX E

Some Nuisance Particulates[q]

TLV, 30 mppcf or 10mg/m³

Alundum (Al_2O_3)	Kaolin
Calcium carbonate	Limestone
Calcium silicate	Magnesite
Cellulose (paper fiber)	Marble
Portland Cement	Mineral Wool
Corundum (Al_2O_3)	Fiber
Emery	Pentaerythritol
Glass, fibrous[r] or dust	Plaster of Paris
Glycerin Mist	Rouge
Graphite (synthetic)	Silicon
Gypsum	Silicon Carbide
Vegetable oil mists	Starch
(except castor,	Sucrose
cashew nut, or	Tin Oxide
similar irritant	Titanium Dioxide
oils)	Zinc Stearate
	*Zinc oxide dust

q) When toxic impurities are not present, e.g. quartz < 1%

r) <7µm in diameter

APPENDIX F

Some Simple Asphyxiants[s]

Acetylene	Hydrogen
Argon	Methane
Butane	Neon
Ethane	Nitrous oxide
Ethylene	Propane
Helium	

s) As defined on pg. 6.

*1975 Addition.

3.2–10   TLVs® THRESHOLD LIMIT VALUES FOR PHYSICAL AGENTS ADOPTED BY ACGIH FOR 1975[a]

1975 TLV PHYSICAL AGENTS COMMITTEE

Herbert H. Jones, Central Missouri State University, Chairman

Peter A. Breysse, University of Washington

Irving H. Davis, Michigan Dept. of Health

LCDR Joseph J. Drozd, USN

Dr. David A. Fraser, Univ. of North Carolina

Maj. George S. Kush, USAF

Tom Cummins, Ontario Dept. of Health

Dr. Wordie H. Parr, NIOSH

David H. Sliney, USAEHA

Dr. Robert N. Thompson, FAA

Thomas K. Wilkinson, USPHS

Eugene G. Wood, U.S. Dept. of Labor

Ronald D. Dobbin, NIOSH

Lt. Col. Robert T. Wangemann, U.S. Army

Any comments or questions regarding these limits should be addressed to:

Secretary - Treasurer
P.O. Box 1937
Cincinnati, Ohio 45201

[a]This publication is reprinted in its entirety by permission of the American Conference of Governmental Industrial Hygienists, Cincinnati, Ohio.

PREFACE

PHYSICAL AGENTS

These threshold limit values refer to levels of physical agents and represent conditions under which it is believed that nearly all workers may be repeatedly exposed day after day without adverse effect. Because of wide variations in individual susceptibility, exposure of an occasional individual at, or even below, the threshold limit may not prevent annoyance, aggravation of a pre-existing condition, or physiological damage.

These threshold limits are based on the best available information from industrial experience, from experimental human and animal studies, and when possible, from a combination of the three.

These limits are intended for use in the practice of industrial hygiene and should be interpreted and applied only by a person trained in this discipline. They are not intended for use, or for modification for use, (1) in the evaluation or control of the levels of physical agents in the community, (2) as proof or disproof of an existing physical disability, or (3) for adoption by countries whose working conditions differ from those in the United States of America.

These values are reviewed annually by the Committee on Threshold Limits for Physical Agents for revisions or additions, as further information becomes available.

Notice of Intent — At the beginning of each year, proposed actions of the Committee for the forthcoming year are issued in the form of a "Notice of Intent." This notice provides not only an opportunity for comment, but solicits suggestions of physical agents to be added to the list. The suggestions should be accompanied by substantiating evidence.

As Legislative Code — The Conference recognizes that the Threshold Limit Values may be adopted in legislative codes and regulations. If so used, the intent of the concepts contained in the Preface should be maintained and provisions should be made to keep the list current.

Reprint Permission — This publication may be reprinted provided that written permission is obtained from the Secretary-Treasurer of the Conference and that this Preface be published in its entirety along with the Threshold Limit Values.

THRESHOLD LIMIT VALUES

HEAT STRESS

These Threshold Limit Values refer to heat stress conditions under which it is believed that nearly all workers may be repeatedly exposed without adverse health effects. The TLVs shown in Table 1 are based on the assumption that nearly all acclimatized, fully clothed workers with adequate water and salt intake should be able to function effectively under the given working conditions without exceeding a deep body temperature of 38°C (WHO technical report series #412, 1969 *Health Factors Involved in Working Under Conditions of Heat Stress*; F. N. Dukes-Dobos and A. Henschel: *"Development of Permissible Heat Exposure Limits for Occupational Work."* ASHRAE Journal, Vol. 15: No. 9, September 1973, pp. 57–62.)

Since measurement of deep body temperature is impractical for monitoring the workers' heat load, the measurement of environmental factors is required which most nearly correlate with deep body temperature and other physiological responses to heat. At the present time Wet Bulb-Globe Temperature Index (WBGT) is the simplest and most suitable technique to measure the environmental factors. WBGT values are calculated by the following equations:

1. Outdoors with solar load:
   $$WBGT = 0.7WB + 0.2GT + 0.1DB$$

2. Indoors or Outdoors with no solar load:
   $$WBGT = 0.7WB + 0.3GT$$
   where:

$WBGT$ = Wet Bulb-Globe Temperature Index
WB = Natural Wet-Bulb Temperature
DB = Dry-Bulb Temperature
GT = Globe Thermometer Temperature

The determination of WBGT requires the use of a black globe thermometer, a natural (static) wet-bulb thermometer, and a dry-bulb thermometer.

TABLE 1

Permissible Heat Exposure Threshold Limit Values
(Values are given in °C. WBGT)

Work — Rest Regimen	Work Load		
	Light	Moderate	Heavy
Continuous work	30.0	26.7	25.0
75% Work — 25% Rest, Each hour	30.6	28.0	25.9
50% Work — 50% Rest, Each hour	31.4	29.4	27.9
25% Work — 75% Rest, Each hour	32.2	31.1	30.0

Higher heat exposures than shown in Table 1 are permissible if the workers have been undergoing medical surveillance and it has been established that they are more tolerant to work in heat than the average worker. Workers should not be permitted to continue their work when their deep body temperature exceeds 38.0°C.

APPENDIX G

HEAT STRESS

I. *Measurement of the Environment*

The instruments required are a dry-bulb, a natural wet-bulb, a globe thermometer, and a stand. The measurement of the environmental factors shall be performed as follows:

A. The range of the dry and the natural wet bulb thermometer shall be − 50°C to 50°C with an accuracy of ± 0.5°C. The dry bulb thermometer must be shielded from the sun and the other radiant surfaces of the environment without restricting the airflow around the bulb. The wick of the natural wet-bulb thermometer shall be kept wet with distilled water for at least 1/2 hour before the temperature reading is made. It is not enough to immerse the other end of the wick into a reservoir of distilled water and wait until the whole wick becomes wet by capillarity. The wick shall be wetted by direct application of water from a syringe 1/2 hour before each reading. The wick shall extend over the bulb of the thermometer, covering the stem about one additional bulb length. The wick should always be clean and new wicks should be washed before using.

B. One globe thermometer, consisting of a 15 cm. (6-inch) diameter hollow copper sphere, painted on the outside with a matte black finish or equivalent shall be used. The bulb or sensor of a thermometer (range −5°C to 100°C with an accuracy of ± 0.5°C) must be fixed in the center of the sphere. The globe thermometer shall be exposed at least 25 minutes before it is read.

C. One stand shall be used to suspend the three thermometers so that they do not restrict free air flow around the bulbs, and the wet-bulb and globe thermometer are not shaded.

D. It is permissible to use any other type of temperature sensor that gives identical reading to a mercury thermometer under the same conditions.

E. The thermometers must be so placed that the readings are representative of the condition where the men work or rest, respectively.

The methodology outlined above is more fully explained in the following publications:

1. "Prevention of Heat Casualties in Marine Corps Recruits, 1955–1960, with Comparative Incidence Rates and Climatic Heat Stresses in other Training Categories," by Captain David Minard, MC, USN, Research Report No. 4, Contract No. MR005.01–0001.01, Naval Medical Research Institute, Bethesda, Maryland, 21 February 1961.

2. "Heat Casualties in the Navy and Marine Corps, 1959–1962, with Appendices on the Field Use of the Wet Bulb-Globe Temperature Index," by Captain David Minard, MC, USN, and R. L. O'Brien, HMC, USN. Research Report No. 7, Contract No. MR 005.01-0001.01, Naval Medical Research Institute, Bethesda, Maryland, 12 March 1964.

3. Minard, D.: Prevention of Heat Casualties in Marine Corps Recruits. Military Medicine *126(4)*: 261–272, 1961.

II. *Work Load Categories*

The heat produced by the body and the environmental heat together determine the total heat load. Therefore, if work is to be performed under hot environmental conditions, the workload category of each job shall be established and the heat exposure limit pertinent to the work load evaluated against the applicable standard in order to protect the worker from exposure beyond the permissible limit.

A. The work load category may be established by ranking each job into light, medium, and heavy categories on the basis of type of operation, where the work load is ranked into one of said three categories, i.e.

(1) light work (up to 200 Kcal/hr or 800 Btu/hr): e.g., sitting or standing to control machines, performing light hand or arm work,

(2) moderate work (200–350 Kcal/hr or 800–1400 Btu/hr): e.g., walking about with moderate lifting and pushing,

(3) heavy work (350–500 Kcal/hr or 1400–2000 Btu/hr): e.g., pick and shovel work,

the permissible heat exposure limit for that work load shall be determined from Table 1.

B. The ranking of the job may be performed either by measuring the worker's metabolic rate while performing his job or by estimating his metabolic rate by the use of the scheme shown in Table 2. Tables available in the literature listed below and in other publications as well may also be utilized. When this method is used the permissible heat exposure limit can be determined by Figure 1.

1. Per-Olaf Astrand and Kaare Rodahl: "Textbook of Work Physiology" McGraw-Hill Book Company, New York, San Francisco, 1970.

2. "Ergonomics Guide to Assessment of Metabolic and Cardiac Costs of Physical Work." Amer. Ind. Hyg. Assoc. J. *32:* 560, 1971.

3. Energy Requirements for Physical Work. Purdue Farm Cardiac Project. Agricultural Experiment Station. Research Progress Report No. 30, 1961.

4. J. V. G. A. Durnin and R. Passmore: "Energy, Work and Leisure." Heinemann Educational Books, Ltd., London, 1967.

TABLE 2

Assessment of Work Load

Average values of metabolic rate during different activities.

A. Body position and movement	Kcal./min.
Sitting	0.3
Standing	0.6
Walking	2.0–3.0
Walking up hill	add 0.8
	per·meter (yard) rise

B. Type of Work	Average Kcal./min.	Range Kcal./min.
Hand work		
light	0.4	0.2–1.2
heavy	0.9	
Work with one arm		
light	1.0	0.7–2.5
heavy	1.8	
Work with both arms		
light	1.5	1.0–3.5
heavy	2.5	

TABLE 2 (continued)

B. Type of Work	Average Kcal./min.	Range Kcal./min.
Work with body		
light	3.5	2.5–15.0
moderate	5.0	
heavy	7.0	
very heavy	9.0	

Light hand work: writing, hand knitting

Heavy hand work: typewriting

Heavy work with one arm: hammering in nails (shoemaker, upholsterer)

Light work with two arms: filing metal, planing wood, raking of a garden
Moderate work with the body: cleaning a floor, beating a carpet

Heavy work with the body: railroad track laying, digging, barking trees

Sample Calculation: Using a heavy hand tool on an assembly line

A. Walking along   2.0 Kcal./min.

B. Intermediate value between heavy work with two arms and light work with the body   3.0 Kcal./min.

                                                5.0 Kcal./min.

C. Add for basal metabolism   1.0 Kcal./min.

                   Total   6.0 Kcal./min.

Adapted from Lehmann, G.E., A. Muller and H. Spitzer: Der Kalorienbedarf bei gewerblicher Arbeit. Arbeitsphysiol. *14:* 166, 1950.

III. *Work-Rest Regimen*

The permissible exposure limits specified in Table 1 and Figure 1 are based on the assumption that the WBGT value of the resting place is the same or very close to that of the work place. If the resting place is air conditioned and its climate is kept at or below 24°C (75°F.) WBGT, the allowable resting time may be reduced by 25%. The permissible exposure limits for continuous work are applicable where there is a work-rest regimen of a 5-day work week and an 8-hour work day with a short morning and afternoon break (approximately 15 minutes) and a longer lunch break (approximately 30 minutes). Higher exposure limits are permitted if additional resting time is allowed. All breaks, including unscheduled pauses and administrative or operational waiting periods during work may be counted as rest time when additional rest allowance must be given because of high environmental temperatures.

It is a common experience that when the work on a job is self-paced, the workers will spontaneously limit their hourly work load to 30–50% of their maximum physical performance capacity. They do this either by setting an appropriate work speed or by interspersing unscheduled breaks. Thus the daily average of the workers' metabolic rate seldom exceeds 330 kcal/hr. However, within an 8-hour work shift there may be periods where the workers' hourly average metabolic rate will be higher.

IV. *Water and Salt Supplementation*

During the hot season or when the worker is exposed to artificially generated heat, drinking water shall be made available to the workers in such a way that they are stimulated to frequently drink small amounts, i.e., one cup every 15–20 minutes (about 150 ml or 1/4 pint).

The water shall be kept reasonably cool (10°–15°C or 50.0°–60.0°F) and shall be placed close to the workplace so that the worker can reach it without abandoning the work area.

The workers should be encouraged to salt their food abundantly during the hot season and particularly during hot spells. If the workers are unacclimatized, salted drinking water shall be made available in a concentration of 0.1% (1g NaCl to 1.0 liter or 1 level tablespoon of salt to 15 quarts of water). The added salt shall be completely dissolved before the water is distributed, and the water shall be kept reasonably cool.

V. *Other Considerations*

A. Clothing: The permissible heat exposure TLVs are valid for light summer clothing as customarily worn by workers when working under hot environmental conditions. If special clothing is required for performing a particular job and this clothing is heavier or it impedes sweat evaporation or has higher insulation value, the worker's heat tolerance is reduced, and the permissible heat exposure limits indicated in Table 1 and Figure 1 are not applicable. For each job category where special clothing is required, the permissible heat exposure limit shall be established by an expert.

B. Acclimatization and Fitness: The recommended heat stress TLVs are valid for acclimated workers who are physically fit.

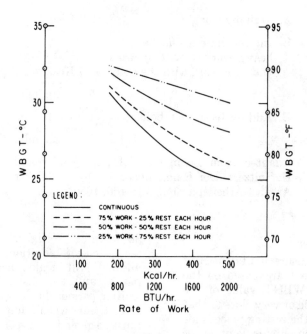

Figure 1 — Permissible Heat Exposure
Threshold Limit Value

IONIZING RADIATION

See U.S. Department of Commerce National Bureau of Standards, Handbook 59, "Permissible Dose from External Sources of Ionizing Radiation," September 24, 1954, and addendum of April 15, 1958. A report, Basic Radiation Protection Criteria, published by the National Committee on Radiation Protection, revises and modernizes the concept of the NCRP standards of 1954, 1957 and 1958; obtainable as NCRP Rept. No. 39, P.O. Box 4867, Washington, D.C. 20008.

LASERS*

The threshold limit values are for exposure to laser radiation under conditions to which nearly all workers may be exposed without adverse effects. The values should be used as guides in the control of exposures and should not be regarded as fine lines between safe and dangerous levels. They are based on the best available information from experimental studies.

Limiting Apertures

The TLVs expressed as radiant exposure or irradiance in this section may be averaged over an aperture of 1 mm except for TLVs for the eye in the spectral range of 400-1400 nm, which should be averaged over a 7 mm limiting aperture (pupil); and except for all TLVs for wavelengths between 0.1-1 mm where the limiting aperture is 10 mm. No modification of the TLVs is permitted for pupil sizes less than 7 mm.

The TLVs for "extended sources" apply to sources which subtend an angle greater than α (Table 5) which varies with exposure time. This angle is *not* the beam divergence of the source.

Correction Factors A and B (C_A and C_B) for Eye Exposure

All TLVs in Tables 3 and 4 are to be used as given for wavelengths 400 nm to 700 nm.

*See Notice of Intended Change, Page 260.

At all wavelengths greater than 1.06 μm and less than 1.4 μm the TLVs are to be increased by a factor of 5. TLV at wavelengths between 700 nm and 1.06 μm are to be increased by a uniformly extrapolated factor as shown in Figure 2. For certain exposure durations at wavelengths between 700-800 nm, correction factor C_B is applied.

Repetitively Pulsed Lasers

Since there are few experimental data for multiple pulses, caution must be used in the evaluation of such exposures. The protection standards for irradiance or radiant exposure in multiple pulse trains have the following limitations:

(1) The exposure from any single pulse in the train is limited to the protection standard for a single comparable pulse.

(2) The average irradiance for a group of pulses is limited to the protection standard as given in Tables 3, 4, or 6 of a single pulse of the same duration as the entire pulse group.

(3) When the Instantaneous Pulse Repetition Frequency (PRF) of any pulses within a train exceeds one, the protection standard applicable to each pulse is reduced as shown in Figure 6 for pulse durations less than 10^{-5} second. For pulses of greater duration, the following formula should be followed:

$$\text{Standard} \left(\begin{array}{c} \text{single pulse} \\ \text{in train} \end{array} \right) = \frac{\text{Standard (pulse } n\tau)}{n}$$

where:

n = number of pulses in train
τ = duration of a single pulse in the train
Standard ($n\tau$) = protection standard of one pulse having a duration equal to $n\tau$ seconds

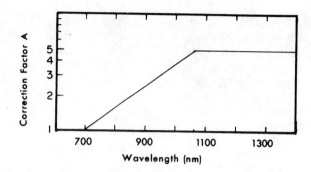

Figure 2. TLV correction factors for
laser wavelengths (eye).

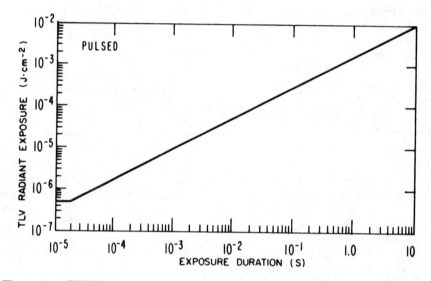

Figure 3a. TLV for intrabeam (direct) viewing of laser beam (400–700 nm).

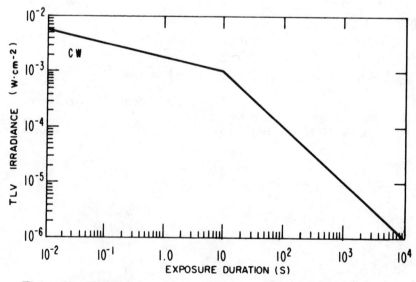

Figure 3b. TLV for intrabeam (direct) viewing of CW laser beam
(400–700 nm).

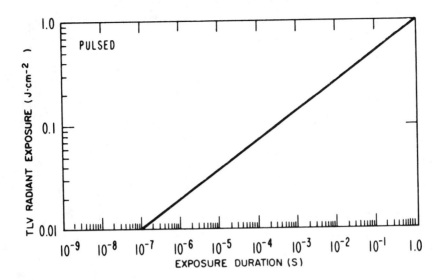

Figure 4a. TLV for laser exposure of skin and eyes for far-infrared radiation (wavelengths greater than 1.4 μm).

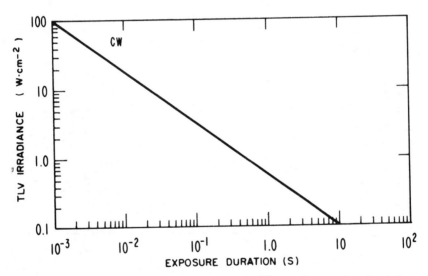

Figure 4b. TLV for CW laser exposure of skin and eyes for far-infrared radiation (wavelengths greater than 1.4 μm).

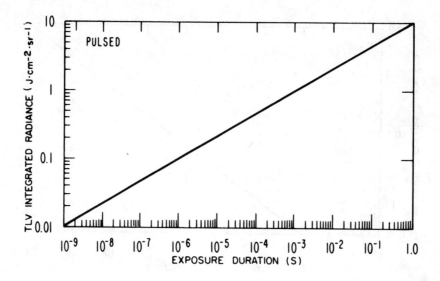

Figure 5a. TLV for extended sources or diffuse reflections of laser radiation (400–700 nm).

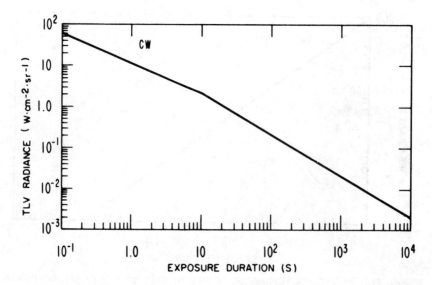

Figure 5b. TLV for extended sources or diffuse reflections of laser radiation (400–700 nm), CW.

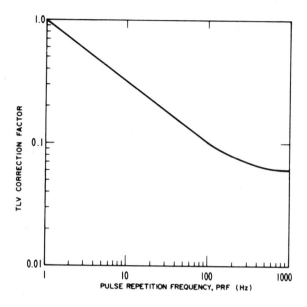

Figure 6.   Multiplicative correction factor for repetitively pulsed lasers having pulse durations less than 10^{-5} second. TLV for a single pulse of the pulse train is multiplied by the above correction factor. Correction factor for PRF greater than 1000 H$_z$ is 0.06.

TABLE 3

Threshold Limit Value for Direct Ocular Exposures
(Intrabeam Viewing) from a Laser Beam

Spectral Region	Wave Length	Exposure Time, (t) Seconds	TLV	
UVC	200 nm to 280 nm	10^{-3} to 3×10^4	3	mJ • cm^{-2}
UVB	280 nm to 302 nm	"	3	"
	303 nm	"	4	"
	304 nm	"	6	"
	305 nm	"	10	"
	306 nm	"	16	"
	307 nm	"	25	"
	308 nm	"	40	"
	309 nm	"	63	"
	310 nm	"	100	"
	311 nm	"	160	"
	312 nm	"	250	"
	313 nm	"	400	"
	314 nm	"	630	"
	315 nm	"	1.0 J • cm^{-2}	

TABLE 3 (Cont.)

UVA	315 nm to 400 nm	10 to 10^3	1.0 J·cm^{-2}
	" "	10^3 to 3×10^4	1.0 mW·cm^{-2}
Light	400 nm to 700 nm	10^{-9} to 1.8×10^{-5}	5×10^{-7} J·cm^{-2}
	" "	1.8×10^{-5} to 10	$\left[\dfrac{1.8t}{\sqrt[4]{t}} \right]$ mJ·cm^{-2}
	" "	10 to 10^4	10 mJ·cm^{-2}
	" "	10^4 to 3×10^4	10^{-6} W·cm^{-2}
Infrared A	700 nm to 1.06 μm	10^{-9} to 18×10^{-5}	0.5 C_A μJ·cm^{-2}
"	700 nm to 1.06 μm	18×10^{-5} to 10	$(1.8t/\sqrt[4]{t})$ C_A mJ·cm^{-2}
"	700 nm to 1.06 μm	10 to 100	10 C_A mJ·cm^{-2}
"	1.06 μm to 1.4 μm	10^{-9} to 1.0×10^{-4}	5 μJ·cm^{-2}
"	1.06 μm to 1.4 μm	1.0×10^{-4} to 10	$\left(\dfrac{9t}{\sqrt[4]{t}} \right)$ mJ·cm^{-2}
"	1.06 μm to 1.4 μm	10 to 100	50 mJ·cm^{-2}
"	700 nm to 800 nm	100 to $\left(\dfrac{10^4}{C_B} \right)$	10 C_A mJ·cm^{-2}
"	700 nm to 800 nm	$\left(\dfrac{10^4}{C_B} \right)$ to 3×10^4	C_A C_B μW·cm^{-2}
"	800 nm to 1.06 μm	100 to 3×10^4	0.1 C_A mW·cm^{-2}
Infrared B & C	1.4 μm to 10^3 μm	10^{-9} to 10^{-7}	10^{-2} J·cm^2
"	" "	10^{-7} to 10	0.56 $\sqrt[4]{t}$ J·cm^{-2}
"	" "	10 to 3×10^4	0.1 W·cm^{-2}

$C_A = e^{(\lambda - 700/224)}$
$C_B = (\lambda - 699)$

TABLE 4

Threshold Limit Values for Viewing a Diffuse Reflection of a Laser Beam or an Extended Source Laser

Spectral Region	Wave Length	Exposure Time, (t) Seconds	TLV
UV	200 nm to 400 nm	10^{-3} to 3×10^4	Same as Table 3
Light	400 nm to 700 nm	10^{-9} to 10	10· $\sqrt[3]{t}$ J·cm^{-2}·sr^{-1}
	" "	10 to 10^4	20 J·cm^{-2}·sr^{-1}
	" "	10^4 to 3×10^4	2×10^{-3} W·cm^{-2}·sr^{-1}
Infrared A	700 nm to 1.06 μm	10^{-9} to 10	10 C_A $\sqrt[3]{t}$ J·cm^{-2}·sr^{-1}
"	700 nm to 1.06 μm	10 to 100	20 C_A J·cm^{-2}·sr^{-1}
"	700 nm to 800 nm	100 to $\left(\dfrac{10^4}{C_B} \right)$	20 C_A J·cm^{-2}·sr^{-1}
"	700 nm to 800 nm	$\left(\dfrac{10^4}{C_B} \right)$	0.2 C_A C_B W·cm^{-2}·sr^{-1}
"	800 nm to 1.06 μm	100 to 3×10^4	0:2 C_A W·cm^{-2}·sr^{-1}
"	1.06 μm to 1.4 μm	10^{-9} to 10	$50 \times \sqrt[3]{t}$ J·cm^{-2}·sr^{-1}
"	" "	10 to 10^2	100 J·cm^{-2}·sr^{-1}
"	" "	10^2 to 3×10^4	1.0 W·cm^{-2}·sr^{-1}
Infrared B & C	1.4 μm to 1 mm	10^{-9} to 3×10^4	Same as Table 3

$C_A = e^{(\lambda - 700/224)}$
$C_B = (\lambda - 699)$

TABLE 5

Limiting Angle of Extended Source
Which May Be Used for Applying Extended Source TLVs

Exposure Duration(s)	Angle α (mrad)
10^{-9}	8.0
10^{-8}	5.4
10^{-7}	3.7
10^{-6}	2.5
10^{-5}	1.7
10^{-4}	2.2
10^{-3}	3.6
10^{-2}	5.7
10^{-1}	9.2
1.0	15
10	24
10^{2}	24
10^{3}	24
10^{4}	24

TABLE 6

Threshold Limit Value for Skin Exposure from a Laser Beam

Spectral Region	Wave Length	Exposure Time, (t) Seconds	TLV
UV	200 nm to 400 nm	10^{-3} to 3×10^{4}	Same as Table 3
Light & Infrared A	400 nm to 1400 nm	10^{-9} to 10^{-7}	$2 \times 10^{-2}\,\mathrm{J \cdot cm^{-2}}$
"	" "	10^{-7} to 10	$1.1\ \sqrt[4]{t}\ \mathrm{J \cdot cm^2}$
"	" "	10 to 3×10^{4}	$0.2\,\mathrm{W \cdot cm^{-2}}$
Infrared B & C	1.4 μm to 1 mm	10^{-9} to 3×10^{4}	Same as Table 3

NOTE: To aid in the determination of TLV's for exposure durations requiring
calculations of fractional powers Figures 3, 4 and 6 may be used.

MICROWAVES*

These Threshold Limit Values refer to microwave energy in the frequency range of 100 MHz to 100 GHz and represent conditions under which it is believed that nearly all workers may be repeatedly exposed without adverse effect.

These values should be used as guides in the control of exposure of microwaves and should not be regarded as a fine line between safe and dangerous levels.

Recommended Values

The Threshold Limit Value for occupational microwave energy exposure where power densities are known and exposure time controlled is as follows:

1. For average power density levels up to but not exceeding 10 milliwatts per square centimeter, total exposure time shall be limited to the 8-hour workday (continuous exposure).
2. For average power density levels from 10 milliwatts per square centimeter up to but not exceeding 25 milliwatts per square centimeter, total exposure time shall be limited to no more than 10 minutes for any 60 minute period during an 8-hour workday (intermittent exposure).
3. For average power density levels in excess of 25 milliwatts per square centimeter, exposure is not permissible.

NOTE: For repetitively pulsed sources the average power density may be calculated by multiplying the peak power density by the duty cycle. The duty cycle is equal to the pulse duration in seconds times the pulse repetition rate in hertz.

NOISE**

These threshold limit values refer to sound pressure levels and durations of exposure that represent conditions under which it is believed that nearly all workers may be repeatedly exposed without adverse effect on their ability to hear and understand normal speech. The medical profession has defined hearing impairment as an average hearing threshold level in excess of 25 decibels (ANSI-S3.6-1969) at 500, 1000,

and 2000 Hz, and the limits which are given have been established to prevent a hearing loss in excess of this level. The values should be used as guides in the control of noise exposure and, due to individual susceptibility, should not be regarded as fine lines between safe and dangerous levels.

Continuous or Intermittent

The sound level shall be determined by a sound level meter, conforming as a minimum to the requirements of the American National Standard Specification for Sound Level Meters, S1.4 (1971) Type S2A, and set to use the A-weighted network with slow meter response. Duration of exposure shall not exceed that shown in Table 7.

These values apply to total duration of exposure per working day regardless of whether this is one continuous exposure or a number of short-term exposures but does not apply to impact or impulsive type of noise.

When the daily noise exposure is composed of two or more periods of noise exposure of different levels, their combined effect should be considered, rather than the individual effect of each. If the sum of the following fractions:

$$\frac{C_1}{T_1} + \frac{C_2}{T_2} + \cdots \frac{C_n}{T_n}$$

exceeds unity, then, the mixed exposure should be considered to exceed the threshold limit value, C_1 indicates the total duration of exposure at a specific noise level, and T_1 indicates the total duration of exposure permitted at that level. All on-the-job noise exposures of 80 dBA or greater shall be used in the above calculations.

Impulsive or Impact Noise

It is recommended that exposure to impulsive or impact noise should not exceed 140 decibels peak sound pressure level. Impulsive or impact noise is considered to be those variations in noise levels that involve maxima at intervals of greater than one per second. Where the intervals are less than one second, it should be considered continuous.

*See Notice of Intent to Study, Page 262.
**See Notice of Intended Changes, Page 262.

Table 7
Threshold Limit Values

Duration per day Hours	Sound Level dBA[a]
16	80
8	85
4	90
2	95
1	100
1/2	105
1/4	110
1/8	115*

*No exposure to continuous or inter-mittent in excess of 115 dBA

a) Sound level in decibels as measured on a sound level meter, conforming as a minimum to the requirements of the American National Standard Specification for Sound Level Meters, S1.4 (1971) Type S2A, and set to use the A-weighted network with slow meter response.

It should be recognized that the application of the TLV for noise will not protect all workers from the adverse effects of noise exposure. A hearing conservation program with audiometric testing is necessary when workers are exposed to noise at or above the TLV levels.

ULTRAVIOLET RADIATION*

These threshold limit values refer to ultraviolet radiation in the spectral region between 200 and 400 nm and represent conditions under which it is believed that nearly all workers may be repeatedly exposed without adverse effect. These values for exposure of the eye or the skin apply to ultraviolet radiation from arcs, gas, and vapor discharges, fluorescent, and incandescent sources, and solar radiation, but do not apply to ultraviolet lasers.* These levels should not be used for determining exposure of photosensitive individuals to ultraviolet radiation. These values should be used as guides in the control of exposure to continuous sources where the exposure duration shall not be less than 0.1 sec.

*See Laser TLVs.

These values should be used as guides in the control of exposure to ultraviolet sources and should not be regarded as a fine line between safe and dangerous levels.

Recommended Values:

The threshold limit value for occupational exposure to ultraviolet radiation incident upon skin or eye where irradiance values are known and exposure time is controlled are as follows:

1. For the near ultraviolet spectral region (320 to 400 mm) total irradiance incident upon the unprotected skin or eye should not exceed 1 mw/cm² for periods greater than 10^3 seconds (approximately 16 minutes) and for exposure times less than 10^3 seconds should not exceed one J/cm².

2. For the actinic ultraviolet spectral region (200 — 315 nm), radiant exposure incident upon the unprotected skin or eye should not exceed the values given in Table 8 within an 8-hour period.

3. To determine the effective irradiance of a broadband source weighted against the peak of the spectral effectiveness curve (270 nm), the following weighting formula should be used:

$$E_{eff} = \sum E_\lambda S_\lambda \Delta_\lambda$$

where:

E_{eff} = effective irradiance relative to a monochromatic source at 270 nm

E_λ = spectral irradiance in W/cm²/nm

S_λ = relative spectral effectiveness (unitless)

Δ_λ = band width in nanometers

4. Permissible exposure time in seconds for exposure to actinic ultraviolet radiation incident upon the unprotected skin or eye may be computed by dividing 0.003 J/cm² by E_{eff} in W/cm². The exposure time may also be determined using Table 9 which provides exposure times corresponding to effective irradiances in μW/cm².

TABLE 8

Relative Spectral Effectiveness
by Wavelength

Wavelength, nm	TLV, mJ/cm²	Relative Spectral Effectiveness S_λ
200	100	0.03
210	40	0.075
220	25	0.12
230	16	0.19
240	10	0.30
250	7.0	0.43
254	6.0	0.5
260	4.6	0.65
270	3.0	1.0
280	3.4	0.88
290	4.7	0.64
300	10	0.30
305	50	0.06
310	200	0.015
315	1000	0.003

TABLE 9

Permissible Ultraviolet Exposures

Duration of Exposure Per Day	Effective Irradiance, E_{eff}, $\mu W/cm^2$
8 hrs...................	0.1
4 hrs...................	0.2
2 hrs...................	0.4
1 hr....................	0.8
1/2 hr..................	1.7
15 min.................	3.3
10 min.................	5
5 min..................	10
1 min..................	50
30 sec.................	100
10 sec.................	300
1 sec..................	3,000
0.5 sec................	6,000
0.1 sec................	30,000

All the preceding TLV's for ultraviolet energy apply to sources which subtend an angle less than 80°. Sources which subtend a greater angle need to be measured only over an angle of 80°.

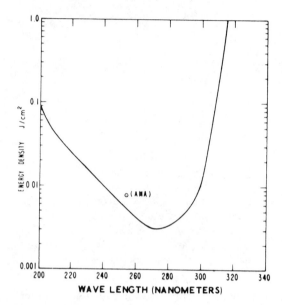

Figure 7. Threshold Limit Values
for Ultraviolet Radiation

Conditioned (tanned) individuals can tolerate skin exposure in excess of the TLV without erythemal effects. However, such conditioning may not protect persons against skin cancer.

NOTICE OF INTENDED CHANGES
(for 1975)

These physical agents, with their corresponding values, comprise those for which either a limit has been proposed for the first time, or for which a change in the "Adopted" listing has been proposed. In both cases, the proposed limits should be considered trial limits that will remain in the listing for a period of at least one year. If, after one year no evidence comes to light that questions the appropriateness of the values herein, the values will be reconsidered for the "Adopted" list.

LASERS

The Threshold Limit Values for Lasers would be changed for ocular exposure to visible and IR-A laser radiation for exposure durations greater than 10 s. Also to be changed would be TLVs for skin exposure to IR-A laser radiation. All other laser TLVs would remain unchanged. The sections of Tables 10, 11, and 12 that would have changes are as follows:

Table 10 Changes
Threshold Limit Values for Direct Ocular Exposures (Intrabeam Viewing) from a Laser Beam

Spectral Region	Wavelengths	Exposure Duration (t) in Seconds	TLV
Light	400 nm to 700 nm	10^{-9} to 1.8×10^{-5}	5×10^{-7} J $\cdot$ cm^{-2}
	400 nm to 700 nm	1.8×10^{-5} to 10	1.8 (t/ $\sqrt[4]{t}$) mJ $\cdot$ cm^{-2}
	400 nm to 549 nm	10 to 10^4	10 mJ $\cdot$ cm^{-2}
	550 nm to 700 nm	10 to T_1	1.8 (t/ $\sqrt[4]{t}$) mJ $\cdot$ cm^{-2}
	550 nm to 700 nm	T_1 to 10^4	10 C_B mJ $\cdot$ cm^{-2}
	400 nm to 700 nm	10^4 to 3×10^4	C_B μW $\cdot$ cm^{-2}
IR-A	700 nm to 1059 nm	10^{-9} to 1.8×10^{-5}	5 $C_A \times 10^{-7}$ J $\cdot$ cm^{-2}
	700 nm to 1059 nm	1.8×10^{-5} to 10^3	1.8 C_A (t/ $\sqrt[4]{t}$) mJ $\cdot$ cm^{-2}
	1060 nm to 1400 nm	10^{-9} to 10^{-4}	5×10^{-6} J $\cdot$ cm^{-2}
	1060 nm to 1400 nm	10^{-4} to 10^3	9(t/ $\sqrt[4]{t}$) mJ $\cdot$ cm^{-2}
	700 nm to 1400 nm	10^3 to 3×10^4	320 C_A μW $\cdot$ cm^{-2}

C_A — See Fig. 2, Laser TLV listing.
$C_B = 1$ for $\lambda = 400$ to 550 nm; $C_B = 10^{[0.015 (\lambda - 550)]}$ for $\lambda = 550$ to 700 nm.
$T_1 = 10$ s for $\lambda = 400$ to 550 nm; $T_1 = 10 \times 10^{[0.02 (\lambda - 550)]}$ for $\lambda = 550$ to 700 nm.

Table 11 Changes
Threshold Limit Values for Viewing a Diffuse Reflection of a Laser Beam or an Extended Source Laser

Spectral Region	Wavelengths	Exposure Duration (t) in Seconds	TLV
Light	400 nm to 700 nm	10^{-9} to 10	10 $\sqrt[3]{t}$ J $\cdot$ cm^{-2} $\cdot$ sr^{-1}
	400 nm to 549 nm	10 to 10^4	21 J $\cdot$ cm^{-2} $\cdot$ sr^{-1}
	550 nm to 700 nm	10 to T_1	3.83 (t/ $\sqrt[4]{t}$) J $\cdot$ cm^{-2} $\cdot$ sr^{-1}
	550 nm to 700 nm	T_1 to 10^4	21/C_B J $\cdot$ cm^{-2} $\cdot$ sr^{-1}
	400 nm to 700 nm	10^4 to 3×10^4	2.1/$C_B \times 10^{-3}$ W $\cdot$ cm^{-2} $\cdot$ sr^{-1}
IR-A	700 nm to 1400 nm	10^{-9} to 10	10 C_A $\sqrt[3]{t}$ J $\cdot$ cm^{-2} $\cdot$ sr^{-1}
	700 nm to 1400 nm	10 to 10^3	3.83 C_A (t/ $\sqrt[4]{t}$) J $\cdot$ cm^{-2} $\cdot$ sr^{-1}
	700 nm to 1400 nm	10^3 to 3×10^4	0.64 C_A W $\cdot$ cm^{-2} $\cdot$ sr^{-1}

C_A, C_B, and T_1 are the same as in footnote to Table 10.

Table 12 Changes

Spectral Region	Wavelengths		Exposure Duration (t) in Seconds	TLV
Light & IR-A	400 nm to 1400 nm		10^{-9} to 10^{-7}	2 $C_A \times 10^{-2}$ J $\cdot$ cm^{-2}
	"	"	10^{-7} to 10	1.1 C_A $\sqrt[4]{t}$ J $\cdot$ cm^{-2}
	"	"	10 to 3×10^4	0.2 C_A W $\cdot$ cm^{-2}

$C_A = 1.0$ for $\lambda = 400$–700 nm; see Figure 2 Laser TLV list for greater wavelength values.

NOISE

These Threshold Limit Values refer to sound pressure levels and durations of exposure that represent conditions under which it is believed that nearly all workers may be repeatedly exposed without adverse effect on their ability to hear and understand normal speech. The medical profession has defined hearing impairment as an average hearing threshold level in excess of 25 decibels (ANSI-S3.6-1969) at 500, 1000, and 2,000 H_z and the limits which are given have been established to prevent a hearing loss in excess of this level. The values should be used as guides in the control of noise exposure and, due to individual susceptibility, should not be regarded as fine lines between safe and dangerous levels.

IMPULSIVE OR IMPACT NOISE

It is recommended that exposure to impulsive or impact noise should not exceed those listed in Table 13. Impulsive or impact noise is considered to be those variations in noise levels that involve maxima at intervals of greater than one per second. Where the intervals are less than one second, it should be considered continuous.

NOTICE OF INTENT TO ESTABLISH THRESHOLD LIMIT VALUES

LIGHT

These Threshold Limit Values refer to visible radiation in the wavelength range of 400 to 700 nm and represent conditions under which it is believed that nearly all workers may be exposed without adverse effect.

Table 13
Threshold Limit Values
Impulsive or Impact Noise

Sound Level dB*	Number of Impulses or Impacts per day
140	100
130	1000
120	10,000

*decibels peak sound pressure level.

These values should be used as guides in the control of exposure to light and should not be regarded as a fine line between safe and dangerous levels.

Recommended Values:

The Threshold Limit Value for occupational exposure to light where luminances are known and exposure durations exceeding 10 seconds in any eight hour workday is as follows:

1. The average luminance of objects continuously viewed shall not exceed 1 candela cm^{-2} (1 candela/cm^2=2920 fL)

This TLV is not to be used for short exposure durations or pulsed light sources.

NOTICE OF INTENT TO STUDY

These agents comprise those which the Physical Agents Committee of ACGIH proposes to study during this year to determine the feasibility of establishing proposed TLVs in 1976. Comments and suggestions, accompanied by substantitive evidence, are solicited.

1. *Radiofrequency Radiation.* Specifically, that portion of the spectrum from 10 MHz to 100 MHz.

2. *Microwave Radiation.* Specifically from 100 GHz to 300 GHz.

3. *Magnetic Fields.* Both pulsed and continuous.

4. *Laser Radiation.* Specifically ultraviolet radiation for pulsed exposures, and repetitively pulsed light and infrared-A laser exposures.

5. *Ultrasonic Energy.* Specifically, acoustic energy at frequencies above 10 kHz.

3.3   NOISE PRODUCTION AND CONTROL

3.3—1   MOST PREVALENT SOUNDS IN HOSPITALS, ARRANGED IN ORDER OF ANNOYANCE

Rank position	Sound
1	Radio or television set
2	Staff talk in corridors
3	Other patients in distress and recovery-room sounds
4	Voice paging
5	Talk in other rooms
6	Babies or children crying
7	Telephones
8	Pantry, kitchen, utility room

Source: Noise in Hospitals, PHS Pub. No. 930-D-11, Public Health Service, U.S. Department of Health, Education, and Welfare, Washington, D.C., 1963.

3.3—2   HOSPITAL NOISE—INTERNAL SOURCES

Power plant — boilers, induced draft fans, pumps, combustion noises

Electrical equipment — auxiliary generators, transformers, switch gear

Heating ventilation and air conditioning equipment — air supply and air exhaust openings.

Transportation systems — elevators, dumbwaiters, pneumatic tubes, conveyor systems, linen chutes, rubbish chutes

Mechanical equipment — sterilizers, autoclaves, ultrasonic cleaners, dishwashing machines, ice machines, refrigerators, freezers

Plumbing fixtures — water flowing from faucet, stream of water hitting metal basin, frequent water hammer

Furnishings — cabinet doors and drawers, metal units, chart cases, racks, beds, bed curtains, furniture

Housekeeping facilities — floor-washing machines, floor polishers, vacuum cleaners, metal trash containers

Communication systems — voice paging systems, telephones, carts used for food, medication, linens, supplies, trash collection

Adapted from G. S. Michaelsen, Noise production and control, in *Environmental Health and Safety in Health-Care Facilities,* R. G. Bond et al., Eds., Macmillan, New York, © 1973.

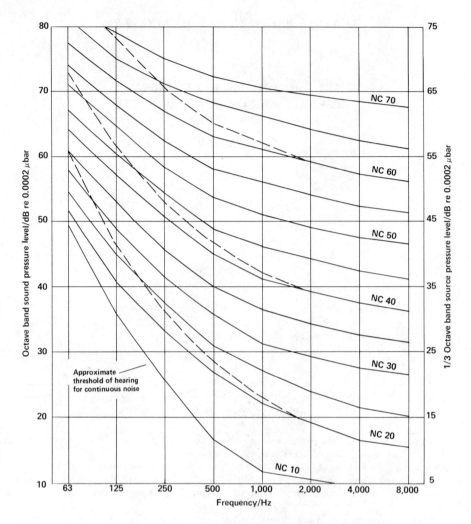

3.3−3   Noise-criterion curves for specifying the design level in terms of the maximum permissible sound-pressure level for each frequency band.

Source: American Society of Heating, Refrigerating, and Air Conditioning Engineers, *ASHRAE Guide and Data Book: Systems,* ASHRAE, New York, 1970. With permission.

3.3–4   RANGES OF INDOOR DESIGN GOALS FOR AIR-CONDITIONING-SYSTEM SOUND CONTROL

Type of area	Range of NC criteria curves[a]	Type of area	Range of NC criteria curves[a]
Residences		Churches and schools	
Private homes (rural and suburban)	20–30	Sanctuaries	20–30
Private homes (urban)	25–35	Libraries	30–40
Apartment houses, 2- and 3-family units	30–40	Schools and classrooms	30–40
Hotels		Laboratories	35–45
Individual rooms or suites	30–40	Recreation halls	35–50
Ballrooms, banquet rooms	30–40	Corridors and halls	35–50
Halls and corridors, lobbies	35–45	Kitchens	40–50
Garages	40–50	Public buildings	
Kitchens and laundries	40–50	Public libraries, museums, courtrooms	30–40
Hospitals and clinics		Post offices, general banking areas, lobbies	35–45
Private rooms	25–35	Washrooms and toilets	40–50
Operating rooms, wards	30–40	Restaurants, cafeterias, lounges	
Laboratories, halls, and corridors ⎰	35–45	Restaurants	35–45
Lobbies and waiting rooms ⎱		Cocktail lounges	35–50
Washrooms and toilets	40–50	Night clubs	35–45
Offices		Cafeterias	40–50
Board rooms	20–30	Stores, retail	
Conference rooms	25–35	Clothing stores ⎰	35–45
Executive offices	30–40	Department stores (upper floors) ⎱	
Supervisors, offices, reception rooms	30–45	Department stores (main floor) ⎰	40–50
General open offices, drafting rooms	35–50	Small retail stores ⎱	
Halls and corridors	35–55	Supermarkets	40–50
Tabulation and computation rooms	40–60	Sports activities, indoor	
Auditoriums and music halls		Coliseums	30–40
Concert and opera halls ⎰	20–25	Bowling alleys, gymnasiums	35–45
Studios for sound reproduction ⎱		Swimming pools	40–55
Legitimate theaters, multipurpose halls	25–30	Transportation (rail, bus, plane)	
Movie theaters, television audience studios		Ticket sales offices	30–40
Semioutdoor amphitheaters	30–35	Lounges and waiting rooms	35–50
Lecture halls, planetariums			
Lobbies	35–45		

[a] NC curves are shown in Figure 3.3–3.

Source: American Society of Heating, Refrigerating, and Air Conditioning Engineers, *ASHRAE Guide and Data Book: Systems,* ASHRAE, New York, 1970, 497. With permission.

3.3–5   Common outdoor noise sources.

Source: *Airborne, Impact, and Structure-Borne Noise – Control in Multifamily Dwellings,* National Bureau of Standards and U.S. Department of Housing and Urban Development, Washington, D.C., 1967.

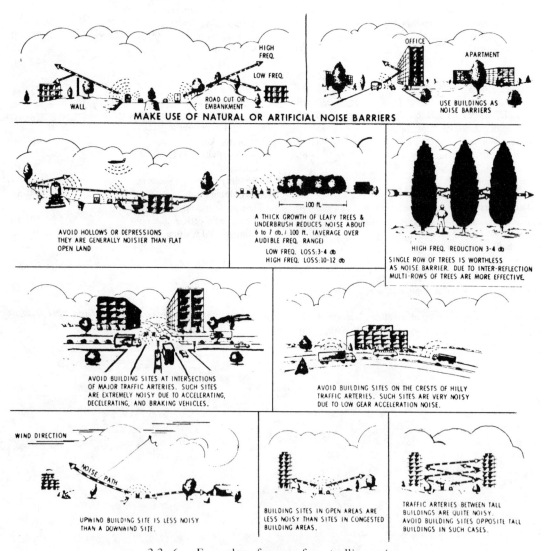

3.3–6   Examples of ways of controlling noise.

Source: *Airborne, Impact, and Structure-Borne Noise – Control in Multifamily Dwellings,* National Bureau of Standards and U.S. Department of Housing and Urban Development, Washington, D.C., 1967.

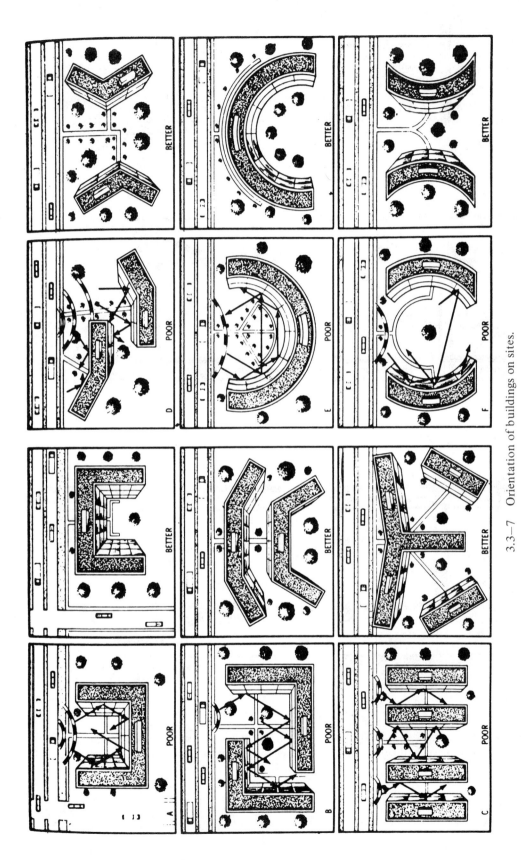

3.3—7 Orientation of buildings on sites.

Source: *Airborne, Impact, and Structure-Borne Noise – Control in Multifamily Dwellings*, National Bureau of Standards and U.S. Department of Housing and Urban Development, Washington, D.C., 1967.

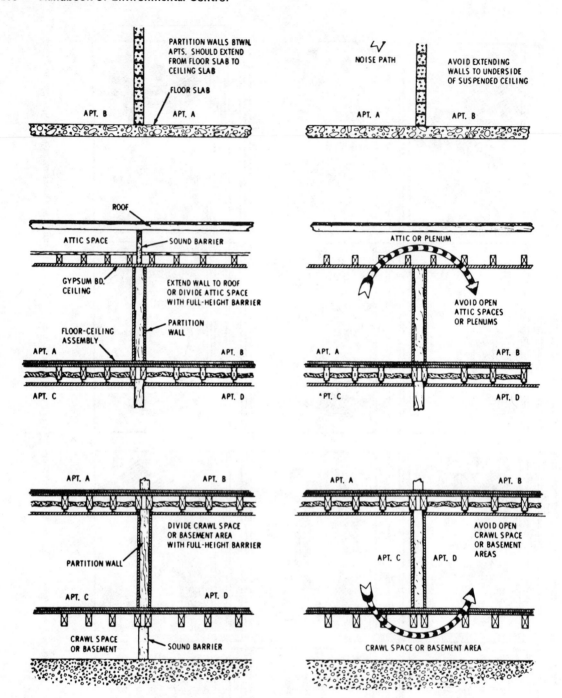

3.3—8   Do's and don'ts in patient-room construction.

Source: *Airborne, Impact, and Structure-Borne Noise — Control in Multifamily Dwellings,* National Bureau of Standards and U.S. Department of Housing and Urban Development, Washington, D.C., 1967.

3.3–9 AVERAGE NOISE LEVELS
IN DECIBELS: A SCALE

	Before carpeting	After carpeting
Corridors	54	48
Patient's room	49	41

Source: R.G. Bond et al., Eds., *Environmental Health and Safety in Health-Care Facilities,* Macmillan, New York, © 1973, 143. With permission.

3.3–10 RECOMMENDED MAXIMUM AVERAGE SOUND-PRESSURE LEVELS[a]

A. For Machines with Maximum Dimension of Less than 6 ft

Sound-pressure levels are to be measured at 6 ft from each machine in at least five positions on each side and above and below (if machine does not sit on floor). Readings shall be averaged arithmetically, except that all readings more than 10.0 dB below the maximum reading shall be disregarded in taking the average.

Sound-pressure level RE 0.0002 μbar[b]

Frequency band, CPS	20–75	75–150	150–300	300–600	600–1200	1200–2400	2400–4800	4800–9600
Schedule I	60	51	43	37	32	30	28	27
Schedule II	63	55	47	41	37	35	33	32
Schedule III	75	68	64	64	64	64	64	64
Schedule IV	100	95	91	87	84	82	82	82

B. For Machines with Dimension of 6 ft or More

Sound-pressure levels are to be measured at 6 ft from each machine in at least five positions on each side and above and below (if machine does not sit on floor). Readings shall be averaged arithmetically, except that all readings more than 10.0 dB below the maximum reading shall be disregarded in taking the average.

Sound-pressure level RE 0.0002 μbar[b]

Frequency band, CPS	20–75	75–150	150–300	300–600	600–1200	1200–2400	2400–4800	4800–9600
Schedule I	55	47	40	32	27	25	25	25
Schedule II	58	50	42	36	32	30	30	30
Schedule III	71	63	60	60	60	60	60	60
Schedule IV	95	90	86	82	89	77	77	77

[a] If hospital rooms do not have acoustical treatment of ¾-in. tile or equivalent on ceiling, subtract 8 dB from all maximum sound levels.
[b] 0 dB = 2 × 10⁻⁴ μbar.

Source: Noise in Hospitals, PHS Pub. No. 930-D-11, Public Health Service, U.S. Department of Health, Education and Welfare, Washington, D.C., 1963.

3.3–11   RECOMMENDED MAXIMUM SOUND-POWER LEVELS[a]

Sound-power level in decibels RE 10^{-12} W[b]

Frequency band, CPS	20–75	75–150	150–300	300–600	600–1200	1200–2400	2400–4800	4800–9600
Schedule I	57	51	45	43	38	36	34	33
Schedule II	62	66	51	50	46	45	44	43
Schedule III	72	68	66	70	70	70	70	70
Schedule IV	100	98	95	94	92	90	89	88

[a] If hospital rooms do not have acoustical treatment of ¾-in. tile or equivalent on ceiling, subtract 8 dB from all maximum sound levels.

[b] 0 dB = 10^{-12} W.

Source: Noise in Hospitals, PHS Pub. No. 930-D-11, Public Health Service, U.S. Department of Health, Education, and Welfare, Washington, D.C., 1963.

3.3–12   SOUND-RATING SCHEDULE FOR PREDOMINANT HOSPITAL NOISE SOURCES

Noise source (equipment)	Schedule
Air conditioning, 6 ft from duct opening, patient room	I
Air conditioning, 6 ft from duct opening, corridor	II
Boiler room equipment	IV
Dishwasher	III
Elevator, corridor 6 ft	II
Elevator door mechanism	II
Elevator machinery, mechanical equipment room	IV
Flush valves, corridor toilet	III
Flush valves, private toilet	I
Ice machine, corridor	II
Ice machine, pantry or kitchen	III
Mechanical ventilation equipment (in separate room) (noise through casing; air ducted out or test room)	IV
Refrigerator, nurses' station	II
Refrigerator, pantry or kitchen	III
Roof ventilators, direct, 1 ft from intake	II
Roof ventilators, ducted to toilets, other areas	II
Room air conditioner	II
Room fans	I
Sterilizers (steam cycle)	III
Transformers, in vault or equipment room	II
Transformers, out of doors, within 50 ft of building	II
Water cooler, corridor	II

Source: Noise in Hospitals, PHS Pub. 930-D-11, Public Health Service, U.S. Department of Health, Education, and Welfare, Washington, D.C., 1963.

3.4   RADIATION PROTECTION

3.4-1   REFERENCES ON RADIATION PROTECTION AND HANDLING AND DISPOSAL OF RADIOACTIVE MATERIALS IN OR FROM HOSPITALS

A Manual of Radioactivity Procedures, National Bureau of Standards Handbook 80, U.S. Government Printing Office, Washington, D.C., November 20, 1961.

Basic Radiation Protection Criteria, NCRP Report No. 39, National Council on Radiation Protection and Measurements, Washington, D. C., January 15, 1971.

Control and Removal of Radioactive Contamination in Laboratories, National Bureau of Standards Handbook 48, U.S. Government Printing Office, Washington, D.C., December 15, 1951.

Dental X-ray Protection, NCRP Report No. 35, National Council on Radiation Protection and Measurements, Washington, D.C., March 9, 1970.

Maximum Permissible Amounts of Radioisotopes in the Human Body and Maximum Permissible Concentrations in Air and Water, National Bureau of Standards Handbook 52, U.S. Government Printing Office, Washington, D.C., March 20, 1953.

Maximum Permissible Body Burdens and Maximum Permissible Concentrations of Radionuclides in Air and Water for Occupational Exposure, National Bureau of Standards Handbook 69, U.S. Government Printing Office, Washington, D.C., June 5, 1959 (supercedes Handbook 52).

Measurement of Absorbed Dose of Neutrons, and of Mixtures of Neutrons and Gamma Rays, National Bureau of Standards Handbook 75, U.S. Government Printing Office, Washington, D.C., February 3, 1961.

Measurement of Neutron Flux and Spectra for Physical and Biological Applications, National Bureau of Standards Handbook 72, U.S. Government Printing Office, Washington, D.C., July 15, 1960.

Medical X-ray and Gamma-Ray Protection for Energies up to 10 MeV. Equipment Design and Use, NCRP Report No. 33, National Council on Radiation Protection and Measurements, Washington, D.C., February 1, 1968.

Medical X-ray and Gamma-Ray Protection for Energies up to 10 MeV. Structural Shielding Design and Evaluation. NCRP Report No. 34, National Council on Radiation Protection and Measurements, Washington, D.C., March 2, 1970.

Precautions in the Management of Patients Who Have Received Therapeutic Amounts of Radionuclides, NCRP Report No. 37, National Council on Radiation Protection and Measurements, Washington, D.C., October 1, 1970.

Protection Against Neutron Radiation, NCRP Report No. 38, National Council on Radiation Protection and Measurements, Washington, D.C., January 4, 1971.

Protection Against Radiation from Brachytherapy Sources, NCRP Report No. 40, National Council on Radiation Protection and Measurements, Washington, D.C., March 1, 1972.

Radiological Monitoring Methods and Instruments, National Bureau of Standards Handbook 51, U.S. Government Printing Office, Washington, D.C., April 7, 1952.

Recommendations for the Disposal of Carbon-14 Wastes, National Bureau of Standards Handbook 53, U.S. Government Printing Office, Washington, D.C., October 26, 1953.

Recommendations for Waste Disposal of Phosphorus-32 and Iodine-131 for Medical Users, National Bureau of Standards Handbook 49, U.S. Government Printing Office, Washington, D.C., November 2, 1951.

Review of the Current State of Radiation Protection Philosophy, NCRP Report No. 43, National Council on Radiation Protection and Measurements, Washington, D.C., January 15, 1975.

Safe Handling of Radioactive Material, National Bureau of Standards Handbook 92, U.S. Government Printing Office, Washington, D.C., March 9, 1964 (supercedes Handbook 42).

3.4—2   DOSE-LIMITING RECOMMENDATIONS

Maximum permissible dose equivalent for
occupational exposure

Combined whole body occupational exposure

Prospective annual limit—paragraphs 229, 233[a]	5 rems in any one year
Retrospective annual limit—paragraphs 230, 233	10—15 rems in any one year
Long-term accumulation to age N years —paragraph 231	$(N - 18) \times 5$ rems
Skin—paragraphs 234, 235	15 rems in any one year
Hands — paragraphs 236, 237	75 rems in any one year (25/quarter)
Forearms—paragraphs 236, 237	30 rems in any one year (10/quarter)
Other organs, tissues and organ systems —paragraphs 238, 239	15 rems in any one year (5/quarter)
Fertile women (with respect to fetus) —paragraphs 240, 241	0.5 rem in gestation period

Dose limits for the public, or occasionally
exposed individuals

Individual or occasional—paragraphs 245, 246, 253, 254	0.5 rem in any one year
Students—paragraphs 255, 256	0.1 rem in any one year

Population dose limits

Genetic—paragraphs 247, 248	0.17 rem average per year
Somatic—paragraphs 250, 251	0.17 rem average per year

Emergency dose limits—life saving

Individual (older than 45 years if possible) —paragraph 258	100 rems
Hands and forearms—paragraph 258	200 rems, additional (300 rems, total)

Emergency dose limits—less urgent

Individual—paragraph 259	25 rems
Hands and forearms—paragraph 259	100 rems, total

Family of radioactive patients

Individual (under age 45)—paragraphs 267, 268	0.5 rem in any one year
Individual (over age 45)—paragraphs 267, 268	5 rems in any one year

[a] Numbers refer to paragraphs in the original source.

Source: Basic Radiation Protection Criteria, NCRP Report No. 39, National Council on Radiation Protection and Measurements, Washington, D. C., January 15, 1971. With permission.

3.4—3   REPRESENTATIVE DOSE-EFFECT RELATIONSHIPS IN MAN FOR WHOLE BODY IRRADIATION[a]

Nature of effect	Representative absorbed dose of whole body X or gamma radiation (rads)
Minimal dose detectable by chromosome analysis or other specialized analyses, but not hemogram	5—25
Minimal acute dose readily detectable in a specific individual (e.g., one who presents himself as a possible exposure case)	50—75
Minimal acute dose likely to produce vomiting in about 10% of people so exposed	75—125
Acute dose likely to produce transient disability and clear hematological changes in a majority of people so exposed.	150—200
Median lethal dose for single short exposure	300

[a] The dose entries in this table should be taken as representative compromises only of a surprisingly variable range of values that would be offered by well-qualified observers asked to complete the right hand column. This comes about in part because whole body irradiation is not a uniquely definable entity. Mid-line absorbed doses are used (see paragraph 139 of original source); the data are a mixed derivative of experience from radiation therapy (often associated with "free-air" exposure dosimetry) and a few nuclear industry accident cases (often with more up to date dosimetry). Also, the interpretation of such qualitative terms as "readily detectable" is a function of the conservatism of the reporter.

Source: Basic Radiation Protection Criteria, NCRP Report No. 39, National Council on Radiation Protection and Measurements, Washington, D.C., January 15, 1971. With permission.

3.4—4 EXAMPLES OF BIOLOGICAL RESPONSE IN HUMAN ORGANS AFTER EXTERNAL PARTIAL BODY IRRADIATION

Organ	Dose schedule	Single dose or extrapolated equivalent (rads)	Effect in relevant organs
Ovary	–	200	Temporary amenorrhea, sterility
	1500 rads/10 days	800[a]	Permanent menopause, sterility
Testis	–	50	Temporary sterility
	1500 rads/10 days	800[a]	Permanent sterility
Bone marrow	25–75 rads in each of 5–10 days. (Portion of bone marrow segments requires higher doses.)	200[a]	Hematopoiesis inhibited in irradiated volume. Usually compensated by marrow activity in unexposed sites.
Kidney	2000 rads/30 days 3000 rads/40 days	800[a]	Nephritis, hypertension
Stomach	1500 rads/20 days 2500 rads/30 days	1000[a]	Atrophic mucosa, anacidity
Liver	3000 rads/30 days 4000 rads/42 days	1500[a]	Hepatitis
Brain and spinal cord	5000 rads/30 days 6000 rads/42 days	2200[a]	Necrosis atrophy
Lung	4000 rads/30 days 6000 rads/42 days	2200[a]	Pneumonitis, fibrosis
Rectum	8000 rads/56 days	2700[a]	Atrophy. Limit of tolerance, most cases.
Bladder	10,000 rads/56 days	3400[a]	Atrophy. Limit of tolerance, most cases.
Ureter	12,000 rads/56 days	4000[a]	Atrophy. Limit of tolerance, most cases.

[a] Extrapolated equivalent calculated from the empirical relation $D_o = D_o t^{-0.27}$, where D_o is the extrapolated equivalent single dose, when an actual dose of D_o is spread over the time t days. This relationship essentially assumes an equal daily dose schedule. Other formulations (e.g., the Ellis formula[1]) give successful empirical results for some erratic fractionation schedules. Such refinements are not needed here; as in Table 3.4—3 these entries are meant to be descriptive, rather than definitive.

Source: Basic Radiation Protection Criteria, NCRP Report No. 39, National Council on Radiation Protection and Measurements, Washington, D.C., January 15, 1971. With permission.

REFERENCE

1. F. Ellis, *Clin. Radiol.*, 20, 1, 1969.

3.4–5 SKIN EFFECTS, SINGLE EXPOSURE[a]

Exposure	Early effect	Chronic effect
50 R	Chromosomal changes only.	None. (Possible slight risk of neoplastic alterations.)
500 R	Transitory erythema. Transitory epilation.	Usually none. Risk of altered function increased
2,500 R	Temporary ulceration. Permanent epilation.	Atrophy, telangiectasis. Altered pigmentation.
5,000 R	Permanent ulceration (unless area very small).	Chronic ulcer, substantial risk of carcinogenesis.
50,000 R	Ordinarily necrotizing, but recovery possible when radiation has extremely low penetration.	Permanent destruction to a depth dependent upon radiation energy.

[a] The skin is tissue that has been studied extensively, and the observed chronic and late effects exemplify what may happen in other tissues. Again, the exposure numbers are representative rather than precise.

Source: Basic Radiation Protection Criteria, NCRP Report No. 39, National Council on Radiation Protection and Measurements, Washington, D.C., January 15, 1971. With permission.

3.4–6 MPC VALUES FOR OCCUPATIONAL AND NONOCCUPATIONAL EXPOSURE FOR SELECTED RADIONUCLIDES IN SOLUBLE FORM

Nuclides Most Frequently Used in Hospitals

Radionuclide	MPC in air, μCi/ml		MPC in water, μCi/ml	
	Occupational	Public	Occupational	Public
^{51}Cr	1×10^{-5}	4×10^{-7}	5×10^{-2}	2×10^{-3}
^{57}Co	3×10^{-6}	1×10^{-7}	2×10^{-2}	5×10^{-4}
^{58}Co	8×10^{-7}	3×10^{-8}	4×10^{-3}	1×10^{-4}
^{60}Co	3×10^{-7}	1×10^{-8}	1×10^{-3}	5×10^{-5}
^{198}Au	3×10^{-7}	1×10^{-8}	2×10^{-3}	5×10^{-5}
^{125}I	5×10^{-9}	8×10^{-11}	4×10^{-5}	2×10^{-7}
^{131}I	9×10^{-9}	1×10^{-10}	6×10^{-5}	3×10^{-7}
^{55}Fe	9×10^{-7}	3×10^{-8}	2×10^{-2}	8×10^{-4}
^{59}Fe	1×10^{-7}	5×10^{-9}	2×10^{-3}	6×10^{-5}
^{32}P	7×10^{-8}	2×10^{-9}	5×10^{-4}	2×10^{-5}
^{99m}Tc	4×10^{-5}	1×10^{-6}	2×10^{-1}	6×10^{-3}
^{113}Sn	4×10^{-7}	1×10^{-8}	2×10^{-3}	9×10^{-5}
^{133}Xe	1×10^{-5}	3×10^{-7}	–	–

Source: Standards for Protection Against Radiation, Title 10, Code of Federal Regulations, Part 20, as amended, December 31, 1968.

3.4–7   SURFACE CONTAMINATION VALUES

$\mu Ci/cm^2$ Except as Indicated

Item	Radiation	ORNL[a]	BCP[b]	AECL[c]
Skin, hands	α	6.7×10^{-7}	10^{-5}	10^{-6}
	$\beta\text{-}\gamma$	0.3 mrad/hr	10^{-4}	10^{-4}
Skin, general	α	6.7×10^{-7}	10^{-5}	10^{-6}
	$\beta\text{-}\gamma$	0.06 mrad/hr	10^{-4}	10^{-4}
Clothing, in noncontaminated zones[d]	α	6.7×10^{-7}	10^{-5}	10^{-6}
	$\beta\text{-}\gamma$	0.25 mrad/hr	10^{-4}	5×10^{-4}
Surfaces in noncontaminated zone areas[d]	α	1.3×10^{-7}	10^{-5}	10^{-6}
	$\beta\text{-}\gamma$	9×10^{-7}	10^{-4}	10^{-5}

[a] Oak Ridge National Laboratory, U.S.A.
[b] British Codes of Practice.
[c] Atomic Energy of Canada Ltd.
[d] For surfaces and clothing exclusively used in controlled contamination zones, the values may be ten times those indicated in the table.

Source: R. G. Bond et al., Eds., *Environmental Health and Safety in Health-Care Facilities,* Macmillian, New York, © 1973, 155. With permission.

REFERENCE

International Commission on Radiological Protection, *Handling and Disposal of Radioactive Materials in Hospitals and Medical Research Establishments,* ICRP Pub. No. 5, Pergamon Press, Oxford, 1965.

3.4–8   TYPICAL POPULATION EXPOSURES FROM NATURAL AND MAN-MADE SOURCES OF RADIATION

Source	Exposure range[a]
Background in the U.S.	
At sea level, outdoors	0.09–0.20 R/year
(varies with soil composition)	
Average genetic dose equivalent	0.125 R/year
At 5000-ft altitude, outdoors	0.11–0.22 R/year
(e.g., in Denver, Colorado)	
In wooden houses	0.09–0.20 R/year
In brick or concrete buildings	0.11–0.50 R/year
(varies with structural material)	
Fallout, additional to natural background	0.01–0.02 R/year
^{40}K in the human body	0.02 R/year
Man-made exposures	
Radioluminous wrist watches, TV, etc.	Up to 4 R/year
Average genetic dose equivalent	<0.005 R/year
Chest X-ray[b]	0.01–1.0 R/film
Dental X-ray[b]	0.2–1.0 R/film
Fluoroscopic examination[b] GI tract	5.0–50 R/exam
All diagnostic medical procedures, average genetic dose equivalent	0.055 R/year
Estimated per capita dose in the population from all sources	0.200 R/year

[a] In this table the roentgen (R) is taken to be equal to the rad or rem. No distinction is made between these quantities because the error resulting from this simplifying assumption is very small in comparison with the accuracy of the numerals provided.

[b] Assumes normal good practice.

Source: R. G. Bond et al., Eds., *Environmental Health and Safety in Health-Care Facilities,* Macmillan, New York, © 1973, 156. With permission.

REFERENCES

1. Radiation Protection in Educational Institutions, NCRP Report No. 32, National Council on Radiation Protection and Measurements, Washington, D.C., July 1, 1966.
2. Basic Radiation Protection Criteria, NCRP Report No. 39, National Council on Radiation Protection and Measurements, Washington, D.C., January 15, 1971.
3. F.W. Spiers, *Radioisotopes in the Human Body,* Academic Press, New York, 1969, 286.

3.4–9   TYPICAL SHIELDING MATERIALS FOR RADIONUCLIDES

Shielding material[a]

Radiation	Permanent	Temporary	Additional clothing
α	Unnecessary	Unnecessary	Unnecessary
β	Lead,[b] copper, iron aluminum, concrete, wood	Iron, aluminum, plastics, wood, glass, water	Leather, rubber, plastic, cloth
γ- X-rays	Lead,[c] iron, copper, lead glass, heavy aggregate concrete, aluminum, ordinary concrete, plate glass, wood, water, paraffin	Lead, iron, lead glass, aluminum, concrete blocks, wood, water	Lead fabrics (but not for "hard" γ)

[a] Arrangement of materials in general order of increased thickness required.
[b] Care must be taken with high atomic number materials to see that the bremsstrahlung (X-rays) generated do not add significantly to the resulting dose after the β rays are absorbed.
[c] For close work, lead is frequently backed with another shield, such as 1/8-in. iron or aluminum, to absorb secondary photoelectrons.

Source: *Safe Handling of Radioactive Materials,* National Bureau of Standards Handbook 92, U.S. Government Printing Office, Washington, D.C., March 9, 1964.

3.4–10   SPECIFIC GAMMA-RAY CONSTANTS AND HALF-VALUE THICKNESSES FOR VARIOUS NUCLIDES

Nuclide	Energy, MeV	Specific gamma-ray constant, mR/hr/mc at 1 m	Lead	Iron	Water	Concrete
^{42}K	1.53	0.15[a]	0.46	0.72	4.7	2.3
^{46}Sc	1.00 (average)	1.10	0.35	0.58	3.8	1.8
^{51}Cr	0.32	0.018[a]	0.077	0.34	2.3	1.1
^{59}Fe	1.20 (average)	0.63	0.41	0.64	4.2	2.0
^{60}Co	1.25 (average)	1.33	0.42	0.65	4.2	2.1
^{65}Zn	1.11	0.30	0.39	0.61	4.0	1.9
^{86}Rb	1.08	0.058[a]	0.38	0.60	3.9	1.9
^{137}Cs-Ba	0.67	0.31	0.24	0.48	3.1	1.5
^{140}Ba-La	1.60	1.40	0.48	0.73	4.7	2.3
^{198}Au	0.41	0.23	0.12	0.38	2.6	1.2

Half-value thickness, inches (Lead, Iron, Water, Concrete)

[a] Gamma rays emitted in a small fraction of the disintegrations.

Source: *Safe Handling of Radioactive Materials,* National Bureau of Standards Handbook 92, U.S. Government Printing Office, Washington, D.C., March 9, 1964.

3.4–11   RADIATIONS EMITTED BY CERTAIN RADIONUCLIDES

Radionuclide	Radiation[a] Type	Radiation[a] Energy[b], MeV	Range[1] Air, cm	Range[1] Aluminum, mm
Naturally Occurring				
Radium-226	α	4.78, 4.60	3.4	–
	γ	0.18	–	–
Radon-222	α	5.5	4.1	–
Polonium-218-Ra A	α	6.0	4.7	–
Lead-214-Ra B	β	0.67, 1.03	–	0.9
	γ	0.053–0.35	–	–
Bismuth-214-Ra C	α	5.5	4.1	–
	β	3.26	–	5.9
	γ	0.6–2.4	–	–
Polonium-214-Ra C′	α	7.69	7.0	–
Thallium-210-Ra C″	β	2.3	–	4.0
	γ	0.296–2.43	–	–
Lead-210-Ra D	β	0.06	–	0.02
	γ	0.047	–	–
Bismuth-210-Ra E	β	1.16	–	1.9
Polonium-210–Ra F	α	5.3	3.9	–
Man-made				
Californium-252	Neutrons	Fission	–	–
	α	6.12, 6.08	4.9	–
	γ	Prompt fission plus fission product	–	–
Cesium-137	β	0.514, 1.176	–	2.0
	γ	0.662	–	–
Chromium-51	Electrons	0.315	–	0.3
	γ	0.320	–	–
Cobalt-60	β	0.314	–	0.4
	γ	1.173, 1.332	–	–
Gold-198	β	0.962	–	1.4
	γ	0.412	–	–
Iodine-125	Electrons	0.004, 0.030	–	–
	γ	0.035	–	–
	Te X-rays		–	–
Iridium-192	β	0.67	–	0.9
	γ	0.296–0.612 (various)	–	–
Tantalum-182	β	0.522	–	0.7
	γ	0.068–1.23 (various)	–	–

[a] From data obtained from Reference 2.

[b] For alpha emission, the energy with the greater abundance is listed first.

Source: Protection Against Radiation From Brachytherapy Sources, NCRP Report No. 40, National Council on Radiation Protection and Measurements, Washington, D.C., March 1, 1972. With permission.

REFERENCES

1. Bureau of Radiological Health, *Radiological Health Handbook,* U.S. Public Health Service, Consumer Protection and Environmental Health Service, Environmental Control Administration, Rockville, Md., 1970.
2. **C. M. Lederer et al.,** *Table of Isotopes,* 6th ed., John Wiley & Sons, New York, 1967.

3.4–12   SELECTED GAMMA-RAY SOURCES[1-3]

Radionuclide	Atomic number	Half-life	Gamma energy, MeV	Gamma-ray exposure rate at 1 m, R/Ci h	Half-value layer[a] Lead, centimeters	Half-value layer[a] Concrete, inches	Tenth-value layer[a] Lead, centimeters	Tenth-value layer[a] Concrete, inches
Cesium-137	55	30.0 years	0.662	0.33	0.65	1.9	2.1	6.2
Chromium-51	24	27.8 days	0.320	0.015	0.2		0.7	
Cobalt-60	27	5.26 years	1.17, 1.33	1.30	1.2	2.5	4.0	8.1
Gold-198	79	2.7 days	0.412	0.23	0.33	1.6	1.1	5.3
Iodine-125	53	60.2 days	0.035	0.123[a]	0.04			
Iridium-192	77	74.2 days	0.296–0.612	0.48	0.3	1.7	2.0	5.8
Radium-226 (with daughters: 0.5-mm Pt filter)	88	1620 years	0.047–2.4	0.825	1.4	2.7	4.6	9.2
Tantalum-182	73	115.1 days	0.068–1.23	0.68	1.2	2.6	4.0	8.6

[a] Approximate values obtained with large attenuation.

Source: Protection Against Radiation from Brachytherapy Sources, NCRP Report No. 40, National Council on Radiation Protection and Measurements, Washington, D.C., March 1, 1972.

REFERENCES

1. Precautions in the Management of Patients Who Have Received Therapeutic Amounts of Radionuclides, NCRP Report No. 37, National Council on Radiation Protection and Measurements, Washington, D.C., October 1, 1970.
2. C.M. Lederer et al., *Table of Isotopes*, 6th ed., John Wiley & Sons, New York., 1967.
3. E.H. Quimby et al., *Radioactive Nuclides in Medicine and Biology*, 3rd ed., Lea & Febiger, Philadelphia, 1970.
4. E.W. Smith, *Nucleonics*, 24, 33, 1966.

3.4–13   PROTECTION REQUIREMENTS FOR VARIOUS GAMMA-RAY SOURCES

Thicknesses of Lead or Concrete Required to Reduce Radiation Exposure to 100 mR at Distances Specified

Millicurie hours	Distance from source 1 ft; 30 cm	Distance from source 3.2 ft; 1 m	Distance from source 6.5 ft; 2 m	Millicurie hours	Distance from source 1 ft; 30 cm	Distance from source 3.2 ft; 1 m	Distance from source 6.5 ft; 2 m
A. Cesium-137[a] – Thicknesses of Lead Required (in Centimeters)				B. Cesium-137 – Thicknesses of Concrete Required (in Inches)			
100	1.1	–	–	100	3.4	–	–
300	2.1	–	–	300	6.3	–	–
1,000	3.2	1.1	–	1,000	9.6	3.2	–
3,000	4.2	2.1	0.9	3,000	12.5	6.1	2.6
10,000	5.3	3.2	1.9	10,000	15.7	9.3	5.7

[a] *TVL* = 2.1 cm lead, or 6.2 in. concrete.

3.4–13 PROTECTION REQUIREMENTS FOR VARIOUS GAMMA-RAY SOURCES (continued)

Millicurie hours	Distance from source 1 ft; 30 cm	3.2 ft; 1 m	6.5 ft; 2 m		Millicurie hours	Distance from source 1 ft; 30 cm	3.2 ft; 1 m	6.5 ft; 2 m
C. Cobalt-60[b] – Thicknesses of Lead Required (in Centimeters)					**H. Iridium-192 – Thicknesses of Concrete Required (in Inches)**			
100	4.5	0.4	–		100	4.3	–	–
300	6.5	2.6	–		300	7.1	1.1	–
1,000	8.6	4.4	2.4		1,000	10.0	4.1	0.6
3,000	10.0	6.4	4.0		3,000	12.8	6.9	3.4
10,000	12.6	8.4	6.0		10,000	16.0	9.8	6.4
D. Cobalt-60 – Thicknesses of Concrete Required (in Inches)					**I. Radium[e] – Thicknesses of Lead Required (in Centimeters)**			
100	9.3	0.9	–		100	4.6	–	–
300	13.1	4.8	–		300	6.5	1.4	–
1,000	17.4	9.2	4.3		1,000	9.0	4.2	1.4
3,000	21.2	12.9	8.7		3,000	11.1	6.4	3.6
10,000	25.5	17.0	12.2		10,000	13.5	8.8	6.0
E. Gold-198[c] – Thicknesses of Lead Required (in Centimeters)					**J Radium – Thicknesses of Concrete Required (in Inches)**			
100	0.5	–	–		100	8.7	–	–
300	1.0	–	–		300	13.0	3.6	–
1,000	1.5	0.4	–		1,000	18.0	8.4	2.9
3,000	2.1	0.9	0.3		3,000	22.0	12.8	7.2
10,000	2.7	1.5	0.8		10,000	27.0	17.6	12.0
F. Gold-198 – Thicknesses of Concrete Required (in Inches)					**K. Tantalum-182[f] – Thicknesses of Lead Required (in Centimeters)**			
100	2.3	–	–		100	3.5	–	–
300	4.6	–	–		300	5.4	1.2	–
1,000[j]	7.4	2.0	–		1,000	7.4	3.3	1.0
3,000	9.9	4.4	1.4		3,000	9.4	5.2	2.8
10,000	12.7	7.2	4.2		10,000	11.4	7.3	4.9
G. Iridium-192[d] – Thicknesses of Lead Required (in Centimeters)					**L. Tantalum-182 – Thicknesses of Concrete Required (in Inches)**			
100	1.5	–	–		100	7.4	–	–
300	2.4	0.4	–		300	11.5	2.7	–
1,000	3.4	1.4	0.2		1,000	16.0	7.2	2.0
3,000	4.4	2.4	1.2		3,000	20.1	11.3	6.1
10,000	5.5	3.4	2.2		10,000	24.6	15.7	10.6

[b] TVL = 4.0 cm lead, or 8.1 in. concrete.
[c] TVL = 1.1 cm lead, or 5.3 in. concrete.
[d] TVL = 2.0 cm lead, or 5.8 in. concrete.
[e] TVL = 4.6 cm lead, or 9.2 in. concrete.
[f] TVL = 4.0 cm lead, or 8.6 in. concrete.

Modified from Protection Against Radiation from Selected Gamma Sources, NCRP Report No. 24, National Bureau of Standards Handbook 73, U.S. Government Printing Office, Washington, D.C., 1960, in Protection Against Radiation from Brachytherapy Sources, NCRP Report No. 40, National Council on Radiation Protection and Measurements, Washington, D.C., March 1, 1972.

3.4–14 APPROXIMATE GAMMA-RAY DOSE RATES TO THE HAND FOR 1 Ci IN A SEALED SOURCE[a]

Nuclide	β max (principal), MeV	γ (principal), MeV	Γ, R/mCi-h at 1 cm	Surface dose rate,[b] r/min	Dose rate at 1-cm tissue depth, R/min	Dose rate at 3-cm tissue depth, R/min
^{137}Cs	0.51, 1.2	0.662	3.26	513	28	3.7
^{60}Co	0.31	1.17, 1.33	13.00	2075	114	16.0
^{192}Ir	0.67	0.468	4.80	813	43	5.5
^{226}Ra	0.4–3.2	0.047–2.4	8.25	1310	72	9.7

[a] Industrial source housings are usually of stainless steel, and for the purpose of the calculations the activity is considered to be a point source. In considering these dose estimates, there is assumed a capsule of outside diameter ¼ in., with a wall of stainless steel (type 304) which is 1/32 in. thick.

[b] The total surface dose rate for the ^{226}Ra source is 1900 R/min based on a 45% increase due to electron production in the stainless steel wall.[1] For the other nuclides given in the table, the increase in surface dose rate due to electron production in the stainless steel wall is estimated to be between 25–45%.

Source: Protection Against Radiation from Brachytherapy Sources, NCRP Report No. 40, National Council on Radiation Protection and Measurements, Washington, D.C., March 1, 1972. With permission.

REFERENCE

1. R.J. Shalek and M. Stovall in *Radiation Dosimetry, Vol. 3, Sources, Fields, Measurements, and Applications,* F. Attix and E. Tochilin, Eds., Academic Press, New York, 1969, 743.

3.4–15 SURFACE DOSE RATE CALCULATIONS FOR RADIUM NEEDLES AND TUBES[a]

Weight, mg	Active length, cm	Outside diameter, mm	Wall thickness, mm Pt	Surface gamma dose rate, R/hr	Total surface[b] dose rates R/hr
Needles					
1	1	1.7	0.5	393	656
2	2	1.7	0.5	393	656
3	3	1.7	0.5	393	656
Tubes					
5	1.15	2.7	1.0	700	1170
10	1.30	2.8	1.0	1240	2070
15	1.45	2.9	1.0	1665	2780
25	1.10	3.5	1.0	3660	6110

[a] Numerically integrated to correct for oblique filtration.

[b] Includes 67% increase in surface dose rate due to electron production in platinum wall.[1]

Source: Protection Against Radiation from Brachytherapy Sources, NCRP Report No. 40, National Council on Radiation Protection and Measurements, Washington, D.C., March 1, 1972. With permission.

REFERENCE

1. R.J. Shalek and M. Stovall in *Radiation Dosimetry, Vol. 3, Sources, Fields, Measurements, and Applications,* F. Attix and E. Tochilin, Eds., Academic Press, New York, 1969, 743.

3.4—16 RADIOACTIVITY LEVELS FOR DISCHARGE OF RADIOACTIVE PATIENTS FROM HOSPITAL

Radionuclide	No restrictions		All persons in household over 45 years of age. Restriction as in Section 4.1.2(d)[a]		Some members of household under 45 years of age. Restrictions as in Section 4.1.2 (d)[b]	
	Exposure rate at 1 m, mR/hr	Activity at discharge, mCi	Exposure rate at 1 m, mR/hr	Activity at discharge, mCi	Exposure rate at 1 m, mR/hr	Activity at discharge, mCi
Cr-51	0.5	35	5	350	1.5	100
Au-198	5.3	23	53	230	16	70
I-125	0.2	8–80[c]	2	20–800[c]	0.6	25–250[c]
I-131	1.8	8	18	80	11	50
Rn-222	3.8	4.6	38	46	15	18
Ir-192	0.2	0.4	2	4	0.6	1.2
Ta-182	0.1	0.2	1	2	0.3	0.6

[a] These levels are in general higher than any likely to be encountered.

[b] These values are rather arbitrarily selected on a basis of the probability of the situation. They represent complete integrated doses of between 1.5 and 2.5 R.

[c] These values cover a large range, due to the variable attenuation of the 35 keV X-rays in the patient.

Source: Precautions in the Management of Patients Who Have Received Therapeutic Amounts of Radionuclides, NCRP Report No. 37, National Council on Radiation Protection and Measurements, Washington, D.C., October 1, 1970. With permission.

3.4—17   APPROXIMATE TIMES FOR PERMISSIBLE EXPOSURES (FOR PERSONS UNDER 45 YEARS OF AGE) AT INDICATED DISTANCES FROM PATIENTS WITH INDICATED EXPOSURE RATES AT 1 m, OR INDICATED RADIONUCLIDE CONTENT, AT TIME OF HOSPITAL DISCHARGE

Radioactive nuclide I	Exposure rate at 1 m at time of discharge from hospital, mR/hr II	Approximate activity at time of discharge, mCi III	"No contact" (greater than 2 m distance) (weeks following discharge) IV	½ hr/day at 1 m plus 2 hr/day at 2 m (weeks following discharge) V	4 hr/day at 1 m (weeks following discharge) VI
Chromium-51	1.5	100	See column V	1st, 2nd, and 3rd	4th, 5th and 6th
Gold-198	16	70	1st	2nd and 3rd	No restrictions
Iodine-125	0.6	75	1st–4th[a]	5th–10th[a]	11th–18th[a]
Iodine-131	11	50	1st	2nd, 3rd, and 4th	5th–8th
Radon	15	18	1st	2nd and 3rd	4th, 5th, and 6th
Iridium-192	8.0	15	1st–24th	25th–35th	36th–45th[b]
Tantalum-182	5.0	17	1st–27th	28th–44th	45th–60th[b]

Note: These rates at the time of discharge from the hospital are arbitrarily selected on the basis of probability of the situation.

[a] For the very unpenetrating radiation from this nuclide, a fluoroscopic-type leaded rubber apron should provide good protection. On the advice of the Radiation Protection Supervisor, such a garment may be used to permit spending more time near the patient.

[b] It is recognized that these conditions are extremely difficult to maintain. Accordingly, use of permanent implants of these nuclides *should* be limited. The activities given are much greater than ten times the values of Table 3.4–18. They represent, however, clinically possible situations.

Source: Precautions in the Management of Patients Who Have Received Therapeutic Amounts of Radionuclides, NCRP Report No. 37, National Council on Radiation Protection and Measurements, Washington, D.C., October 1, 1970. With permission.

3.4–18 INITIAL EXPOSURE RATE AND CORRESPONDING INITIAL ACTIVITY WHICH RESULT IN A TOTAL INTEGRATED EXPOSURE OF 0.5 R AT 1 m DURING COMPLETE DECAY

Radionuclide	Exposure rate at 1 m, mR/mCi-hr	Initial exposure rate resulting in 0.5 R to total decay, mR/hr at 1 m	Corresponding activity, mCi
Chromium-51	0.015	0.5	34.8
Gold-198	0.23	5.3	23
Iodine-125	0.003–0.03[a]	0.2	8–80[a]
Iodine-131	0.22	1.8	8[b]
Radon-222[c]	0.825	3.8	4.6

Also for Integrated Exposure of 0.5 R in 1 Year

Iridium-192	0.48	0.2	0.4
Tantalum-182	0.68	0.1	0.2

[a] These values cover a large range due to the variable attenuation of the 35 keV X-rays in the patient.

[b] For the patient receiving radioiodine therapy, thyroidal uptake and urinary excretion are maximal before 24 hr. Therefore, thyroidal radioactivity at 24 hr should be used in place of the initial activity for determinations of integrated exposure rate. This will usually be approximately one third of the administered activity.

[c] In equilibrium with short-lived daughters and filtered by 0.5 mm Pt.

Source: Precautions in the Management of Patients Who Have Received Therapeutic Amounts of Radionuclides, NCRP Report No. 37, National Council on Radiation Protection and Measurements, Washington, D.C., October 1, 1970. With permission.

3.4–19 RADIATION DOSE TO HANDS IN PERITONEAL CAVITY

Isotope	No gloves, rem/mCi-hr	Single surgical gloves, rem/mCi-hr	Double autopsy gloves, rem/mCi-hr
Au-198[a]	0.7	0.4	0.1
P-32 or Y-90[b]	0.8	0.6	0.3

[a] The Au-198 values include a factor for the gamma rays.

[b] The same values are given for P-32 and for Y-90. These values would also be applicable for any other beta-ray emitter whose radiation energies are 1.5 MeV or greater. For beta radiation of lower energy the dose rate will be markedly less, especially when gloves are used.

Source: Precautions in the Management of Patients Who Have Received Therapeutic Amounts of Radionuclides, NCRP Report No. 37, National Council on Radiation Protection and Measurements, Washington, D.C., October 1, 1970. With permission.

3.4–20 APPROXIMATE TIME FOR HANDS IN PERITONEAL CAVITY TO RECEIVE 1.5 REM[a]

Total activity on surface, mCi	Au-198		P-32 or Y-90	
	Single surgical gloves, min	Double autopsy gloves, min	Single surgical gloves, min	Double autopsy gloves, min
10	21	64	17	32
20	11	32	8	16
30	7	21	6	11
40	5	16	4	8
50	4	13	3	6
60	4	11	3	5
70	3	9	2	5
80	3	8	2	4
90	2	7	2	4
100	2	6	2	3

[a] A dose of 1.5 rems is selected as being the average permissible weekly dose because the annual occupational permissible dose for hands and forearms is 75 rems. For occupationally exposed personnel, 25 rems is permissible if the procedure is not expected to occur more often than once in any 13 consecutive weeks, and no other exposure is to be received in this period. The time for 25 rems is approximately 15 times that in the table.

Source: Precautions in the Management of Patients Who Have Received Therapeutic Amounts of Radionuclides, NCRP Report No. 37, National Council on Radiation Protection and Measurements, Washington, D.C., October 1, 1970. With permission.

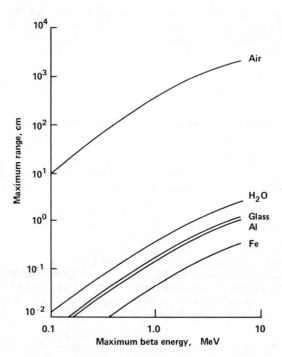

3.4–21   Penetration ability of beta radiation.

Source: *Safe Handling of Radioactive Materials,* National Bureau of Standards Handbook 92, U.S. Government Printing Office, Washington, D.C., March 9, 1964.

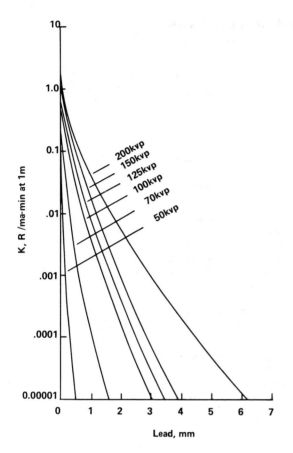

3.4–22 Attenuation in lead of X-rays generated at 50 to 200 kVp.

Source: *Safety Standard for Non-medical X-ray and Sealed Gamma-ray Sources. Part I. General,* National Bureau of Standards Handbook 93, U.S. Government Printing Office, Washington, D.C., 1964.

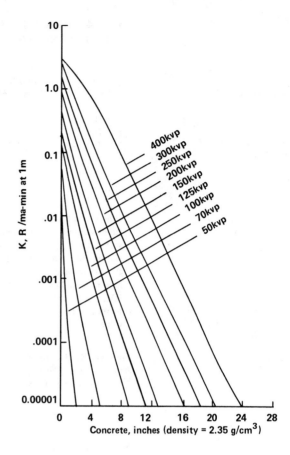

3.4–23 Attenuation in concrete of X-rays generated at 50 to 400 kVp.

Source: *Safety Standard for Non-medical X-ray and Sealed Gamma-ray Sources. Part I. General,* National Bureau of Standards Handbook 93, U.S. Government Printing Office, Washington, D.C., 1964.

3.4–24   MINIMUM BEAM FILTRATION FOR DIAGNOSTIC X-RAY MACHINES

Operating kVp	Minimum total filter,[a] mm Al	Beam HVL,[b] mm Al
<50	0.5	
50–70	1.5	
>70	2.5	–
49	–	Not less than 0.6
70	–	Not less than 1.6
90	–	Not less than 2.6

[a] Inherent plus added filter. Beryllium window X-ray tubes with no added filtration emit low-energy X-rays at very high exposure rates. A keyed filter interlock switch should be used on beryllium window machines so that the operator can tell by a glance at the control panel how much added filter, if any, is present.

[b] HVL: half-value layer, the amount of filtration required to reduce the radiation intensity to one half its original value.

Source: R. G. Bond et al., Eds., *Environmental Health and Safety in Health-Care Facilities,* Macmillian, New York, © 1973, 161. With permission.

REFERENCE

Medical X-ray and Gamma-ray Protection for Energies Up to 10 MeV (Equipment Design and Use), NCRP Report No. 33, National Council on Radiation Protection and Measurements, Washington, D.C., 1968.

3.4–25   EFFECT OF TUBE POTENTIAL, DISTANCE, AND FILTRATION ON AIR EXPOSURE RATE AT PANEL OF FLUOROSCOPES[a]

Potential kVp	Source to panel distance		Equivalent total aluminum filtration (in roentgens per milliampere minute)				
	Centimeters	Inches	1 mm	2 mm	2.5 mm[b]	3 mm	4 mm
70	30	12	5.3	2.7	2.2*	1.8	1.3
	38	15	3.5	1.7	1.4†	1.2	0.8
	46	18	2.4	1.2	1.0	0.8	0.6
80	30	12	7.0	3.9	3.2*	2.6	2.0
	38	15	4.6	2.5	2.1†	1.7	1.3
	46	18	3.2	1.8	1.4	1.2	0.9
90	30	12	9.0	5.2	4.3*	3.6	2.8
	38	15	5.8	3.3	2.8†	2.3	1.8
	46	18	4.0	2.3	1.9	1.6	1.2
100	30	12	11.0	6.6	5.5*	4.7	3.7
	38	15	7.0	4.2	3.5†	3.0	2.3
	46	18	4.9	2.9	2.5	2.1	1.6
110	30	12	13.1	8.0	6.8*	5.9	4.6
	38	15	8.4	5.1	4.4†	3.8	3.0
	46	18	5.8	3.5	3.0	2.6	2.0
120	30	12	14.7	9.3	8.0*	7.0	5.5
	38	15	9.5	6.0	5.1†	4.5	3.6
	46	18	6.5	4.1	3.6	3.1	2.5
130	38	15	–	6.8	5.9†	5.2	4.2
	46	18	–	4.7	4.1	3.6	2.9
140	38	15	–	7.6	6.6†	5.9	4.8
	46	18	–	5.3	4.6	4.1	3.3
150	38	15	–	8.5	7.5†	6.7	5.4
	46	18	–	5.8	5.2	4.6	3.7

[a] Typical exposure rates produced by equipment with medium length cables, derived from the references by interpolation and extrapolation. Filtration includes that of the tabletop and the X-ray tube with its inherent and added filter. As used above, panel means either panel or tabletop.

[b] When the fluoroscope is operated at 80 kVp, the exposure rate measured in air at the position where the beam enters the patient *shall not* exceed 3.2 R/mA-in. and *should not* exceed 2.1 R/mA-min. For other fluoroscopic tube potentials, the exposure per unit charge *shall not* exceed the values marked with an asterisk in the table and *should not* exceed the values marked with a dagger.

Source: Medical X-ray and Gamma-ray Protection for Energies Up to 10 MeV, NCRP Report No. 33, National Council on Radiation Protection and Measurements, Washington, D.C., reprinted June 1, 1971. With permission.

REFERENCES

J. Hale, *Radiology,* 86, 147, 1966.
E.D. Trout and J.P. Kelley, *Radiology,* 82, 977, 1964.

3.4–26 HALF-VALUE LAYERS AS A FUNCTION OF FILTRATION AND TUBE POTENTIAL FOR DIAGNOSTIC UNITS[a]

Total filtration, mm Al	Peak potential (kVp); typical half-value layers in millimeters of aluminum									
	30	40	50	60	70	80	90	100	110	120
0.5	0.36[b]	0.47[b]	0.58	0.67	0.76	0.84	0.92	1.00	1.08	1.16
1.0	0.55	0.78	0.95	1.08	1.21	1.33	1.46	1.58	1.70	1.82
1.5	0.78	1.04	1.25[b]	1.42[b]	1.59[b]	1.75	1.90	2.08	2.25	2.42
2.0	0.92	1.22	1.49	1.70	1.90	2.10	2.28	2.48	2.70	2.90
2.5	1.02	1.38	1.69	1.95	2.16	2.37[b,c]	2.58[b,c]	2.82[b,c]	3.06[b,c]	3.30[b,c]
3.0	—	1.49	1.87	2.16	2.40	2.62	2.86	3.12	3.38	3.65
3.5	—	1.58	2.00	2.34	2.60	2.86	3.12	3.40	3.68	3.95

[a] For full-wave rectified potential. Derived from Reference 1 by interpolation and extrapolation.
[b] Recommended minimum half-value layer for radiographic units. See Section 3.2.2 (a) of original source.
[c] Recommended minimum half-value layer for fluoroscopes. See Section 3.1.2 (b) of original source.

Source: Medical X-ray and Gamma-ray Protection for Energies Up to 10 MeV, NCRP Report No. 33, National Council on Radiation Protection and Measurements, Washington, D.C., reprinted June 1, 1971. With permission.

REFERENCE

1. J. Hale, *Radiology*, 86, 147, 1966.

3.4–27 EXPOSURE RATE THROUGH FLUOROSCOPIC SCREEN WITHOUT PATIENT[a]

X-ray tube potential, kVp	Source to table top distance		Lead equivalent of screen protective barrier[b] (typical exposure rate: mR/hr per R/min at tabletop)		
	Inches	Centimeters	1.5 mm	1.8 mm	2.0 mm
80	12	30	10	4.5	2.5
	15	38	13	6	3.5
	18	46	15	7	4
90	12	30	12	6	3.5
	15	38	16	7.5	4.5
	18	46	19	9	5.5
100	12	30	15	7	4.5
	15	38	20	9	5.5
	18	46	23	11	7
110	12	30	19	9	5.5
	15	38	24	12	7
	18	46	29	14	8.5
120	12	30	23	11	7
	15	38	30	14	9
	18	46	35	17	10
130	15	38	35	17	10
	18	46	42	20	12
140	15	38	41	19	12
	18	46	49	23	14
150	15	38	46	20	12
	18	46	55	24	15

Notes: Total filtration: 3 mm aluminum equivalent Table top to screen distance: 14 in. Screen to chamber distance: 2 in. Medium length high tension cables.

[a] Adapted from Reference 1 by interpolation and extrapolation. Actual exposure rate values may differ from the typical values given above by ±15% depending upon length of high tension cables.
[b] See Sections 3.1.1 (d) and 3.1.2 (c) of original source.

Source: Medical X-ray and Gamma-ray Protection for Energies Up to 10 MeV, NCRP Report No. 33, National Council on Radiation Protection and Measurements, Washington, D.C., reprinted June 1, 1971. With permission.

REFERENCE

1. E.D. Trout and J.P. Kelley, *Radiology*, 82, 977, 1964.

3.4–28   SCATTERED RADIATION EXPOSURE RATE AT SIDE OF FLUOROSCOPY TABLE[a]

Peak potential, kVp	Exposure rate in mR per mAh	
	No screen drape	With screen drape[b]
80	300	4.5
90	400	8.0
100	650	17.5
110	850	32.
120	1050	42.
130	1330	60.
140	1550	77.
150	1750	98.

Notes: Phantom: 100-lb sack of flour 15-in source-panel distance 10 × 10 cm field at tabletop Phantom edge 6 in. from side of table

[a] After J.S. Krohmer, unpublished data.
[b] 0.3-mm lead equivalent.

Source: Medical X-ray and Gamma-ray Protection for Energies up to 10 MeV, NCRP Report No. 33, National Council on Radiation Protection and Measurements, Washington, D.C., reprinted June 1, 1971. With permission.

3.4–29   AVERAGE EXPOSURE RATES PRODUCED BY DIAGNOSTIC X–RAY EQUIPMENT[a]

Distance from source to point of measurement		Tube potential, roentgens per 100 mA-sec						
Inches	Centimeters	50 kVp	60 kVp	70 kVp	80 kVp	90 kVp	100 kVp	125 kVp
12	30	1.8	2.8	4.2	5.8	8.0	9.8	15.2
18	46	0.8	1.3	1.8	2.5	3.4	4.2	6.7
24	61	0.4	0.7	1.1	1.4	1.9	2.3	3.8
39	100	0.2	0.3	0.4	0.5	0.7	0.9	1.4
54	137	0.1	0.1	0.2	0.3	0.4	0.5	0.7
72	183	0.1	0.1	0.1	0.2	0.2	0.3	0.4

[a] Measured in air with total filtration equivalent to 2.5 mm aluminum.

Source: Medical X-ray and Gamma-ray Protection for Energies Up to 10 MeV, NCRP Report No. 33, National Council on Radiation Protection and Measurements, Washington, D.C., reprinted June 1, 1971. With permission.

3.4–30   THICKNESS OF LEAD REQUIRED TO REDUCE USEFUL BEAM TO 5%[a]

Beam quality

Potential	Half-value layer, millimeters	Required lead thickness, millimeters
60 kVp	1.2 Al	0.10
100 kVp	1.0 Al	0.16
100 kVp	2.0 Al	0.25
100 kVp	3.0 Al	0.35
140 kVp	0.5 Cu	0.7
200 kVp	1.0 Cu	1.0
250 kVp	3.0 Cu	1.7
400 kVp	4.0 Cu	2.3
1000 kVp	3.2 Pb	20.5
2000 kVp	6.0 Pb	43.0
2000 kVcp	14.5 Pb	63.0
3000 kVcp	16.2 Pb	70.0
6000 kV	17.0 Pb	74.0
8000 kV	15.5 Pb	67.0
Cobalt 60	10.4 Pb	47.0

[a] Approximate values for broad beams. Transmission data for brass, steel, and other material for potentials up to 2000 kVp may be found in Reference 1. Measurements on 1000 and 2000 kVp made with resonant-type therapy units. Data for 6000 kV taken from Reference 2, for a linear accelerator. Data for 2000 and 3000 kVcp and 8000 kV derived by interpolation from graph presented in Reference 3. The third column refers to lead or to the required equivalent lead thickness of lead-containing materials (e.g., lead rubber, lead glass, etc.).

Source: Medical X-ray and Gamma-ray Protection for Energies Up to 10 MeV, NCRP Report No. 33, National Council on Radiation Protection and Measurements, Washington, D.C., reprinted June 1, 1971. With permission.

REFERENCES

1. E.D. Trout and R.M. Gager, *Am. J. Roentgenol.,* 63, 396, 1950.
2. C.J. Karzmark and C. Tatiana, *Br. J. Radiol.,* 41, 33, 1968.
3. J.H. Bly and E.J. Burrill, High Energy Radiography in the 6 to 30 MeV Range. Symposium on Nondestructive Testing in the Missile Industry, Special Technical Publication No. 278, American Society for Testing and Materials, Philadelphia, 1960.

3.4–31   PERMISSIBLE EXPOSURE LEVELS FOR UNLIMITED EXPOSURE TO VARIOUS TYPES OF RADIATION

Radiation and conditions	Permissible level at the cornea		Wavelength, nm
Laser, ruby, Q-switched, pulsed, 7-mm pupil	1×10^{-8}	J/cm^2	694.3
Laser, ruby, non-Q-switched, pulsed, 7-mm pupil	1×10^{-7}	J/cm^2	694.3
Laser, ruby, CW, 7-mm pupil	1×10^{-6}	W/cm^2	694.3
Laser, argon, CW, 7-mm pupil	2×10^{-3}	W/cm^2	476.5–514.5
Laser, neodymium, pulsed	10^{-6}	J/cm^2	1060
Infrared, pulsed	0.1	J/cm^2	–
Infrared, CW	1.0	J/cm^2	–
Ultraviolet, CW	1×10^{-7}	W/cm^2	253.7
Microwaves	10^{-2}	W/cm^2	–

Source: R.G. Bond et al., Eds., *Environmental Health and Safety in Health-Care Facilities,* Macmillan, New York, © 1973. 174. With permission.

REFERENCES

C.H. Powell and L. Goldman, *Arch. Environ. Health,* 18, 448, 1968.
R.J. Rockwell, *Arch. Environ. Health,* 18, 394, 1969.
L.R. Setter et al., Regulations, Standards, and Guides for Microwave, Ultraviolet Radiation, and Radiation from Lasers and Television Receivers — An Annotated Bibliography, PHS Pub. No. 999-RH-35, Public Health Service, U.S. Department of Health, Education, and Welfare, Washington, D.C., 1969.

3.4–32 DIELECTRIC CONSTANT AND CONDUCTIVITY IN mMho/cm FOR VARIOUS BODY TISSUES AT 37°C

Frequency (Me)

	25	50	100	200	400	700	1000	3000	8500
Dielectric constant ε									
Muscle	103–115	85–97	71–76	56	52–54	52–53	49–52	45–48	40–42
Heart muscle				59–63	52–56	50–55			
Liver	136–138	88–93	76–79	50–56	44–51	42–51	46–47	42–43	34–38
Spleen	>200	135–140	100–101						
Kidney	>200	119–132	87–92	62	53–55	50–53			
Lung				35	35	34			
Brain	>160	110–114	81–83						
Fat		11–13		4.5–7.5	4–7		5.3–7.5	3.9–7.2	3.5–4.5
Bone marrow		6.8–7.7					4.3–7.3	4.2–5.8	4.4–5.4
Conductivity, κ, in mMho/cm									
Muscle		6.80–8.85		9.52–10.5	11.1–11.8	12.7–13.7	12.7–13.3	21.7–23.3	83.3
Heart muscle				8.7–10.5	10–11.8	10.5–12.8			
Liver	4.76–5.41	5.13–5.78	5.59–6.49	6.67–9.09	7.69–9.52	8.7–11.8	9.43–10.2	20–20.4	58.8–66.7
Spleen		6.62–7.81							
Kidney		6.9–11.1		11.1	11.8	1.3–1.32			
Lung		2.22–3.85		6.25	7.14	7.69			
Brain	4.55	4.76–5.26	5.13–5.56						
Fat		0.4–0.59		0.29–0.95	0.36–1.11		0.83–1.49	1.11–2.27	2.7–4.17
Bone marrow		0.2–0.36					0.43–1	1.16–2.25	1.67–4.76

Source: H.P. Schwan, *Advances in Biological and Medical Physics*, Academic Press, New York, 1957, 147. With permission.

3.4–33   SUMMARY OF MAXIMUM RECOMMENDED LEVELS FOR HUMAN EXPOSURE

Country and source	Radiation frequency	Maximum recommended level	Conditions or remarks
U.S.A. (USASI)	10 MHz–100 GHz	10 mW/cm²	Periods of 0.1 hr
		1 mW hr/cm²	Averaged over any 0.1-hr period
U.S. Army and Air Force	–	10 mW/cm²	Continuous exposure
		10–100 mW/cm²	Maximum exposure time in minutes at $W(mW/cm^2) = 6000W^{-2}$
		100 mW/cm²	No occupancy
Great Britain (Post Office Regulation)	30 MHz–30 GHz	10 mW/cm²	Continuous 8-hr exposure, average power density
NATO (1956)		0.5 mW/cm²	
Canada	10 MHz–100 GHz	1 mW hr/cm²	Averaged over any 0.1-hr period
		10 mW/cm²	Periods of 0.1 hr
Poland	>300 MHz	10 μW/cm²	8-hr exposure/day
		100 μW/cm²	2–3 hr/day
		1 mW/cm²	15–20 min/day
German Socialist Republic	–	10 mW/cm²	
U.S.S.R.	0.1–1.5 MHz	20 V/m	Alternating magnetic fields
		5 A/m	
	1.5–30 MHz	20 V/m	
	30–300 MHz	5 V/m	
	>300 MHz	10 μW/cm²	6 hr/day
		100 μW/cm²	2 hr/day
		1 mW/cm²	15 min/day
Czechoslovak Socialist Republic	0.01–300 MHz	10 V/m	8 hr/day
	>300 MHz	25 μW/cm²	8 hr/day, CW operation
		10 μW/cm²	8 hr/day, pulsed

Source: S.F. Cleary, *CRC Crit. Rev. Environ. Control,* 1, 257, 1970.

Section 4

Safety

_____

4. SAFETY

4.1 PATIENT AND PERSONNEL SAFETY

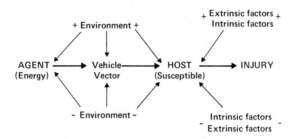

4.1–1a   Causal accidental-injury relationship.

Source: R.G. Bond et al., Eds., Environmental Health and Safety in Health-Care Facilities, Macmillan, New York, © 1973, 179. With permission.

4.1–1b   INJURY RESULTS FROM EXCHANGE OF ENERGY IN OR ON THE BODY THAT EXCEEDS THE BODY'S INJURY THRESHOLD

The Agent in the Injury Sequence Can Be Considered as Energy in One of the Forms Given Below

Energy form	Nature of injury	Examples
Mechanical	Impact and/or contact with fixed or moving bodies	Patient fall, being struck by swinging door, accidental puncture with syringe needle, contact with power-operated equipment
Chemical	External and/or internal chemical action in body tissue	Dermatitis from kitchen detergents, skin contact with strong acids or alkalis, inhalation of vapors, skin contact with chemicals in chemical laboratories
Thermal	Contact with open flame or heated surface	Contact with steam radiators, hot water bottles, heat lamps, cooking utensils, oxygen tank fire
Electrical	Contact with electric current or electric arcs	Current being used or carried in electric wiring, electrically powered hand-operated equipment, electric life-support or monitoring equipment
Radiation	Contact with or exposure to ionizing, sonic, or light radiation	Diagnostic and therapeutic X-rays, radioisotopes, microwave blood warmers, sonic instruments and bottles, lasers

Source: R.G. Bond et al., Eds., *Environmental Health and Safety in Health-Care Facilities,* Macmillan, New York, © 1973, 178. With permission.

4.1–2 EFFECTS OF 60-CYCLE ELECTRICAL CURRENT FLOWING THROUGH THE TRUNK (FROM HAND TO HAND OR FOOT TO HAND) FOR 1 sec OR MORE

Milliamperes	Effect
0.5–2	Tingling perception
2–10	Muscular contraction
10–20	Painful shock (unable to let go)
20–50	Severe muscle contraction
100–300	Ventricular fibrillation
Over 600	Respiratory paralysis, burns

Source: R.G. Bond et al., Eds., *Environmental Health and Safety in Health-Care Facilities,* Macmillan, New York, © 1973, 193. With permission.

4.1–3 ACCIDENT RATE IN RELATION TO NUMBER OF ADMISSIONS, MOUNT CARMEL MERCY HOSPITAL, DETROIT

	1962	1963	Total
Number of admissions	21,818	21,375	43,193
Number of patients who had accidents	363	423	786
Patients having accidents per thousand admissions	16.6	19.8	18.2
Number of accidents	397	471	868
Accidents per thousand admissions	18.2	22.0	20.0

Source: J.D. Fagin and Sister Mary Vita, *Hospitals,* 39, 60, 1965. With permission.

4.1–4 OTHER DATA ON ACCIDENTS

Rate per 1000 admissions	Reference
20	Fagin and Sister Mary Vita[1]
28.5	Weil and Parrish[2]
24.5	Williams[3]
18.5	Snell[4]
20.4	Petrovsky[5]

REFERENCES

1. J.D. Fagin and Sister Mary Vita, *Hospitals,* 39, 60, 1965.
2. T.P. Weil and H.M. Parrish, *Hospitals,* 32, 43, 1958.
3. W.R. Williams, *Hospitals,* 22, 39, 1948.
4. W.E. Snell, *Lancet,* ii, 1202, 1956.
5. C.C. Petrovsky, *Med. J. Aust.,* 2, 1123, 1967.

4.1–5   AGE INCIDENCE OF PATIENTS HAVING ACCIDENTS, MOUNT CARMEL MERCY HOSPITAL, DETROIT

Age group	Number of patients	Number of patients having accidents per thousand admissions
Newborn	3	0.36
1 week–1 year	5	5.49
1–10 years	63	12.1
11–20	43	11.9
21–30	40	4.99
31–40	50	7.95
41–50	67	12.1
51–60	112	21.8
61–70	198	39.6
71–80	174	62.0
Over 80 years	31	41.9
Total	786	

Source: J.D. Fagin and Sister Mary Vita, *Hospitals,* 39, 60, 1965. With permission.

4.1–6   TIME OF OCCURRENCE OF ACCIDENT

A. No significant difference noted in monthly or seasonal incidence

B. Day of week

Sunday	92
Monday	151
Tuesday	125
Wednesday	122
Thursday	133
Friday	133
Saturday	107
Indeterminate	5
Total	868

C. Time of day (in relation to nursing shift)

Day shift	349
Night shift	258
Swing shift	240
Indeterminate	21
Total	868

Source: J.D. Fagin and Sister Mary Vita, *Hospitals,* 39, 60, 1965. With permission.

4.1–7   LOCATION OF ACCIDENT

Patient's room or ward	592
Bed	233
Bedside	227
Other	132
Bathroom	108
Operating and recovery rooms	62
Hallway	48
X-ray department	14
Physiotherapy	7
Emergency department	4
Utility room	4
Obstetrical suite	3
Phone booth	2
Waiting room	2
Nurses' station	1
Stairway	1
Pharmacy	1
Nursery	1
Pediatrics playroom	1
Unspecified	17
Total	868

Source: J.D. Fagin and Sister Mary Vita, *Hospitals,* 39, 60, 1965. With permission.

4.1–8   NATURE OF ACCIDENT, MOUNT CARMEL
MERCY HOSPITAL, DETROIT

A. Falls	602
1. From bed	250
Siderails up	130
Siderails down	45
Patient in restraints	14
Condition of siderails not specified	75
2. From wheelchair, cart, chair, or couch	94
3. In patient's room other than above	118
4. In bathroom	83
From toilet	14
In tub	7
Other	62
5. In hallway or on stairs	32
6. Using crutches	4
7. From operating table	2
8. From window (suicide)	1
9. Other	18
B. Medication or treatment error	61
C. Hit by or bumped into object	61
D. Injured by hospital equipment other than furniture	60
E. Burned by hot object or liquid	23
F. Damage to teeth during anesthesia	20
G. Cut by sharp objects (including broken thermo-meters and other broken glass)	17
H. Self-inflicted injuries	12
I. Injury to dentures	10
J. Injured by another person	8
K. Chemical burns	3
L. Other	5
Total	882

Source: J.D. Fagin and Sister Mary Vita, *Hospitals,* 39, 60, 1965. With permission.

4.1–9   CONTRIBUTION OF HOSPITAL
PERSONNEL TO ACCIDENT

Incorrect medication or treatment	61
Patient injured while being assisted by staff	28
Teeth broken by insertion of endotracheal tube	13
Bed rail improperly placed	10
Patient bumped by personnel	5
Insertion of broken thermometer	3
Other instances of negligence	13
No contributory negligence reported	735
Total	868

Source: J.D. Fagin and Sister Mary Vita, *Hospitals,* 39, 60, 1965. With permission.

4.1—10   TYPE OF INJURY TO PATIENT, SEVERITY, AND PART OF BODY INJURED

Type		Severity		Part injured	
None	447	None	447	None or	475
Contusion	101	Mild or	353	unspecified	
Pain only	70	unspecified		Head	96
Abrasion	69	Moderate	62	Upper extremity	94
Laceration	61	Severe	4	Lower extremity	88
Burn or scald	23	Fatal	2	Face	37
Epistaxis	3			Back	31
Fracture	2			Abdomen	9
Other	88			Thorax	7
Unspecified	17			Neck	5
				Other	48

Source: J.D. Fagin and Sister Mary Vita, *Hospitals,* 39, 60, 1965. With permission.

4.1—11   OBJECTS INVOLVED IN ACCIDENTS

None or unspecified	260
Bed	85
Siderails	59
Chair	55
Medication	45
Floor	42
Wheelchair	35
Door	25
Toilet	22
Tubing	21
Bedpan	16
Radiator	16
Endotracheal tube	13
Table	11
Food tray	9
Bath tub	9
Nightstand	9
Airway	6
X-ray equipment	6
Transfusion apparatus	5
Traction apparatus	5
Thermometer	4
Drinking cup	4
Crutches	4
Safety belt	3
Laryngoscope	3
Screen	3
Signal cord	3
Glass	3
Other	87
Total	868

Source: J.D. Fagin and Sister Mary Vita, *Hospitals,* 39, 60, 1965. With permission.

4.1–12   SERVICE CLASSIFICATION OF PATIENTS INVOLVED IN ACCIDENTS AND RATIO IN RELATION TO SERVICE

Service	Number of patients	Ratio per 1000 patients
Medical	452	52.2
Surgical	258	17.9
Pediatric	46	6.15
Obstetrical	27	2.19
Newborn	3	0.36
Total	786	

Source: J.D. Fagin and Sister Mary Vita, *Hospitals,* 39, 60, 1965. With permission.

4.1–13   PSYCHOLOGICAL CONDITION OF PATIENT AT TIME OF ACCIDENT

Mental or emotional state	Number of patients
Alert and rational	353
Disturbed in some manner	427
Not adequately described	88
Type of disturbed behavior	
Confused or irrational	138
Restless, depressed, or anxious	86
Dizzy or weak	57
Anesthetized	56
Lethargic or sleeping	49
Fainted	44
Uncooperative	34
Drugged	10
Senile	9
Convulsing	6
Highly febrile or toxic	5
Intoxicated	2

Source: J.D. Fagin and Sister Mary Vita, *Hospitals,* 39, 60, 1965. With permission.

4.1–14 TYPES OF DRUGS ADMINISTERED WITHIN 12 hr BEFORE ACCIDENT

Drugs	Number of instances
Sedatives	337
Soporifics	238
Narcotics	175
Antihypertensives	48
Anesthetics	37
Antihistamines	14
Stimulants	2

Source: J.D. Fagin and Sister Mary Vita, *Hospitals*, 39, 60, 1965. With permission.

4.1–15 ACTIVITY STATUS OF PATIENT AT TIME OF ACCIDENT

Ambulant	258
Absolute bed rest	219
Bathroom privileges	187
Up in wheelchair	102
Up with assistance	13
Up with crutches	11
Up in chair only	5
Up with walker	3
In restraints	2
Other or unspecified	68
Total	868

Source: J.D. Fagin and Sister Mary Vita, *Hospitals*, 39, 60, 1965. With permission.

4.1–16 DISTRIBUTION OF INPATIENT ACCIDENTS, LAUNCESTON GENERAL HOSPITAL, AUSTRALIA

January 1, 1954 – January 1, 1966

Cause of accident	Number of accidents	Percent of total
Bedside falls	832	57.5
All other falls	396	27.4
Errors in medication	74	5.1
Burns	34	3.0
Injuries by furniture, equipment, or doors	25	1.7
Thermometer broken in mouth or rectum	18	1.2
Cut with glass or razor	7	
Given food while on cease-nourishment order	7	
Transfusion mishaps	7	
Bed or clothing set alight	6	
Broken injection needle *in situ*	3	4.2
Fights with other patients or staff	11	
Other accidents[a]	27	
Total	1447	100.1

[a] This includes error in pack count, four cases; pack left *in situ*, one case; drain tube lost *in situ*, one case; swallowed gastroenterostomy tube, one case; eye wrongly prepared for operation, one case; bowel ruptured with sigmoidoscope, one case.

Source: C.C. Petrovsky, *Med. J. Aust.*, 2, 1123, 1967. With permission.

4.1–17   ADMISSIONS AND TOTAL NUMBER OF FALLS CLASSIFIED BY AGE GROUPS AND PERCENTAGES

Age group	Total number of falls	Percentage of patients admitted	Percentage of total falls
Children	136	21.8	11.1
Middle	344	55.7	28.0
Pensionable	748	22.6	60.9
Total	1228	100.0	100.0

Source: C.C. Petrovsky, *Med. J. Aust.,* 2, 1123, 1967. With permission.

4.1–18   MEDICAL DISORDERS OF FALL PATIENTS

Disorder	Number of patients	Percent of total
Cardiovascular	209	17.0
Neurological and neurosurgical	203	16.5
Orthopedic	111	9.0
Malignant disease	109	8.9
Pulmonary	120	9.8
Psychiatric	87	7.1
Senility (uncomplicated)	68	5.5
Gastrointestinal	55	4.5
Genitourinary	58	4.7
Metabolic disorders and anemias	50	4.1
Gynecological	10	0.8
Eye	17	1.4
All other	131	10.7
Total	1228	100.0

Source: C.C. Petrovsky, *Med. J. Aust.,* 2, 1123, 1967. With permission.

4.1–19   FALLS CLASSIFIED BY TYPE AND RESULT

Type of fall	Number of falls	Result of fall			
		No injury	Minor injury	Moderate injury	Severe injury
Bedside	832	479 (57.6%)	303 (36.4%)	28 (3.4%)	22 (2.6%)
Other	396	172 (43.4%)	180 (45.5%)	22 (5.6%)	22 (5.6%)
Total	1228	651 (53.0%)	483 (39.3%)	50 (4.1%)	44 (3.6%)

Source: C.C. Petrovsky, *Med. J. Aust.,* 2, 1123, 1967. With permission.

4.1–20   CHARACTERIZATION OF SEVERE
INJURY IN FALL PATIENTS

Type of fracture	Number of patients	
Single fractures		
Femur	15 ⎫	6 deaths
Refractured femur	6 ⎭	
Humerus	3	2 deaths
Nose	2	
Skull (parietal bone)	3	
Clavicle	1	
Supracondylar of arm	2	
Lumbar vertebra	1	
Multiple fractures		
Skull and nose (with sutures)	1	
Femur and Colles'	2	1 death
Femur and olecranon	1	1 death
Ribs	1	
Humerus and scapula	1	
Radius and ulna	2	

Source: C.C. Petrovsky, *Med. J. Aust.,* 2, 1123, 1967.
With permission.

4.1–21   ANNUAL ACCIDENTS PER
1 MILLION hr WORKED

Department	On-job accidents per 1 million hr worked
Nursing service	1.82
Dietary	4.00
Housekeeping	5.72
Clinical laboratory	1.18
Building-grounds	2.50
Laundry	5.73
Business office	0.66
X-ray	2.56
Data processing	2.20
Average	2.70

Source: C. Lewis, *Arch. Environ. Health,* 11, 16, 1965.
With permission. Copyright 1965, American Medical
Association.

4.1–22   SUMMARY OF LASER SAFETY IN BIOMEDICAL INSTALLATIONS

Some installations	Laser techniques	Hazards	Protection
Cytology	Microbeam radiation	Eye	Protective glasses (external laser sources) Microscope shields Closed television circuit
Embryology	Microbeam radiation	Eye	Protective glasses
Microbiology	Microbeam radiation	Eye	Protective glasses
	Light-scattering techniques	Air pollution	Closed circuit television Plume traps Enclosed systems Exhaust systems
Chemistry and biochemistry	Raman spectroscopy	Eye	Protective glasses
	Microemission spectroscopy	Skin	Skin protection
	Light-scattering techniques	Air pollution	Enclosed systems Plume traps Protective screens Enclosed systems Exhaust systems
Air pollution studies	Atmospheric propagation	Eye Chronic skin exposure	Protective glasses
Animal research Clinical research	All lasers	Eye Skin	Prepared and planned installations
Ophthalmology	Ruby, argon photocoagulators	Eye	Protective glasses
Outpatient clinics	Ruby, Nd,[a] argon, krypton, CO_2, YAG, He-Ne	Air pollution	Skin protection Plume traps Exhaust systems
Inpatient operation room	Ruby, Nd, argon, krypton, CO_2, YAG, He-Ne		Enclosed systems

[a] Neodymium.

Source: L. Goldman, *Arch. Environ. Health*, 20, 193, 1970. With permission. Copyright 1970, American Medical Association.

4.1–23   INTERIOR LIGHTING REQUIREMENTS – HOSPITALS

Table A

While for convenience of use this table sometimes lists locations rather than tasks, the recommended footcandle values have been arrived at for specific visual tasks. The tasks selected for this purpose have been the most difficult ones which commonly occur in the various areas.

In order to assure these values at all times, higher initial levels should be provided as required by the maintenance conditions.

Where tasks are located near the perimeter of a room special consideration should be given to the arrangement of the luminaires in order to provide the recommended level of illumination on the task.

The illumination levels shown in the table are intended to be minimum on the task irrespective of the plane in which it is located. In some instances, denoted by a dagger (†), the values shown will be for equivalent sphere illumination, E_s. The commonly used lumen method of illumination calculation, which gives results only for a horizontal work plane, cannot be used to calculate or predetermine E_s values. The ratio of vertical to horizontal illumination will generally vary from one third for luminaires having narrow distribution to one half for luminaires of wide distribution. For a more specific determination one of the point calculation methods should be used. Where the levels thus achieved are inadequate, special luminaire arrangements should be used or supplemental lighting equipment employed.

Supplementary luminaires may be used in combination with general lighting to achieve these levels. The general lighting should be not less than 20 footcandles and should contribute at least one tenth the total illumination level.

Many of the following values have appeared, or in the future will appear, in other reports of the Illuminating Engineering Society, some of which are jointly sponsored with other agencies and organizations.

Area	Footcandles on tasks[a]	Dekalux[b] on tasks[a]
Anesthetizing and preparation room	30	32
Autopsy and morgue		
Autopsy room	100	100
Autopsy table	1000	1080
Museum	50	54
Morgue, general	20	22
Central sterile supply		
General, work room	30	32
Work tables	50	54
Glove room	50	54
Syringe room	150	160
Needle sharpening	150	160
Storage areas	30	32
Issuing sterile supplies	50	54
Corridor		
General in nursing areas – daytime	20	22
General in nursing areas – night (rest period)	3	3.2
Operating, delivery, recovery, and laboratory suites and service areas	30	32

[a] Minimum on the task at any time for young adults with normal and better than 20/30 corrected vision. For general notes see beginning of table.

[b] Dekalux is an SI unit equal to 0.929 footcandles; 1 dekalux = 10 lux.

4.1–23   INTERIOR LIGHTING REQUIREMENTS – HOSPITALS (continued)

Area	Footcandles on tasks[a]	Dekalux[b] on tasks[a]
Cystoscopic room		
General	100	110
Cystoscopic table	2500	2690
Dental suite		
Operatory, general	70	75
Instrument cabinet	150	160
Dental entrance to oral cavity	1000	1080
Prosthetic laboratory bench	100	110
Recovery room, general	5	5.4
Recovery room, local for observation	70	75
EEG (encephalographic) suite		
Office (see Offices, Table B)		
Work room, general	30	32
Work room, desk or table	100	110
Examining room	30	32
Preparation rooms, general	30	32
Preparation rooms, local	50	54
Storage, records, charts	30	32
Electromyographic suite		
Same as EEG but provisions for reducing level in preparation area to 1		
Emergency operating room		
General	100	110
Local	2000	2150
Emergency outpatient suite		
Same as Examination and treatment room		
Minor surgery	2000	2150
EKG, BMR, and specimen room		
General	30	32
Specimen table	50	54
EKG machine	50	54
Examination and treatment room		
General	50	54
Examining table	100	110
Exits, at floor	5	5.4
Eye, ear, nose, and throat suite		
Darkroom (variable)	0–10	0–11
Eye examination and treatment	50	54
Ear, nose, throat room	50	54
Flower room	10	11
Formula room		
Bottle washing	30	32
Preparation and filling	50	54
Fracture room		
General	50	54
Fracture table	200	220
Splint closet	50	54
Plaster sink	50	54
Intensive care nursing areas		
General	30	32
Local	100	110
Laboratories		
General	50	54
Close work areas	100	110

4.1–23   INTERIOR LIGHTING REQUIREMENTS – HOSPITALS (continued)

Area	Footcandles on tasks[a]	Dekalux[b] on tasks[a]
Linens (see Laundries, Table B)		
Sorting soiled linen	30	32
Central (clean) linen room	30	32
Sewing room, general	30	32
Sewing room, work area	100	110
Linen closet	10	11
Lobby (or entrance foyer)		
During day	50	54
During night	20	22
Locker rooms	20	22
Medical records room	100[†]	110[†]
Nurses station		
General – day	70[†]	75[†]
General – night	30	32
Desk for records and charting	70[†]	75[†]
Table for doctor's making or viewing reports	70[†]	75[†]
Medicine counter	100[†]	110[†]
Nurses gown room		
General	30	32
Mirror for grooming	50	54
Nurseries, infant		
General	30	32
Examining, local at bassinet	100	110
Examining and treatment table	100	110
Nurses station and work space (see Nurses station)		
Obstetrical suite		
Labor room, general	20	22
Labor room, local	100	110
Scrub-up area	30	32
Delivery room, general	100	110
Substerilizing room	30	32
Delivery table	2500	2690
Clean-up room	30	32
Recovery room, general	30	32
Recovery room, local	100	110
Patients' rooms (private and wards)		
General	20	22
Reading	30	32
Observation (by nurse)	2	2.2
Night light, maximum at floor (variable)	0.5	0.5
Examining light	100	110
Toilets	30	32
Pediatric nursing unit		
General, crib room	20	22
General, bedroom	10	11
Reading	30	32
Playroom	30	32
Treatment room, general	50	54
Treatment room, local	100	110
Pharmacy		
Compounding and dispensing	100	110
Manufacturing	50	54
Parenteral solution room	50	54
Active storage	30	32
Alcohol vault	10	11

4.1–23   INTERIOR LIGHTING REQUIREMENTS – HOSPITALS (continued)

Area	Footcandles on tasks[a]	Dekalux[b] on tasks[a]
Radioisotope facilities		
Radiochemical laboratory, general	30	32
Uptake or scanning room	20	22
Examining table	50	54
Retiring room		
General	10	11
Local for reading	30	32
Solarium		
General	20	22
Local for reading	30	32
Stairways	20	22
Surgical suite		
Instrument and sterile supply room	30	32
Clean-up room, instrument	100	110
Scrub-up area (variable)	200	220
Operating room, general (variable)	200	220
Operating table	2500	2690
Recovery room, general	30	32
Recovery room, local	100	110
Anesthesia storage	20	22
Substerilizing room	30	32
Therapy, physical		
General	20	22
Exercise room	30	32
Treatment cubicles, local	30	32
Whirlpool	20	22
Lip reading	150	160
Office (see Offices, Table B)		
Therapy, occupational		
Work area, general	30	32
Work tables or benches, ordinary	50	54
Work tables or benches, fine work	100	110
Toilets	30	32
Utility room		
General	20	22
Work counter	50	54
Waiting rooms, or areas		
General	20	22
Local for reading	30	32
X-ray suite		
Radiographic, general	10	11
Fluoroscopic, general (variable)	0–50	0–54
Deep and superficial therapy	10	11
Control room	10	11
Film viewing room	30	32
Darkroom	10	11
Light room	30	32
Filing room, developed films	30	32
Storage, undeveloped films	10	11
Dressing rooms	10	11

4.1–23 INTERIOR LIGHTING REQUIREMENTS – HOSPITALS (continued)

Table B

Area	Footcandles on tasks[a]	Dekalux[b] on tasks[a]
Foodservice facilities		
Dining areas		
Cashier	50	54
Intimate type		
Light environment	10	11
Subdued environment	3	3.2
For cleaning	20	22
Leisure type		
Light environment	30	32
Subdued environment	15	16
Quick service type		
Bright surroundings	100	110
Normal surroundings	50	54
Food displays – twice the general levels but not under	50	54
Kitchen, commercial		
Inspection, checking, preparation, and pricing	70	75
Laundries		
Washing	30	32
Flat work ironing, weighing, listing, marking.	50	54
Machine and press finishing, sorting	70	75
Fine hand ironing	100	110
Library		
Reading areas		
Reading printed material	30[†]	32[†]
Study and note taking	70[†]	75[†]
Conference areas	30[†]	32[†]
Seminar rooms	70[†]	75[†]
Book stacks (30 in. above floor)		
Active stacks	30[c]	32[c]
Inactive stacks	5[c]	5.4[c]
Book repair and binding	70	75
Cataloging	70[†]	75[†]
Card files	100[†]	110[†]
Carrels, individual study areas	70[†]	75[†]
Circulation desks	70[†]	75[†]
Rare book rooms – archives		
Storage areas	30	32
Reading areas	100[†]	110[†]
Map, picture, and print rooms		
Storage areas	30	32
Use areas	100[†]	110[†]
Audiovisual areas		
Preparation rooms	70	75
Viewing rooms (variable)	70	75
Television receiving room (shield viewing screen)	70	75

[a] Minimun on the task at any time for young adults with normal and better than 20/30 corrected vision. For general notes see beginning of table.

[b] Dekalux is an SI unit equal to 0.929 footcandles; 1 dekalux = 10 lux.

[†] Equivalent sphere illumination. See general notes at beginning of table.

[c] Vertical.

4.1–23   INTERIOR LIGHTING REQUIREMENTS – HOSPITALS (continued)

Area	Footcandles on tasks[a]	Dekalux[b] on tasks[a]
Audio listening areas		
General	30	32
For note taking	70[†]	75[†]
Record inspection table	100[d]	110[d]
Microform areas		
Files	70[†]	75[†]
Viewing areas	30	32
Locker rooms	20	22
Offices		
Drafting rooms		
Rough layout drafting	150[†]	160[†]
Accounting offices		
Auditing, tabulating, bookkeeping, business machine operation, computer operation	150[†]	160[†]
General Offices		
Reading poor reproductions, business machine operation, computer operation	150[†]	160[†]
Reading handwriting in hard pencil or on poor paper, reading fair reproductions, active filing, mail sorting	100[†]	110[†]
Reading handwriting in ink or medium pencil on good quality paper, intermittent filing	70[†]	75[†]
Private offices		
Reading poor reproductions, business machine operation	150[†]	160[†]
Reading handwriting in hard pencil or on poor paper, reading fair reproductions	100[†]	110[†]
Reading handwriting in ink or medium pencil on good quality paper	70[†]	75[†]
Reading high-contrast or well-printed materials	30[†]	33[†]
Conferring and interviewing	30	33
Conference rooms		
Critical seeing tasks	100[†]	110[†]
Conferring	30	33
Note-taking during projection (variable)	30[†]	32[†]
Corridors	20[e]	22[e]
Service space (see also Storage rooms)		
Stairways, corridors	20	22
Elevators, freight and passenger	20	22
Toilets and wash rooms	30	32
Storage rooms or warehouses		
Inactive	5	5.4
Active		
Rough bulky	10	11
Medium	20	22
Fine	50	54

[d] Obtained with a combination of general lighting plus specialized supplementary lighting. Care should be taken to keep within the recommended luminance ratios. These seeing tasks generally involve the discrimination of fine detail for long periods of time and under conditions of poor contrast. The design and installation of the combination system must not only provide a sufficient amount of light, but also the proper direction of light, diffusion, color, and eye protection. As far as possible it should eliminate direct and reflected glare as well as objectionable shadows.

[e] Or not less than one fifth the level in adjacent areas.

Source: J.E. Kaufman, Ed., *IES Lighting Handbook,* 5th ed., Illuminating Engineering Society, New York, 1972. With permission.

4.1–24 INTERIOR LIGHTING REQUIREMENTS – NURSING HOMES

Area	Footcandles on tasks[a]	Dekalux[b] on tasks[a]
Administrative and lobby areas, day	50	54
Administrative and lobby areas, night	20	22
Barber and beautician areas	50	50
Chapel or quiet area, general	5	5.4
Chapel or quiet area, local for reading	30	32
Corridors and interior ramps	20	22
Dining area	30	32
Doorways	10	11
Exit stairways and landings, on floor	5	5.4
Janitor's closet	15	15
Nurse's desk, for charts and records	70[†]	75[†]
Nurse's medicine cabinet	100[†]	110[†]
Nurse's station, general		
Day	50[†]	54[†]
Night	20	22
Occupational therapy	30	32
Patient care room, reading	30	32
Patient care unit (or room), general	20	22
Pharmacy area, general	30	32
Pharmacy, compounding, and dispensing area	100	110
Physical therapy	20	22
Recreation area	50	54
Stairways other than exits	30	32
Toilet and bathing facilities	30	32
Utility room, general	20	22
Utility room, work counter	50	54
Work table, course work	100	110
Work table, fine work	200	220

[a] Minimum on the task at any time for young adults with normal and better than 20/30 corrected vision. For general notes see beginning of Table 4.1–23.

[b] Dekalux is an SI unit equal to 0.929 footcandles; 1 dekalux = 10 lux.

[†] Equivalent sphere illumination. See general notes at beginning of Table 4.1–23.

Source: J.E. Kaufman, Ed., *IES Lighting Handbook,* 5th ed., Illuminating Engineering Society, New York, 1972. With permission.

4.2   LABORATORY SAFETY

4.2–1   LIST OF INCOMPATIBLE CHEMICALS

Substances listed in the left column should be stored and handled so that they cannot accidentally contact corresponding substances listed in the right column.

Alkaline earth metals such as sodium, potassium, cesium, lithium, and magnesium	Carbon dioxide, chlorinated hydrocarbons, and water
The halogens: fluorine, chlorine, bromine, and iodine	Ammonia, acetylene, and hydrocarbons
Acetic acid, hydrogen sulfide, hydrocarbons, sulfuric acid, analine, or any flammable liquid	Oxidizing agents such as chromic acid, nitric acid, peroxides, and permanganates

Source: R.G. Bond et al., Eds., *Environmental Health and Safety in Health-Care Facilities,* Macmillan, New York, © 1973, 203. With permission.

4.2–2   SAFETY REGULATIONS FOR THE USE OF COMPRESSED-GAS CYLINDERS

1. Cylinders shall have the name of the chemical contents appearing in legible form on the cylinder. A color code is not a satisfactory designation.

2. Cylinders shall be held securely in an upright position. String, wire, or similar makeshift materials are not acceptable.

3. Cylinders shall be located so they are not exposed to direct flame or heat in excess of 125°F.

4. Cylinders not in use shall have the valve protective cap securely in place. (Lecture bottles are an exception.)

5. Cylinders shall be moved only in suitable hand carts.

6. Cylinders containing flammable gas shall be used and stored in a ventilated area (ten air changes per hour minimum). No other gases or chemicals shall be stored in the same area.

7. Cylinders containing toxic or corrosive gases shall be used and stored in well-ventilated areas separated from cylinders containing other gases.

8. Cylinders containing corrosive gas shall be returned to the supplier no later than 6 months from time of first use. Cylinder and regulator valves shall be opened and closed at least weekly during periods of nonuse.

9. Cylinders discharging into liquids or closed systems containing other chemicals shall have a trap, check valve, or vacuum breaker between the cylinder and system or liquid.

10. Systems mixing two or more gases shall be provided with necessary control or check valves to prevent contamination of the separate gas sources.

11. Closed systems, or any arrangement that might accidentally become a closed system to which a cylinder is attached, shall have a safety relief valve set at a relief pressure that will prevent damage to any part of the equipment.

12. Cylinder valves and regulators shall have outlets and inlets, respectively, for the specific gas as designated in the American Standard of Compressed Gas Association Pamphlet V-1 or for flush type cylinder valves according to Compressed Gas Association Pamphlet V-3.

13. Use of adaptors between cylinder valve and regulator should be discouraged, but if used, they should be only a type listed in Appendix of Compressed Gas Association Pamphlet V-1.

14. Emergency plans shall be developed to insure control or safe removal of leaking cylinders. Proper protective clothing and equipment for the type of compressed gas being used should be available in the immediate area to allow safe entry should the gas be accidentally released from the cylinder.

Source: Safety Standard, Division of Environmental Health and Safety, University of Minnesota, Minneapolis.

4.2–3 USE OF GLASSWARE

Procedures for General Use of Laboratory Glassware

1. Inspect glassware prior to and after use; damaged glassware is not to be used, sent for washing, or returned for storage; repair or discard.

2. Use brush and dust pan to clean up broken glass; remove glass from sink by using tongs for large pieces and cotton held by tongs for chips and slivers.

3. Discard broken and damaged glass in "Broken Glass" receptacle.

4. After use rinse or purge flammable or toxic residue from glassware.

5. Use only very low pressure (quiet) compressed air in purging; use additional care with small-necked vessels.

6. Use following procedure to break glass tubing (1-cm tubing is maximum diameter that can be broken; larger diameters should be cut).

    a. Scratch deeply with triangular file at break point.

    b. Wrap towel around tubing: grip tubing on both sides of scratch; apply upward pressure with thumbnails directly below scratch.

7. Fire polish rough ends of tubing; if unable to fire polish, bevel edge with file or emery.

8. Bend tubing with smooth curves; sharp angles break easily.

9. Use following procedure to bore cork or stopper:

    a. Select sharp bore of smallest size that will accept glass tube or rod.

    b. Place stopper, small end up, on wood block – never hold in palm of hand.

    c. Protect hand with leather glove, hold stopper with thumb and forefinger.

    d. Twist borer slowly with gentle downward pressure.

    e. For rubber stopper lubricate with water or glycerol.

10. Insert glass tubing through cork or stopper as follows:

    a. Select stopper with hole size for snug but not tight fit.

    b. Lubricate tubing and hole with water or glycerol.

    c. Using tubing manipulator or towel, grasp tubing near entry end.

    d. Protect other hand with leather glove or several folds of towel, hold stopper with thumb and forefinger with palm parallel to direction of force.

    e. Insert tubing with gentle twisting motion completely through stopper (failure to go completely through may result in sealing off of tubing).

11. Insert *thin-walled* glass tubing or devices as follows:

    a. Select smallest size borer that will accept the tubing or device.

    b. Insert borer through the stopper.

    c. Insert tubing through the lumen or borer, then gently extract borer.

12. Separate glass tubing and stoppers immediately after use; protect hands with towel.

13. Cut off, using sharp knife, stoppers or tubing that gentle pressure will not separate.

14. Lift beakers by grasping side just below rim; use two hands for large beakers, one on side and one on bottom.

15. Lift, shake, and carry flasks by placing fingers of one hand around neck and other hand on bottom.

16. Guard against breaking the fragile neck of volumetric flasks by holding stopper with thumb and forefinger and inserting with a gentle twisting motion.

17. Store and use carboys and 5-gal bottles in the crate provided. Avoid handling large unprotected glass containers when filled with liquid. Remember large bottles, Pyrex® included, are susceptible to thermal shock.

18. Do not use soft glass containers, of any size, to receive materials having a temperature in excess of 50°C (122°F).

19. Liquid level in containers provided with a positive closure should leave a 10% air space.

20. Vessels with positive closures should not normally be placed in a steam bath or subjected to direct heat.

21. Beakers, flasks, or bottles should be protected by asbestos-centered gauze when heating by direct flame.

22. Beaker or flasks of over 1 liter must not be heated by flame or placed in direct contact with hot plate.

23. Dry the outside of test tubes before heating with direct flame.

24. Avoid heating soft glass vessels by any means.

25. Glass apparatus set-ups should impart no appreciable stress to any glass part.

26. Flasks should always be supported at the bottom with a neck clamp used for added stability.

27. Large flasks and beakers should be supported by tripod or bench jack, not by ring clamps.

28. Check all tubing for blockages before assembly.

29. Compressed gas introduced into glass apparatus should not be more than 5 psi. The apparatus should have pressure-relief devices.

4.2—3 USE OF GLASSWARE (continued)

30. Ground glass connections should be lubricated before assembly and disassembled immediately after use.

31. Frozen ground-glass joints should not be forced; discard or return to "Glass Blowing."

32. Use a stop-cock puller to remove frozen stop-cocks or ground-glass stoppers. The use of heat is not recommended and must never be used if contents is an unknown or is affected by heat.

33. Store glassware well back on shelf; place large heavy pieces on lower shelves; place glass tubing horizontally behind stop.

34. Laboratory glassware shall never be used for the storage of food or liquids intended for human consumption.

Source: Recommended Practice, Division of Environmental Health and Safety, University of Minnesota, Minneapolis.

4.2—4 PROCEDURES FOR USE OF LABORATORY GLASSWARE SUBJECTED TO PRESSURE OR VACUUM

1. Glassware to be subjected to pressure or vacuum shall be:
   a. Pyrex® glass
   b. Heavy walled if vessel has flat bottom
   c. Carefully inspected for flaws before use
2. Any separate part of a vacuum system, 1 liter or larger, shall be wrapped with tape or covered with wire mesh.
3. Wrap all dewars with tape.
4. Desiccators used in vacuum service shall be completely enclosed in shield.
5. Protect vacuum set-ups by barrier or shields between system and all personnel.
6. Interpose a cold trap of proper size and temperature between vacuum system and pump. Check trap frequently for plugging.
7. Exhaust vacuum pumps into fume hood or special vent stack.
8. Label all systems: DANGER—Equipment Under Vacuum.
9. Protect all parts of vacuum system from physical shock.
10. When using Gooch crucible in vacuum system, select size that cannot slip through holder. Use similar care in selecting corks or stoppers.
11. Pressure bottles should be Pyrex and must be wrapped in cloth or placed in a bottle bag. Use water bath to heat and cool in air with bag on.
12. Before distillation of ether or acetals destroy peroxides.
13. Use round-bottomed flasks or kettles of Pyrex for distillations. Use tripod or bench jack to support flask or kettle, with neck clamp for added security.
14. Place flammable liquid distillations in fume hood. Use electric mantle, water, or steam bath for heat. Never use open flame or hot plate.
15. Place boiling stones or an ebullator in distillation flask or bottle.
16. Provide a vent in every distillation system. Check to see it does not plug.
17. In starting distillation:
   a. Check joints and connections for tightness.
   b. Check that coolant is flowing.
   c. Check that receiver of adequate capacity is in place.
18. In stopping a distillation:
   a. Shut off heat
   b. Turn off coolant only after vapors disappear from condenser.
   c. On vacuum stills, hold vacuum until kettle contents have cooled below boiling
   d. Do not stopper hot flasks or kettles
19. Distillations under reduced pressure should have inert gas introduced by ebullator.

Source: Recommended Practice, Division of Environmental Health and Safety, University of Minnesota, Minneapolis.

4.2–5   PROCEDURES FOR USING GLASSWARE-CLEANING SOLUTIONS

1. Wear a face shield or splashproof safety goggles to avoid injury to eyes by splashing.
2. Use long rubber gloves and aprons to prevent accidental exposure to the hands and body.
3. Use tongs or special perforated containers for handling objects that have been immersed in the cleaning solution.
4. Remove completely and immediately dispose of any solution that is spilled on bench tops, working surfaces, or floors.
5. Immediately wash off cleaning solution that is accidentally spilled on the skin with large quantities of water. Do not hesitate to use the emergency shower if necessary. After washing thoroughly, secure medical care.
6. Wash hands promptly and thoroughly after the cleaning task is completed.

Source: Chromic-Sulfuric Acid Cleaning Solution, Division of Environmental Health and Safety, University Health Service, University of Minnesota, Minneapolis.

4.3   FIRE SAFETY

4.3–1   DEFINITIONS OF TERMS USED IN PRACTICE OF FIRE SAFETY

Combustible – any material that will burn when exposed to fire, even if only under special circumstances, is considered to be combustible.

Fire prevention – effort directed toward the elimination of fires that are injurious to man and his environment.

Fire load – the amount of combustible material in a given location (usually expressed in units of weight per unit area).

Fire protection – the control of the effects of fire on man and his environment, including techniques for isolation and extinguishment.

Fire resistance – the relative ability of a material to resist the effects of fire (usually expressed in units of time), generally applied to building materials.

Fire retardant – a lower degree of fire resistance usually obtained by special flame-retarding treatments and techniques of building materials and fabrics.

Flame spread – the relative hazard of building materials, which is determined by the rate of flame propagation on the surface of the material. Cement-asbestos board has a flame spread rating of 0, and Douglas fir, of 100. These ratings are used as standards against which other materials are compared.

Ignition temperature – the temperature at which self-sustained combustion will take place in the absence of a spark or open flame.

Flammable range – the percent by volume of either flammable vapor or gas concentrations in ambient air at which combustion will take place. The lower limit indicates the least concentration in air, and the upper limit the greatest concentration that will support combustion.

Flash point – the lowest temperature at which a flammable liquid gives off enough vapors to form a flammable mixture with air. This temperature is usually much lower than the ignition temperature.

Flammable – a higher burning rate than would be found in most combustible materials. Flammable materials usually ignite very readily at very low ignition temperatures.

Noncombustible and nonflammable – materials that will not burn, or will burn at very slow rates.

Source: R.G. Bond et al., Eds., *Environmental Health and Safety in Health-Care Facilities,* Macmillan, New York, © 1973, 211. With permission.

4.3–2   CAUSES OF 245 FATAL AND NONFATAL HOSPITAL FIRE INJURIES

	Fatal injuries	Nonfatal injuries
Bodily burns from flames	42	36
(Contributing factors – trapped by explosion or rapid spread of fire; no automatic protection; bedridden or other physical handicap; mentally deranged or under restraint; smoking in presence of oxygen; left unattended in oxygen or croup tent; mishandling of oxygen or flammable liquids)		
Suffocated or overcome by smoke or fire gases	98	39
(Contributing factors – inadequate exit facilities; trapped by explosion or rapid spread of fire; no automatic protection; bedridden or other physical handicap; mentally deranged or under restraint; fire fighting; attempted rescue)		
Explosions	13	7
(Ignition of combustible anesthetic agents by static sparks, cautery, etc.)		
Attempting to escape	–	2
(Jumping; falls from roof, etc.)		
Falling walls or other parts of building	–	3
Other direct injuries (falls, etc.)	1	4
Total	154	91[a]

[a] Includes 32 fire fighters.

Source: Reprinted with permission from *Occupancy Fire Record – Hospitals,* FR61-1, copyright 1961, National Fire Protection Association, Boston, 1961.

4.3–3   WHEN HOSPITAL FIRES OCCURRED

	Fires	Percent[a]
6:01 A.M.–5:00 P.M.	151	53.4
5:01 P.M.–midnight	76	26.8
Midnight–6:00 A.M.	56	19.8
Unknown	17	–
Total	300	100

[a] Percentages shown are of fires in which time of discovery was known.

Source: Reprinted with permission from *Occupancy Fire Record – Hospitals,* FR61-1, copyright 1961, National Fire Protection Association, Boston, 1961.

4.3–4   FREQUENCY OF FIRE DEPARTMENT RESPONSES TO HOSPITAL ALARMS

13 Hospitals – 4680 Beds; 4-year Study
(1966–1969)

Working shift	Number of responses
7:00 A.M.–3:00 P.M.	731 (51%)
3:00 P.M.–11:00 P.M.	471 (34%)
11:00 P.M.–7:00 A.M.	201 (15%)

Source: Minneapolis Bureau of Fire Prevention, Education Division.

4.3–5   ORIGIN OF HOSPITAL FIRES REPORTED TO THE NATIONAL FIRE PROTECTION ASSOCIATION

Origin of fire	Number of fires
Patients' rooms	63
Oxygen tent (30)	
Bedding, including mattress (24)	
Other (9)	
Employees' quarters	57
Heat or power plant	42
Storeroom	28
Laboratory	27
Operating room	25
Chute (laundry or rubbish)	19
Utility shaft	18
Lounge	15
Kitchen	14
Laundry	13
Incinerator	11
Linen closet	7
Other closets	7
Maintenance area	5
Walls, other concealed spaces	5
Miscellaneous known locations	6
No data	19
Total	381

Source: National Fire Protection Association, *Fire J.,* 64, 14, 1970. With permission.

4.3–6   FIRE DEPARTMENT RESPONSE TO HOSPITAL ALARMS

13 Hospitals — 4680 Beds; 4-year Study (1966–1969)

Fires involving	Number of fires
Sheets, bedcloths, linens, pillows, gowns, etc. (all articles where patient carelessness was possible cause)	574 (42%)
Ashtrays, trash containers, wastebaskets, burns on tables, overstuffed chairs, etc. (all articles where patient, visitor, or employee carelessness was possible cause)	326 (23%)
Electric motors, malfunctioning laundry equipment, light fixtures, laboratory and X-ray equipment, etc. (all tools and equipment which would be the responsibility of the hospital)	285 (21%)
Smoke conditions (visible smoke and odors of possible smoke)	111 (8%)
False alarms (faulty alarm systems or false calls to hospitals)	87 (6%)

Source: Minneapolis Bureau of Fire Prevention, Education Division.

4.3–7   FLASH POINTS AND IGNITION TEMPERATURES OF FLAMMABLE LIQUIDS COMMONLY FOUND IN HOSPITALS

Liquid	Flash point, °F	Ignition temperature, °F	NFPA classification[a]
Acetic acid	109	984	II
Acetone	−4	1000	I
Benzene	2	1044	I
Ethyl alcohol	55	793	I
Ethyl ether	−49	356	I
Formalin (with methanol)	122	806	II
Gasoline	−45	700	I
Isopropyl alcohol	53	750	I
Isopropyl ether	−18	830	I
Kerosene	100	444	II
Lubricating oil	300–450	500–700	III
Toluene	40	997	I
Turpentine	95	488	I

[a] Class I flammable liquids have flash points below 100°F, Class II liquids have flash points between 100 and 140°F, and Class III liquids have flash points over 140°F.

Adapted from National Fire Protection Association, *Fire Protection Guide on Hazardous Materials,* 4th ed., NFPA, Boston, 1972, in R.G. Bond et al., Eds., *Environmental Health and Safety in Health-Care Facilities,* Macmillan, New York, © 1973, 217. With permission.

4.3—8 CAUSES OF HOSPITAL FIRES AND EXPLOSIONS REPORTED TO THE NATIONAL FIRE PROTECTION ASSOCIATION

Cause of fire	Number of fires
Matches and smoking	73
Electrical	
Fixed services	35
Appliances	31
Malfunction of heater	30
Mishandling flammable liquids	23
Spontaneous ignition	20
Anesthesia accidents	18
Oxygen accidents	18
Incendiary, suspicious	12
Kitchen hazards	10
Combustibles too close to heater	8
Welding or cutting	8
Incinerator spark	7
Static, other than anesthesia	5
Miscellaneous known	19
Unknown	64
Total	381

Source: National Fire Protection Association, *Fire J.*, 64, 14, 1970. With permission.

4.3—9 CAUSES OF DEATH IN HOSPITAL FIRES REPORTED TO THE NATIONAL FIRE PROTECTION ASSOCIATION

Cause of death	Number of deaths
Smoke poisoning or asphyxiation	119
Burns	50
Anesthesia explosions	18
Falls or falling objects	2
Total	189

Source: National Fire Protection Association, *Fire J.*, 64, 14, 1970. With permission.

4.3—10 FREQUENCY OF FIRE DEPARTMENT RESPONSES

13 Hospitals — 4680 Beds;
4-year Study
(1966—1969)

Month	Number of responses
January	116
February	130
March	144
April	127
May	84
June	95
July	109
August	104
September	105
October	120
November	124
December	125
Total	1383

Source: Minneapolis Bureau of Fire Prevention, Education Division.

4.3−11   TYPICAL FUEL CONTENTS OF MATERIALS

Material	Fuel load per pound, Btu	Material	Fuel load per pound, Btu
Woods		Fats and waxes	
Douglas fir, untreated	8,400[1]	Animal fats, mean	17,100[2]
Douglas fir, fire retardant	7,050−8,290[1]	Butter fat	16,800[2]
Fir (dry)	9,060[2]	Lard	17,200[2]
Maple, soft, untreated	7,940[1]	Vegetable and fish oils	
Ash (dry)	8,480[2]	Cottonseed	16,920[2]
Beech, 13% moisture	7,510[2]	Linseed	16,860
Elm (dry)	8,510[2]	Olive	17,020[2]
Hardwood, average	8,120[2]	Cod-liver	16,980[2]
Locust	8,640[2]	Flammable liquids	
Oak, 13% moisture	7,180[2]	Acetone	13,500[3]
Pine, 12.3−10.5% moisture	8,080−8,420[2]	Acetylene	21,600[3]
Soft wood, resinous	8,330[2]	Ethyl alcohol	12,900[3]
Average for soft and hard	8,000[3]	Benzene (benzol)	18,000[3]
woods approximately 12% moisture		Bitumen	15,200[3]
Birch, 12% moisture	7,580[2]	Pentane	20,900[3]
Paper		Kerosene	19,800[2]
Average	7,000[3]	Gasoline	20,000[2]
Ash, 7.0%−1.4%	6,710−7,830[2]	Crude oil	19,000[2]
Building paper (asphalt	13,620[1]	Xylene	18,400[3]
impregnated)		Toluene	18,300[3]
Building paper (rosin sized)	7,650[1]	Diethyl ether	16,500[3]
Fibers		Plastics	
Cotton	7,160[3]	Polystyrene	17,420[1]
Silk (raw)	9,200[3]	Natural rubber	17,000[3]
Wool (raw)	9,800[3]		
Flax	6,500[3]		
Metals			
Structural steel, unpainted	230[1]		
Aluminum	30[1]		
Carbon	60[1]		
Magnesium	10,800[1]		

Source: H.E. Nelson, *J. Am. Soc. Saf. Eng.*, 10, 17, 1965. With permission.

REFERENCES

1. Potential Heat of Building Materials in Building Fires, *NBS Tech. Bull.*, November 1960.
2. *Fire Protection Handbook,* 12th ed., National Fire Protection Association, Boston.
3. Fire Gradings of Buildings, Part 1, Joint Committee of the Building Research Board of the Department of Scientific and Industrial Research and of the Fire Offices' Committee, Her Majesty's Stationery Office, London.

4.3−12   TYPICAL FUEL-LOAD OCCUPANCIES

Light	Moderate	High
Office with metal furnishings	Department store sales area	Flammable liquid operations
File rooms (metal files)	Most storerooms	Woodworking shops
Classrooms	Library stacks	Oils
Hospital patient areas	Paper or records on open shelves	Explosives and pyrotechnics
Dining rooms	Most industrial operations	Trash rooms
Conference rooms, auditoriums	Drafting rooms, map making	General storage warehouses
Restrooms, locker rooms	Office with wooden furnishings	Chemical or other laboratories
(metal lockers)	Shops − no flammable liquids or	involving flammable liquids
	production woodworking	

Source: H.E. Nelson, *J. Am. Soc. Saf. Eng.*, 10, 17, 1965. With permission.

4.3–13   HAND-OPERATED EXTINGUISHERS COMMONLY FOUND IN HOSPITALS

Type of extinguisher	Dry chemical	Carbon dioxide	Vaporizing liquid	Foam
Class A fires Wood, cloth, paper, rubbish, etc.	No, but will control small fires	No, but will control small fires	No	Yes
Class B fires Oil, gasoline, grease, paint, etc.	Yes	Yes	Yes	Yes
Class C fires Electrical equipment	Yes	Yes	Yes	No
Nominal capacities and corresponding highest Underwriters' Laboratories classification[a]	2½ lb– 4B:C 5 lb– 8B:C 10 lb–16B:C 20 lb–20B:C 30 lb–20B:C 150 lb–80B:C 350 lb–80B:C	2½ lb– 2B:C 5 lb– 4B:C 10 lb– 8B:C 15 lb–10B:C 20 lb–12B:C 50 lb–16B:C 75 lb–20B:C 100 lb–40 B:C	2½ lb–2B:C	1¼ gal– 1A: 2B 2½ gal– 2A: 6B 5 gal– 4A: 6B 17 gal–10A:12B 33 gal–20A:40B
Maximum effective range	5 lb–10–12 ft 20 and 30 lb–20–25 ft 150 and 350 lb–20–35 ft	3–8 ft	4–6 ft	1¼–5 gal–30–40 ft 17 and 33 gal–up to 50 ft
Approximate weight, fully charged	5 lb–11½ lb 10 lb–22 lb 20 lb–38 lb 30 lb–54½ lb	2 lb–11 5/8 lb 2 1/2 lb–8 1/2–13 lb 3 5/8 lb–11 1/8 lb 4 lb–20 lb 5 lb–16–22 lb 7 1/2 lb–35 lb 10 lb–38 lb 15 lb–48 lb 20 lb–60 lb 25 lb–65 lb	9 lb	1¼ gal } 20 lb– 1½ gal } 25 lb 2½ gal–35 lb 5 gal–70 lb
Subject to freezing	No	No	No	Yes
Yearly inspection	Weigh cartridge	Weigh	Weigh	Discharge
Operation	Push lever	Turn handwheel, pull trigger, or squeeze handle	Push lever	Invert
Means of expelling extinguishing agent	Carbon dioxide or nitrogen under pressure	Under pressure in extinguisher	Under pressure in extinguisher	Chemical reaction to form carbon dioxide
Composition of extinguisher charge	Specially treated sodium bicarbonate in powdered form with important components for producing free-flow and water-repellency	Carbon dioxide	Bromotrifluoro-methane	Solutions of aluminum sulfate and sodium bicarbonate with foam-stabilizing agent

[a] Numerical listings for Class B fire extinguishers are synonymous with area in square feet of flammable liquid fire which can be extinguished by an average operator. An experienced operator can extinguish fires two and one-half times these areas according to U.L. tests. Standard test fires are utilized at rating the Class A fire extinguishers numerically. The ratings are directly proportional to the amount of solution contained, with the 1¼-gal water extinguisher used as the base of 1A. Due to variables involved, 10-gal sizes and larger are reduced 25% from the direct proportion. Extinguishers with a Class C rating have no numerical rating. The main requirement for Class C extinguishers is that they can be used on electrical equipment without electrical hazard to the operator. Class C ratings are given only in conjunction with B ratings and never alone. The ratings of vaporizing liquid extinguishers vary according to the method of discharge — hand pump or stored pressure. Generally, the hand pump units receive lower ratings.

4.3–13 HAND-OPERATED EXTINGUISHERS COMMONLY FOUND IN HOSPITALS (continued)

Type of extinguisher	 Soda and acid	 Pressurized water type	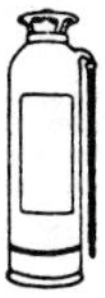 Cartridge water type	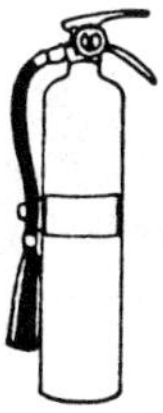 All class dry chemical
Class A fires Wood, cloth, paper, rubbish, etc.	Yes	Yes	Yes	Yes
Class B fires Oil, gasoline, grease, paint, etc.	No	No	Yes	Yes
Class C fires Electrical equipment	No	No	No	Yes
Nominal capacities and corresponding highest Underwriters' Laboratories classification[a]	1¼ gal— 1A 1½ gal— 1A 2½ gal— 2A 17 gal—10A 33 gal—20A	1¼ gal—1A 1½ gal—1A 2½ gal—2A	1 gal— 1A 1¾ gal— 2A:½B 2½ gal— 2A:1B 17 gal—10A 33 gal—20A	5 lb–8B:C 15 lb–1A,12B:C 25 lb–2A,20B:C
Maximum effective range	1¼–2½ gal–30–40 ft 17 and 33 gal–45–60 ft	30–40 ft	1 to 2½ gal–30–40 ft 17 and 33 gal– up to 50 ft	8–16 ft
Approximate weight, fully charged	1¼ gal } 20 lb– 1½ gal } 25 lb 2½ gal–35 lb	2½ gal–35 lb	1 gal } 20 lb– 1¾ gal } 25 lb 2½ gal–35 lb	5 lb–11½ lb 15 lb–31 lb 25 lb–45 lb
Subject to freezing	Yes	Yes	No	No
Yearly inspection	Discharge	Check pressure gauge	Weigh cartridge	Check pressure gauge or cartridge
Operation	Invert	Push lever	Invert and bump on floor	Push lever
Means of expelling extinguishing agent	Chemical reaction to form carbon dioxide	Under pressure in extinguisher	Carbon dioxide cartridge	Nitrogen under pressure
Composition of extinguisher charge	Sodium bicarbonate solution and sulfuric acid	Water	Water	Specially treated dry chemical

Source: R.G. Bond et al., Eds., *Environmental Health and Safety in Health-Care Facilities,* Macmillan, New York, © 1973, 226. With permission.

5. GENERAL SANITATION

5.1 FOOD HYGIENE

5.1–1 FOODBORNE DISEASE OUTBREAKS–1970

Agent	Number	Percent
Bacterial		
Clostridium perfringens	6,952	29.7
Salmonella	4,747	20.4
Staphylococcus	4,699	19.8
Shigella	1,668	7.1
Escherichia coli (coliforms)[a]	1,297	5.5
Vibrio parahemolyticus	168	0.7
Bacillus cereus	49	0.2
C. botulinum	14	0.0
Enterococci	23	0.1
Parasitic		
Trichinella spiralis	41	0.2
Viral		
Hepatitis	107	0.5
Chemical		
Monosodium glutamate	23	0.1
Metals	24	0.1
Other chemicals	248	1.0
Unknown	3,388	14.6
Total	23,448	100.0%

[a] In one outbreak several thousand persons were ill. Because of uncertainty about the number of cases this outbreak was omitted.

Source: *Foodborne Outbreaks, Annual Summary, 1970,* Center for Disease Control, Enteric Diseases Section, Bacterial Diseases Branch, Epidemiology Program, Public Health Service, U.S. Department of Health, Education, and Welfare, Atlanta, 1970.

5.1—2   CHARACTERISTICS OF FOOD INFECTIONS CAUSED BY MICROORGANISMS AND/OR THEIR PRODUCTS

	Salmonella sp.	*Streptococcus* sp.[a]	Infectious hepatitis virus
Incubation period	1. Range: 6—48 hr 2. Average: 19 hr	1. Range: 1—7 days 2. Average: 4 days	1. Range: 7—28 days 2. Average: 21 days
Symptoms	1. Sudden onset 2. Abdominal pains 3. Diarrhea 4. Frequent vomiting 5. Fever nearly always present	1. Fever 2. Sore throat 3. Headache	1. Headache 2. Aching muscles 3. Jaundice 4. Loss of appetite 5. Sometimes nausea
Duration	3—7 days	1—2 weeks	Weeks to many months
Causative agent	More than 1300 types of these bacteria produce illness in man.	Beta-hemolytic streptococci	Virus of infectious hepatitis
Common foods involved	1. Meat 6. Eggs, egg products 2. Poultry 7. Gravies 3. Custards 8. Raw vegetables 4. Shellfish 9. Sauces 5. Soups 10. "Warmed-over" foods	Food contaminated with nasal and oral discharges from persons with active cases	1. Drinking water 2. Reportedly found in milk and orange juice 3. Oysters
How foods are contaminated	1. Fecal contamination by food-service workers 2. Contaminated raw food or contaminated water 3. Food-service worker and other carriers	1. Coughing 2. Sneezing 3. Droplet infection 4. Unsanitary handling	1. Feal contamination by food-service workers 2. Water supply used for washing food utensils 3. Grown in sewage-polluted estuary
Preventive measures	1. Strict personal cleanliness and habits among food-service workers 2. Sufficient cooking and rapid refrigeration of perishable foods 3. Control of insects and rodents	1. Exclude food-service workers with known strep infections 2. Adequate heating and proper re-frigeration of food	1. Exclude food-service workers with known illness 2. Determine safety of water supply

	Staphylococcus aureus	*Clostridium perfringens*	*C. botulinum*
Incubation period	1. Range: 1—6 hr 2. Average: 3 hr	1. Range: 8—22 hr 2. Average: 10 hr	1. Range: 12—36 hr 2. Average: 19 hr
Symptoms	1. Abrupt and often violent onset 2. Nausea 3. Cramps 4. Vomiting 5. Severe diarrhea 6. Prostration 7. Subnormal temperature 8. Lowered blood pressure	1. Sudden abdominal colic followed by diarrhea 2. Nausea and vomiting often absent	1. Weakness 2. Dizziness 3. Headache 4. Constipation 5. Eye and cranial nerve pain 6. Change in voice
Duration	Usually 2—3 days	1 day or less	Prolonged periods: 3 weeks to months; fatality rate very high

[a] Enterococcus group not included.

5.1—2 CHARACTERISTICS OF FOOD INFECTIONS CAUSED BY MICROORGANISMS AND/OR THEIR PRODUCTS (continued)

	Staphylococcus aureus	*Clostridium perfringens*	*C. botulinum*
Causative agent	Poison develops when certain kinds of *Staphylococcus* organisms grow and produce their toxin in food; toxin is heat stable.	*C. perfringens* bacteria grow rapidly under favorable conditions; poison develops in intestinal tract	Poison develops when *C. botulinum* organisms grow in food; they grow without air; poison is usually fatal; toxin is heat labile.
Common food involved	1. Cooked ham, chicken a la king 2. Other meats 3. Cream-filled pastries and other desserts made of dairy products 4. Potato salad; chicken, fish, and other meat salads 5. Hollandaise sauce	1. Meat that has been boiled, steamed, braised, or partially roasted and allowed to cool slowly at room temperature 2. Meat served cool or reheated	Improperly processed or unrefrigerated food of low acid content
How foods are contaminated	By food-service workers (boils, pimples, infected cuts, droplets from nose and mouth)	1. Contaminated after slaughter 2. During processing 3. During transportation 4. During handling	1. Soil and dirt 2. Spores not killed in underheated foods
Preventive measures	1. Keep foods hot or cold. 2. Cool or heat rapidly. 3. Use hands in food preparation only when necessary.	1. Keep food hot or cold. 2. Refrigerate or heat rapidly between 40° and 150°F.	1. Use only commercially prepared food. 2. Discard leaking cans or jars of food. 3. Discard bulged cans and foods with off-odors in freshly opened containers.

Source: R.G. Bond et al., Eds., *Environmental Health and Safety in Health-Care Facilities,* Macmillan, New York,© 1973, 240. With permission.

5.1–3   CHARACTERISTICS OF CERTAIN POISONS THAT CAUSE FOODBORNE ILLNESS

	Zinc-tin-cadmium	Antimony
Onset	1. Range: 3–30 min 2. Average: 15 min	1. Range: few minutes to 1 hr 2. Average: 30 min
Symptoms	1. Abdominal pain 2. Vomiting 3. Diarrhea	Vomiting
Duration	Usually short duration but may be up to 3 or 4 days	1–2 days

	Cyanide (e.g., silver polish)	Arsenic, fluoride, sodium fluoride, lead
Onset	1. Range: minutes to 4 hr 2. Average: 2 hr	1. Range: few minutes to 2 hr 2. Average: 20 min
Symptoms	1. Weakness 2. Headache 3. Confusion 4. Nausea and vomiting 5. Blood pressure usually normal 6. Pulse more rapid than normal	1. Acute poisoning 2. Vomiting 3. Diarrhea 4. Contraction of pupils 5. Partial paralysis of eye muscles, facial muscles, hand, and lower extremities
Duration	2 hr to a few days	Variable

Source: R.G. Bond et al., Eds., *Environmental Health and Safety in Health-Care Facilities,* Macmillan, New York, © 1973, 242. With permission.

5.1—4 CHARACTERISTICS OF CERTAIN POISONOUS ANIMALS AND PLANTS THAT CAUSE FOODBORNE ILLNESS

	Shellfish	Fish	Wheat grain
Onset	5–30 min	20 min–7 or 8 hr	Gradually, after several meals
Specific agent	Plankton (Gonyaulax), food of mussels	Gymnothorax toxin Moray eel *Gymnothorax buroensis* (Bleeker), *G. flavimarginatus* (Ruppell), *G. javanicus* (Bleeker), etc.	Parasitic fungus of rye *Claviceps purpurea* Ergotism
Symptoms	1. Respiratory paralysis 2. Trembling about lips to complete loss of power in the muscles of extremities and neck	1. Vomiting 2. Coma 3. Perspiration 4. Nausea 5. Aphonia 6. Foaming at mouth 7. Convulsions 8. Respiratory paralysis	1. Gangrenous, involves limbs, fingers, toes; occasionally ears and nose 2. Convulsions 3. Depression 4. Weakness and drowsiness 5. Giddiness 6. Painful cramps in limbs 7. Itching of skin

	Rhubarb	Snakeroot	Mushroom
Onset	2–12 hr	Variable	6–15 hr
Specific agent	Oxalic acid (in greens)	Milk from cows pastured on snakeroot *Eupatorium urticaefolium*	*Amanita phalloides* and other species
Symptoms	1. Cramplike pains 2. Failure of blood to clot 3. Possibly death	1. Weakness or prostration 2. Pernicious vomiting 3. Severe constipation 4. Epigastric pain 5. Temperature normal	1. Convulsions 2. Severe abdominal pain 3. Nausea 4. Vomiting 5. Hypoglycemia

Source: R.G. Bond et al., Eds., *Environmental Health and Safety in Health-Care Facilities,* Macmillan, New York, © 1973. 242. With permission.

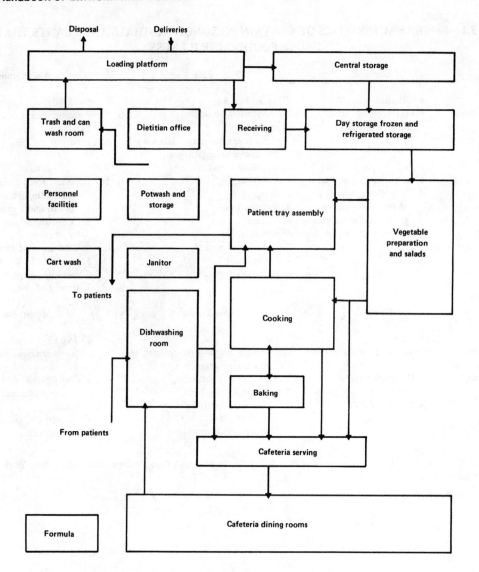

5.1—5   Flow diagram of food from receiving to service.

Source: Hospital Dietary Service: A Planning Guide, The Hill-Burton Program, PHS Pub. No. 930-C-11, Public Health Service, U.S. Department of Health, Education, and Welfare, Washington, D.C., 1967.

5.2 SOLID WASTE HANDLING AND DISPOSAL

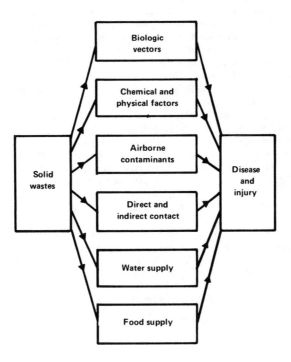

5.2–1 Solid waste human disease relationships (postulated).

Adapted from T.G. Hanks, Solid Waste/Disease Relationships: A Literature Survey, PHS Rep. No. SW-1c, PHS Pub. No. 999-UIH-6, Public Health Service, U.S. Department of Health, Education, and Welfare, Washington, D.C., 1967, in R.G. Bond et al., Eds., *Environmental Health and Safety in Health-Care Facilities,* Macmillan, New York, © 1973, 258. With permission.

5.2–2   DEFINITIONS APPLICABLE TO WASTES FROM HEALTH-CARE FACILITIES

1. Wastes	Useless, unused, unwanted, or discarded materials
2. Refuse	Includes all solid wastes; in practice this category includes garbage, rubbish, ashes, dead animals.
3. Garbage	Designates putrescible wastes resulting from handling, preparation, cooking, and serving of food.
4. Rubbish	This term includes all nonputrescible refuse except ashes; there are two categories of rubbish, combustible and noncombustible.

    a. Combustible: this material is primarily organic; it includes items such as paper, plastics, cardboard, wood, rubber, bedding.

    b. Noncombustible: this material is primarily inorganic and includes tin cans, metals, glass, ceramics, and other mineral refuse.

5. Ashes	Residue from fires used for cooking, heating, and on-site incineration.
6. Biologic wastes	Wastes resulting directly from patient diagnosis and treatment procedures; includes materials of medical, surgical, autopsy, and laboratory origin.

    a. Medical wastes: these wastes are usually produced in patient rooms, treatment rooms, and nursing stations; the operating room may also be a contributor; items include soiled dressings, bandages, catheters, swabs, plaster casts, receptacles, and masks.

    b. Surgical and autopsy (pathologic wastes): these wastes may be produced in surgical suites or autopsy rooms; items that may be included are placenta, tissues and organs, amputated limbs, and similar material.

    c. Laboratory wastes: these wastes are produced in diagnostic or research laboratories; items that may be included are cultures, spinal-fluid samples, dead animals, and animal bedding.

Sources: D. L. Snow et al., *A Report on Hospital Solid Wastes and Their Handling,* Committee on Hospital Facilities, Engineering and Sanitation Section, American Public Health Association, New York, 1955; American Public Works Association, *Municipal Refuse Disposal,* 2nd ed., Public Administration Service, Chicago, 1966. With permission.

5.2–3   AREAS GENERATING WASTES AND THEIR TYPICAL WASTE PRODUCTS

Area	Waste products
Administration	Paper goods
Obstetrics department, including patient rooms in the department	Soiled dressings; sponges; placentas; waste ampules, including silver nitrate capsules; needles and syringes; disposable masks; disposable drapes; sanitary napkins; disposable blood lancets; disposable catheters; disposable enema units; disposable diapers and underpads; disposable gloves
Emergency and surgical departments, including patient rooms	Soiled dressings; sponges; body tissue, including amputations; waste ampules; disposable masks; needles and syringes; drapes; casts; disposable blood lancets; disposable emesis basins; Levine tubes; catheters; drainage sets; colostomy bags; underpads; surgical gloves
Laboratory, morgue, pathology, and autopsy rooms	Contaminated glassware, including pipettes, petri dishes, specimen containers, and specimen slides; body tissue; organs; bones
Isolation rooms other than regular patient rooms	Paper goods containing nasal and sputum discharges; dressings and bandages; disposable masks; leftover food; disposable salt and pepper shakers
Nurses' stations	Ampules; disposable needles and syringes; paper goods
Service areas	Cartons; crates; packing materials; paper goods; metal containers, including tin cans, drums; bottles, including food containers, solution bottles, and pharmaceutical bottles; wastes from public and patient rooms, including paper goods, flowers; waste food materials from the central and floor kitchens; wastes from X-ray, discarded furniture; rags

Source: V.R. Oviatt, *Hospitals,* 42, 73, 1968. With permission.

5.2–4 CLASSIFICATION OF WASTES

Type O: Trash, a mixture of highly combustible waste such as paper, cardboard cartons, wood boxes, and combustible floor sweepings, from commercial and industrial activities. The mixtures contain up to 10% by weight of plastic bags, coated paper, laminated paper, treated corrugated cardboard, oily rags, and plastic or rubber scraps.

This type of waste contains 10% moisture and 5% incombustible solids, and has a heating value of 8500 Btu per pound as fired.

Type 1: Rubbish, a mixture of combustible waste such as paper, cardboard cartons, wood scrap, foliage and combustible floor sweepings, from domestic, commercial, and industrial activities. The mixture contains up to 20% by weight of restaurant or cafeteria waste, but contains little or no treated papers, plastic, or rubber wastes.

This type of waste contains 25% moisture and 10% incombustible solids, and has a heating value of 6500 Btu per pound as fired.

Type 2: Refuse, consisting of an approximately even mixture of rubbish and garbage by weight.

This type of waste is common to apartment and residential occupancy, consisting of up to 50% moisture and 7% incombustible solids, and has a heating value of 4300 Btu per pound as fired.

Type 3: Garbage, consisting of animal and vegetable wastes from restaurants, cafeterias, hotels, hospitals, markets, and like installations.

This type of waste contains up to 70% moisture and up to 5% incombustible solids, and has a heating value of 2500 Btu per pound as fired.

Type 4: Human and animal remains, consisting of carcasses, organs, and solid organic wastes from hospitals, laboratories, abattoirs, animal pounds, and similar sources consisting of up to 85% moisture and 5% incombustible solids, and having a heating value of 1000 Btu per pound as fired.

Type 5: By-product waste, gaseous, liquid or semiliquid, such as tar, paints, solvents, sludge, fumes from industrial operations. Btu values must be determined by the individual materials to be destroyed.

Type 6: Solid by-product waste such as rubber, plastics, wood waste from industrial operations. Btu values must be determined by the individual materials to be destroyed.

Source: Incinerator Institute of America, *I.I.A. Incinerator Standards,* I.I.A., New York, 1968, 3A. With permission.

5.2–5 CLASSIFICATION OF INCINERATORS

Class 1: Portable, packaged, completely assembled, direct-fed incinerators, having not over 5 ft³ storage capacity, or 25 lb per hour burning rate, suitable for type 2 waste.

Class 1A: Portable, packaged or job assembled, direct-fed incinerators 5–15 ft³ primary chamber volume, or a burning rate of 25 lb per hour up to, but not including, 100 lb per hour of Type 0, Type 1, or Type 2 waste, or a burning rate of 25 lb per hour up to, but not including 75 lb per hour of Type 3 waste.

Class II: Flue-fed, single chamber incinerators with more than 2 ft² burning area for Type 2 waste. This type of incinerator is served by one vertical flue functioning both as a chute for charging waste and to carry the products of combustion to atmosphere. This type of incinerator has been installed in apartment houses or multiple dwellings.

Class IIA: Chute-fed multiple chamber incinerators for apartment buildings with more than 2 ft burning area suitable for Type 1 or Type 2 waste (not recommended for industrial installations). This type of incinerator is served by a vertical chute for charging wastes from two or more floors above the incinerator and a separate flue for carrying the products of combustion to atmosphere.

Class III: Direct-fed incinerators with a burning rate of 100 lb per hour and over, suitable for Type 0, Type 1, or Type 2 waste.

Class IV: Direct-fed incinerators with a burning rate of 75 lb per hour or over, suitable for Type 3 waste.

Class V: Municipal incinerators suitable for Type 0, Type 1, Type 2, or Type 3 wastes, or a combination of all four wastes, and are rated in tons per hour or tons per 24 hr.

Class VI: Crematory and pathological incinerators, suitable for Type 4 waste.

Class VII: Incinerators designed for specific by-product wastes, Type 5 or Type 6.

Source: Incinerator Institute of America, *I.I.A. Incinerator Standards,* I.I.A., New York, 1968, 3A. With permission.

5.2—6   CONTRIBUTIONS OF HOSPITAL SOLID WASTE PER PATIENT PRESENTED IN VARIOUS PUBLICATIONS

Year presented	Contribution of solid waste	Reference
1937	7 lb/day/bed[a] comprising 40 to 50% garbage, and the balance rubbish	1
1949	7.5 lb/day/bed, compared with "government estimates" of 8 to 8.5 lb/day/bed.[a] The latter included 4 lb/day/bed[a] of "kitchen garbage."	2
1952	7 lb/day/patient, average generation	3
1955	Design basis for incinerators of 8 lb/day/bed[a] plus 3 lb/day for each person in an "auxiliary section" (e.g., nurse's residence), plus 1 lb/patient in outpatient department. If garbage were excluded, the first value might be reduced to 5 lb/day/bed. Base incinerator design on 7 hr/day.	4

1956	Mean contributions:	lb/day/patient	ft³/day/patient	5
	Garbage	3.28	0.064	
	Noncombustibles	1.10	0.111	
	Combustibles	2.61	0.521	
	Surgical and	0.14	Omitted	
	autopsy waste	7.13	0.696	

Year presented	Contribution of solid waste	Reference
1958	7 lb/day/patient, for general hospitals	6
1961	7 lb/day/patient, or almost 1 ft³/day/patient	7
1961	7 to 8 lb/day/bed	8
1963	9.6 lb/day/bed, or 1.7 ft³/day/bed for a 1000-bed military hospital	9
1964	Estimates of combustible waste generation had risen from 8 lb/day/patient "a few years ago" to 10 lb/day/patient, or more.	10
1965	Recommended values, in absence of better data, were 5 lb/day/bed, or 0.5 ft³/day/bed. A range was reported of 0.1—0.93 ft³/day/bed	11
1965	Estimated increase from approximately 6 lb/day/patient in 1955 to 13 lb/day/patient in 1975	12
1966	19 lb/day/patient at a teaching hospital with 1100 beds (included 8.3 lb/day/patient of garbage and noncombustibles, and 10.7 lb/day/patient of "readily burnable material")	13
1966	Design basis for incinerators of 8 lb/day/bed and 6 hr/day of operation	14
1967	Estimated increase during preceding 10 years from 6—9 lb/day/patient, to 20 lb/day/patient	15
1967	4.3—11.0 lb/day/patient found at hospitals with 200—725 beds.	16
1967	20 lb/day/patient	17
1968	17 lb/day/patient approximately	18
1968	Estimated 12 lb/day/bed	19

[a] Contribution of solid waste listed on a "lb/day/bed" basis may be converted to "lb/day/patient" by using percent occupancy of the hospital, as follows: lb/day/patient = (100/percent occupancy) (lb/day/bed). Since nongovermental, nonprofit hospitals have an average occupancy of 80% the conversion factor would typically be 1.25.

5.2–6   CONTRIBUTIONS OF HOSPITAL SOLID WASTE PER PATIENT PRESENTED IN VARIOUS PUBLICATIONS (continued)

Year presented	Contribution of solid waste	Reference
1968	Mean of 12.3 lb/day/bed found at 7 hospitals (range of 9–19 lb/day/bed)	20
1968	Estimated increase in preceding 10 years from 6–9 lb/day/patient to 12–19 lb/day/patient	21
1969	15 lb/day/bed	22
1969	At least 10 lb/day/bed from "general hospital buildings," excluding food service wastes and loads from auxiliary buildings. Waste contribution was expected to peak at 20 to 25 lb/day/bed.	23

Adapted from A.F. Iglar, *CRC Crit. Rev. Environ. Control*, 1, 507, 1971, in R.G. Bond et al., Eds., *Environmental Health and Safety in Health-Care Facilities*, Macmillan, New York, © 1973, 264. With permission.

REFERENCES

1. W.J. Overton, *Mod. Hosp.*, 48, 87, 1937.
2. O.E. Olson, *Mod. Hosp.*, 73, 116, 1949.
3. American Hospital Association, *Manual of Hospital Housekeeping*, Pub. No. M 16–52, Committee on Housekeeping in Hospitals, A.H.A., Chicago, 1952.
4. Incinerator Institute of Amercia, *The Selection of Incinerators for Hospital Use*, Bull. H., I.I.A., New York, 1955.
5. American Public Health Association, *Am. J. Public Health*, 46, 357, 1956.
6. Sister M. Clarisse, *Tex. Hosp.*, 14, 30, 1958.
7. O. Vance, *Hosp. Manage.*, 92, 48, 1961.
8. Anonymous, *Institutions*, 49, 127, 1961.
9. J.P. McKenna, A Study of the Requirements for Disposing of Waste Materials in the United States Air Force Hospital (at) Lackland, unpublished report submitted in partial fulfillment of requirements for Master of Hospital Administration degree, Baylor University, Waco, Tex., 1963.
10. R.C. Paul, *Hospitals*, 38, 99, 104, 1964.
11. A.G.R. Farr and H.B. Healy, *Hosp. Eng.*, 19, 65, 1965.
12. J. Falick, *Archit. Eng. News*, 7, 46, 1965.
13. J.A. Holbrook, *Mod. Hosp.*, 107, 126, 1966.
14. Incinerator Institute of America: I.I.A. Incinerator Standards, I.I.A., New York, 1966.
15. A.J.J. Rourke, *Mod. Hosp.*, 109, 132, 1967.
16. V.R. Oviatt, Waste Handling, an Old Problem, Tech. Inform. Rep., Institute of Sanitation Management, Hicksville, N.Y., 1967.
17. Anonymous, *Mod. Hosp.*, 109, 214, 1967.
18. Syska and Hennessey, Inc., Hospital systems. Part II. Materials handling, *Tech. Lett.*, 1968.
19. R.W. Davis, *Mod. Hosp.*, 111, 138, 1968.
20. H.J. Baker, A Summary of Solid Waste Handling in Seven Hospitals within the District of Columbia, unpublished course paper, University of Minnesota School of Public Health, Minneapolis, 1968.
21. R.B. Groce, *Tex. Hosp.*, 24, 32, 1968.
22. T.L. Jacobsen, *Hospitals*, 43, 89, 1969.
23. Syska and Hennessey, Inc., Incineration: an engineering approach to the waste disposal crisis, *Tech. Lett.*, 1968.

5.2–7   SOLID WASTE PRODUCTION – LOS ANGELES HOSPITALS

Hospital	Quantity of waste produced, tons/day		
	Disposables	Reusables	Total
Los Angeles County, USC Medical Center	11.60	27.25	38.85
Long Beach General Hospital	0.55	2.57	3.12
Harbor General Hospital	4.53	8.00	12.53
Rancho Los Amigos Hospital	2.77	10.26	13.03
John Wesley Hospital	0.68	2.19	0.37
Olive View Hospital	2.19	4.06	6.25
Mira Loma Hospital	0.37	0.86	1.23

Source: Reprinted from W.E. Small, *Mod. Hosp.*, 117, 100, 1971. © 1971 by McGraw-Hill, Inc. All rights reserved.

5.2–8   BREAKDOWN OF DAILY WASTE PRODUCTION BY TYPES OF WASTE IN POUNDS

Type of waste	LAC-USC Medical Center	Long Beach General Hospital	Harbor General Hospital	Rancho Los Amigos Hospital	John Wesley Hospital	Olive View Hospital	Mira Loma Hospital
Sharps, needles, etc.	75	3	22	40	8	20	5
Pathological and surgical	1,000	Traces	156	4	115	6	Traces
Soiled linen	45,500	3,740	13,600	16,320	2,900	5,630	1,120
Rubbish	16,200	540	6,569	2,760	717	1,722	362
Reusable patient items	Traces	Traces	Traces	Traces	Traces	Traces	Traces
Noncombustibles	1,500	75	465	725	80	250	80
Garbage (nongrindable)[a]	1,800	150	660	875	160	475	110
Food service items	9,000	1,400	2,400	4,200	800	2,500	600
Radiological	Traces	–	Traces	Traces	–	Traces	–
Ash and residue	Traces	–	20	20	50	20	25
Animal carcasses	25	–	220	20	10	23	–
Food waste (grindable)	2,600	330	950	1,100	210	1,860	150
Total production	77,700	6,238	25,062	26,064	5,050	12,506	2,452
Daily production disposable	23,200	1,098	9,062	5,544	1,350	4,376	732
Pounds per bed patient	11.6	3.6	16.7	6.0	7.9	7.8	5.1
Pounds per capita[b]	3.75	2.08	5.57	2.80	3.44	4.32	3.37
Daily production reusable	54,500	5,140	16,000	20,520	3,700	8,130	1,720
Pounds per bed patient	27.2	16.9	29.6	22.1	21.7	14.5	11.9
Pounds per capita[b]	8.75	9.74	9.73	10.20	9.41	8.08	7.93

[a] Predominantly garbage mixed with substantial quantities of paper, plastics, metal, etc.
[b] Per capita production based on equivalent 24-hr population.

Source: Reprinted from W.E. Small, *Mod. Hosp.*, 117, 100, 1971. © 1971 by McGraw-Hill, Inc. All rights reserved.

5.2–9 SOURCES OF SOLID WASTE WITHIN THE HOSPITAL

Source	Mean percent of total solid waste weight
Administrative and other offices	1.60
Central supply	0.84
Teaching facilities	0.03
Construction and demolition	0.22
Dietary facilities	49.00
Emergency	0.82
Extended care	0.54
Grounds	0.36
Hospitality shop	0.94
Housekeeping	0.13
Intensive care	0.46
Isolation	0.26
Laboratories	2.10
Laundry	0.46
Maintenance	0.34
Maternity	3.80
Morgue	0.04
Nursing stations, general	20.00
Outpatient department	0.24
Pediatric care	0.85
Pharmacy	0.74
Physical therapy	0.05
Psychiatric care	0.20
Public areas	0.29
Residences	0.76
Storerooms	1.20
Surgery	4.50
X-ray	1.10
Mixed, other and unknown	8.30

Source: A.F. Iglar, Hospital Solid Waste Management, Doctoral thesis, University of Minnesota, Minneapolis, 1970. With permission.

5.2–10 ENUMERATION OF DISPOSABLE ITEMS IN HOSPITALS

Hospital department	Different items[a]
Nursing	126
Dietary	29
Surgery and obstetrics	24
Laboratory	26
Housekeeping	17
Total	222

[a] Examples: canopies, catheters, diapers.

Source: G.S. Michaelsen and D. Vesley, *Hosp. Manage.*, 101, 23, 1966. With permission.

5.2–11 TYPES OF STORAGE CONTAINERS

Type of container	Number of hospitals
No containers	5
Metal cans and similar receptacles	25
Improvised containers	17
Noncompacting bulk receptacles	37
Stationary compactor receiver receptacles	9
Other	7

Source: A.F. Iglar, Hospital Solid Wastes Management, Doctoral thesis, University of Minnesota, Minneapolis, 1970. With permission.

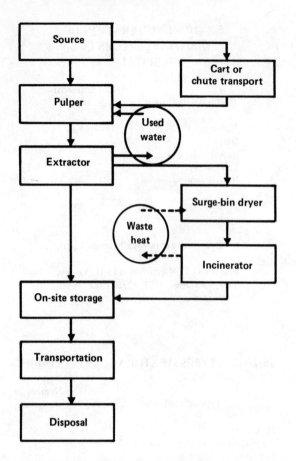

5.2—12   Pulper waste-transport system.

Adapted from V.R. Oviatt, *Hospitals,* 42, 74, 1968, in R.G. Bond et al., Eds., *Environmental Health and Safety in Health-Care Facilities,* Macmillan, New York, © 1973, 272. With permission.

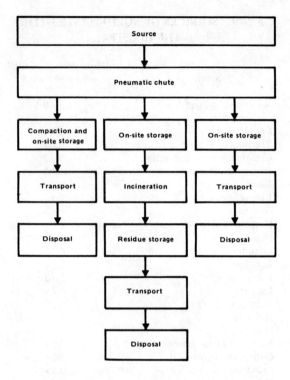

5.2—13   Pneumatic waste-transport system.

Adapted from V.R. Oviatt, *Hospitals,* 42, 74, 1968, in R.G. Bond et al., Eds., *Environmental Health and Safety in Health-Care Facilities,* Macmillan, New York, © 1973, 272. With permission.

5.2–14   DISPOSITION OF HOSPITAL SOLID WASTE

Method of solid waste disposal	N	Percent by weight for hospitals visited	
		Mean percent[a]	Range of percent[b]
Incineration at the hospital	63[c]	35	3–89
Grinding	63[c]	21	1–47
Municipal incineration	7	2.9	3–100
Sanitary landfill	24	15	3–98
Dumping	49	21	6–100
Hog feeding	13	3.7	4–40
Mixed, other, and unknown	4	1.2	2–45

[a] Average percentage disposed of by this method for all hospitals visited. Means include values of zero.

[b] Ranges include only hospitals for which waste was measured for indicated method of disposal.

[c] Excludes certain instances in which method was used, at least occasionally, but no waste was measured.

Source: A.F. Iglar, Hospital Solid Wastes Management, Doctoral thesis, University of Minnesota, Minneapolis, 1970. With permission.

5.2–15   PRIMARY AND SECONDARY STANDARDS PROPOSED

Sulfur oxides primary standards
  80 μg per cubic meter – annual arithmetic mean.
  365 μg per cubic meter – maximum 24-hr concentration not to be exceeded more than once per year.
Sulfur oxides secondary standards
  60 μg per cubic meter – annual arithmetic mean.
  260 μg per cubic meter – maximum 24-hr concentration not to be exceeded more than once per year.
Particulate-matter primary standards
  75 μg per cubic meter – annual geometric mean.
  260 μg per cubic meter – maximum 24-hr concentration not to be exceeded more than once per year.
Particulate-matter secondary standards
  60 μg per cubic meter – annual geometric mean.
  150 μg per cubic meter – maximum 24-hr concentration not to be exceeded more than once per year.
Carbon monoxide primary and secondary standards
  10 mg per cubic meter – maximum 8-hr concentration not to be exceeded more than once per year.
  15 mg per cubic meter – maximum 1-hr concentration not to be exceeded more than once per year.
Photochemical oxidants primary and secondary standards
  125 μg per cubic meter – maximum 1-hr concentration not to be exceeded more than once per year.
Nitrogen oxides primary and secondary standards
  100 μg per cubic meter–annual arithmetic mean. 250 μg per cubic meter – 24-hr concentration not to
    be exceeded more than once per year.
Hydrocarbons primary and secondary standards
  125 μg per cubic meter – maximum 3-hr concentration (6–9 A.M.) not to be exceeded more than
    once per year.

Adapted from *Environ. News,* January 30, 1971, in R. G. Bond et al., Eds., *Environmental Health and Safety in Health-Care Facilities,* Macmillan, New York, © 1973, 275. With permission.

5.2–16   PERCENT OF TOTAL SOLID WASTE WEIGHT COLLECTED DURING PEAK HOUR[a]

Range of total solid waste weight collected during peak hour	Hospitals visited	
Percent	Number	Percent
No data	10	13
1–9	1	1
10–19	19	24
20–29	29	36
30–39	11	14
40–49	9	11
50–59	1	1
Total	80	100

[a] Construction and demolition wastes not included.

Source: A.F. Iglar, Hospital Solid Waste Management, Doctoral thesis, University of Minnesota, Minneapolis, 1970. With permission.

5.2–17 CONCENTRATIONS OF GASES RELEASED DURING THERMAL DEGRADATION OF POLYVINYL CHLORIDE (PVC)

	Concentration (milligrams per cubic meter)				
	100°C			20°C	40°C
Substance	1 hr	2 hr	3 hr	30 days	30 days
Aldehydes	0.68	1.0	1.32	0	0.64
Dibutyl phthalate	31.2	33.2	35.6	6.4	8.8
Carbon dioxide	14,080	16,000	17,600	3,520	7,920
Fatty acids	Trace	6	10	0	0
Carbon monoxide	38	40	40	30	34
Hydrocarbons	96	168	396	552	720
Vinyl chloride	152	162	164	0	0
Hydrogen chloride	8	12	20	0	0

Source: L.A. Popv and V.D. Yablochkin, *Hyg. Sanit. (USSR),* 32, 114, 1967. (Translations for the U.S. Department of Commerce, Clearinghouse for Federal Scientific and Technical Information.)

5.2–18 GARBAGE GRINDER SIZE

Horsepower	Capacity (lb/hr)	Up to number of persons per meal
½	200	125
¾	400	300
1½	1200	1500
3	2000	1500 plus

Source: R. J. Black, in Environmental Aspects of Hospitals: Vol. II, Supportive Departments, PHS Pub. No. 930-C-16, Public Health Service, U.S. Department of Helath, Education, and Welfare, Washington, D.C., 1967, 20.

5.2–19 EFFICACY OF INCINERATOR OPERATIONS IN THE DESTRUCTION OF THE MICROFLORA ASSOCIATED WITH MUNICIPAL SOLID WASTES

Material	Bacterial population[a]	Incinerator design[b]			
		I	II	III	IV
Solid waste	Total cells	7.6×10^7	4.1×10^8	5.6×10^7	3.8×10^8
	Heat resistant[c]	4.2×10^4	6.8×10^4	2.7×10^4	1.7×10^4
	Total coliforms	6.2×10^5	4.8×10^6	5.4×10^5	1.2×10^5
	Fecal coliforms	9.1×10^4	4.0×10^5	1.2×10^5	2.3×10^4
Residue	Total cells	4.4×10^7	1.7×10^6	1.2×10^6	7.1×10^3
	Heat resistant[c]	1.0×10^5	2.0×10^4	3.9×10^3	4.4×10^3
	Total coliforms	1.5×10^4	2.3×10^2	4.1×10^1	5
	Fecal coliforms	2.4×10^3	9	5	<1

[a] Expressed as counts per gram.
[b] See Table 5.2–20 for characteristics.
[c] Expressed as spores per gram.

Source: H.L. Peterson and F. J. Stutzenberger, *Appl. Microbiol.*, 18, 8, 1969. With permission.

5.2–20 INCINERATOR CHARACTERISTICS

Characteristics	Incinerator			
	I	II	III	IV
Design capacity[a]	500	500	1,200	200
Number of furnaces	2	4	4	2
Feed mechanism	Continuous	Batch	Continuous	Batch
Grate	Traveling	Circular	Rotary-kiln	Reciprocating
Operating temperature	1800–2000°F	1800–2000°F	1200–1700°F (primary) 650–925°C	1800–2000°F
	980–1090°C	980–1090°C	1700–2200°F (secondary) 925–1200°C	980–1090°C
Duration of burning (hours)	1.75–2.0	1.5–1.75	0.5–1.5	1.0
Total burning rate (tons/hour)	22	20	50	6.5
Quench water recirculated	No	No quench water	Yes	No quench water
Estimated volume reduction	80–85%	80–85%	80–85%	80–85%

[a] Expressed as tons per 24 hr.

Source: H. L. Peterson and F. J. Stutzenberger, *Appl. Microbiol.*, 18, 8, 1969. With permission.

5.3 LIQUID WASTE COLLECTION AND DISPOSAL; LAUNDRY WASTES

5.3–1 EFFECT OF SEWAGE TREATMENT PROCESSES ON THE REMOVAL OF PATHOGENIC MICROORGANISMS

Investigator	Date	Organism studied	Percentage of organisms removed by					Remarks
			Trickling filter	Activated sludge	Anaerobic digestion	Chlorination	Stabilization ponds	
A. Enteric Bacteria								
Courmont and Rochaix	1922	Typhoid group	—	Present	—	—	—	5- to 6-hr treatment
		Cholera vibrios	—	Not found	—	—	—	Same
Burns and Sierp	1927	Typhoid group	—	96	—	—	—	5- to 6-hr treatment
		Cholera vibrios	—	98	—	—	—	Same
		Total count	—	95–98	—	—	—	Same
		Two paratyphoids	—	97–98	—	—	—	3 hr with 15% sludge
		Shiga-Kruse	—	97.5	—	—	—	—
		Shigella flexneri	—	97	—	—	—	—
		Shigella Y	—	98	—	—	—	—
Pesch and Sauerborn	1929	Typhoid group	—	95	—	—	—	24-hr treatment
Ruchhoft	1934	Salmonella typhosa	—	99.2	25–75	98–99	—	5½-hr aeration and settling
Metcalf and Eddy	1935	Total bacteria	70–85	90–98	—	—	—	After 6–8 days
Mom and Schaeffer	1940	S. typhosa	—	—	Not found	—	—	—
Allen, Brooks, and Williams	1949	Total bacteria	80–93	—	—	—	—	—
		Coliforms	82–95	—	—	—	—	—
		B. coli type 1	83–92	—	—	—	—	—
		Streptococcus fecalis	84–94	—	—	—	—	—
Allen and Brooks	1949	Total bacteria	—	—	—	96–98.7	—	Residual by 0-tolidine after 15-min contact
Day, Horchler, and Marks	1953	Coliforms	—	—	—	99	—	0-tolidine residual of 1–1.2 mg/l for 15 min
Gilcreas and Kelly	1954	Coliform	95	—	—	99+	—	0.2 mg/l residual
Cooke and Kabler	1955	Allescheria boydii	Present	Present	—	—	—	Also present in dry sludge
		Aspergillus fumigatus	—	—	—	—	—	—
		Geotrichum candidum	—	—	—	—	—	—

5.3–1   EFFECT OF SEWAGE TREATMENT PROCESSES ON THE REMOVAL OF PATHOGENIC MICROORGANISMS (continued)

Investigator	Date	Organism studied	Percentage of organisms removed by					Remarks
			Trickling filter	Activated sludge	Anaerobic digestion	Chlorination	Stabilization ponds	
A. Enteric Bacteria (continued)								
Imhoff and Fair	1956	Total bacteria	90–95	90–98	–	98–99	–	–
Mathews	1956	Coliforms	–	91–98	–	–	–	–
Ware and Mellon	1956	Total bacteria	97	70–99	–	–	–	–
McCoy	1957	Coliforms	94–96	–	–	–	–	–
		B. paratyphosi B	84–99+	–	–	–	–	–
McKinney, Langley, and Tomlinson	1958	S. typhosa	–	–	84–92.4	–	–	6 and 20 days, respectively
Bloom, Mack, and Mallmann	1958	Salmonella	–	Present	–	–	–	22 of 35 final effluent positive – 8 species
Towne, Bartsch, and Davis	1957	Coliforms	–	–	–	–	59.5–99.9	–
Merz, Merrell, and Stone	1957	Coliforms	–	–	–	–	50–91	–
Fitzgerald and Rohlich	1958	Coliforms	–	–	–	–	98.2–99+	–
Gillespie	1944	S. typhosa	–	–	–	–	See Remarks	Reduced from 41/ml to not demonstrable
B. Tubercle Bacilli								
Pramer, Heukelekian, and Ragotzkie	1950	Mycobacterium tuberculosis	99	–	90	–	–	With chlorination in filters
Jensen	1954	M. tuberculosis	Survive	Survive	Survive	99+	–	0-tolidine residuals of 1–5 mg/l for 2 hr
Kelly, Clark, and Coleman	1955	M. tuberculosis	Survive	–	–	Survive	–	No residual chlorine
Heukelekian and Albanese	1956	M. tuberculosis	66	88	69	99+	–	0-tolidine residual of 2 mg/l for 30 min or 1 mg/l for 1 hr
Greenberg and Kupka	1957	M. tuberculosis	Survive	Survive	Survive	–	–	Literature review

5.3–1 EFFECT OF SEWAGE TREATMENT PROCESSES ON THE REMOVAL OF PATHOGENIC MICROORGANISMS (continued)

Investigator	Date	Organism studied	Trickling filter	Activated sludge	Anaerobic digestion	Chlorination	Stabilization ponds	Remarks
C. Viruses								
Carlson, Ridenour, and McKhann	1943	Mouse adapted poliovirus	–	Mostly removed	–	–	–	Aerated 6 hr – 6 of 30 mice died
Neefe, Stokes, Baty, and Reinhold	1945	Infectious hepatitis	–	–	–	Survive / Inactivated	–	1 mg/l chlorine residual after 30 min; 15 mg/l chlorine residual after 30 min
Gilcreas and Kelly	1954	Coxsackie A / B. coli B phage	60 / 15	–	–	–	–	–
Kelly, Clark, and Coleman	1955	Coxsackie	Reduced	–	Survive	–	–	More numerous in late summer
Ware and Mellon	1956	B. coli B phage	57–73	–	–	–	–	Greater reduction with added B. coli
Mack, Mallmann, Bloom, and Krueger	1958	Poliovirus I and III / Coxsackie / ECHO	–	Survived	–	–	–	Virus isolated more often from sludge than from liquid
Weidenkopf	1958	Poliovirus I	–	–	–	99	–	For 2.5 min with 5 mg/l residual; For 6.5 min with 1.95 mg/l residual; For 14 min with 0.53 mg/l residual
D. Parasites								
Vassilkova	1936	Tapeworm ova	18–26	–	97	No effect	–	Air drying for 2 years removed all.
Cram	1943	E. histolytica cysts / Ascaris lumbricoides ova	88–99.9	No reduction	Removed	–	–	Viability not reduced after 3 months.
		Ancylostoma caninum ova (dog hookworm) / Toxascaris leonina ova (dog ascaris)	70–76	Does not affect viability	Not effective	–	–	Viable after 64 days at 20°C

5.3–1   EFFECT OF SEWAGE TREATMENT PROCESSES ON THE REMOVAL OF PATHOGENIC MICROORGANISMS (continued)

Investigator	Date	Organism studied	Percentage of organisms removed by					Remarks
			Trickling filter	Activated sludge	Anaerobic digestion	Chlorination	Stabilization 'ponds	
D. Parasites (continued)								
Hamlin	1946	Tapeworm ova	–	Not removed	–	–	–	–
Jones et al.	1947	Schistosoma japonicum	Reduced	Excellent hatching medium	90 in 25–35 days	–	–	–
Jones and Hummel	1947	S. japonicum ova	–	–	–	Killed	–	30 min residual of 3.9–11 mg/l
		miracidia	–	–	–	–	–	30 min residual of 0.2–0.4 mg/l
		S. mansoni miracidia	–	–	–	–	–	30 min residual of 0.2–0.6 mg/l
Reinhold	1949	A. lumbricoides	–	–	Reduced	–	–	Removed by 1-hr settling
Newton, Bennett, and Figgat	1949	T. saginata	62–70	Little effect	Very slow	–	–	Normal eggs recovered after 6-month digestion at 75–85°F
Bhaskaran, Sampathkumaran,	1956	Ascaris	99.8	93–98	45	–	–	In 105 days at 80–92°F
Sur, and Radhakrishnan		Hookworm	–	81.5–96	–	–	–	–
		Trichuris	100	91.8–100	–	–	–	–

Source: P. Kabler, *Sew. Ind. Wastes*, 31, 1375, 1959. With permission.

5.3–2   SUMMARY DATA ON SURVIVAL OF PATHOGENS IN HEALTH-CARE-FACILITY AND OTHER WASTES

Microorganism	Remarks	Reference
1. Mycobacteria	Low in wastes from general hospitals, but high in treated effluents from sanatoria.	1
2. *Mycobacterium tuberculosis*	Samples from septic tank at sanatorium, Poona, India; microscopic examination showed mycobacteria were present in 91.7% of inlet samples, 83.3% of tank samples, and 86.1% of outlet samples	2
3. *M. scotochromogenes, M. kansasii*	Found in poorly treated unchlorinated effluents from childrens' hospitals	1
4. *M. aquae*	Resisted free Cl_2 levels up to 2 to 2.5 mg/l in swimming-pool water	1
5. Tubercle bacilli	Chlorine used for disinfection in ranges of 20–80 mg/l for exposure periods of 15–60 min; practically no organisms survive when product of Cl_2 concentration and time in minutes > 1200	3
6. Tubercle bacilli	Sludge not free of bacteria even after 15 weeks' retention	4
7. Tubercle bacilli	Resistant to Cl_2 when in sludge. Each cubic meter of infected sludge must be mixed with 10 kg quicklime before disposal.	5
8. Enterobacteriaceae (*Klebsiella, Escherichia,* and *Aerobacter*)	Airborne bacteria from one conventional and two extended aeration plants showed an average emission rate of 440 bacteria/ m^2 from aeration tanks; average velocity of dispersion of bacterial cloud was 1 cm/sec; under worst conditions contaminated air extended 100–200 ft downstream of plant	6
9. Typhoid bacilli	Not detectable after 23 days of alkaline methane fermentation	7
10. Paratyphoid bacilli	Maximum survival in digested sludge was 3 months	7
11. *Shigella flexneri 2a*	Epidemic of dysentery in hospital due to defective sewer contaminating water supply following heavy rain	8
12. Enteric viruses	Isolated in 44% of samples of crude sewage and of Imhoff-tank effluent; 28% of samples from biologic filtration and final sedimentation	9
13. Poliovirus	Sewage heated with steam for 1 hr at temperature above 100°C, cooled to 30°C before discharge	10

Source: R.G. Bond et al., Eds., *Environmental Health and Safety in Health-Care Facilities,* Macmillan, New York, © 1973, 285. With permission.

5.3–2   SUMMARY DATA ON SURVIVAL OF PATHOGENS IN HEALTH-CARE-FACILITY AND OTHER WASTES (continued)

REFERENCES

1.   L. Coin et al., Modern microbiological and virological aspects of water pollution, in *Proc. 2nd Int. Conf. Water Pollut. Res.*, Vol. 1, Tokyo, 1965, 1; *Water Pollut. Abstr.*, 40, 13, 1967.

2.   F.L. Saldanha et al., *Indian J. Med. Res.*, 52, 1051, 1964.

3.   S. Zolaczkowski et al., *Acta Microbiol. Pol.*, 8, 181, 1959.

4.   P. Hofmann and J. von Schimmelmann, *Gesundheitsing*, 83, 175, 1962; *Water Pollut. Abstr.*, 37, 112, 1964.

5.   F. von Ammon, *Münch. Beitr. Abwass. Fisch. Flussbiol.*, 8, 174, 1961; *Water Pollut. Abstr.*, 37, 1793,1964.

6.   P.A. Kenline, The Emission, Identification, and Fate of Bacteria Airborne from Activated Sludge and Extended Aeration Sewage Treatment Plants, thesis, University of Cincinnati, 1968.

7..   W. Ahrens, *Wiss. Z. Tech. Hochsch. Dresden*, 6, 1099, 1956/1957; *Water Pollut. Abstr.*, 34, 868, 1961.

8.   I. Nikodemusz and L. Ormay, *Arch. Inst. Pasteur Tunis*, 36, 43, 1959; *Water Pollut. Abstr.*, 34, 181, 1961.

9.   M. Bendinelli et al., *Riv. Ital. Ig.*, 27, 198, 1967; *Water Pollut. Abstr.*, 42, 1103, 1969.

10.   W. Schulz, *Münch. Beitr. Abwass. Fisch. Flussbiol.*, 8, 231, 1961; *Water Pollut. Abstr.*, 37, 1797, 1964.

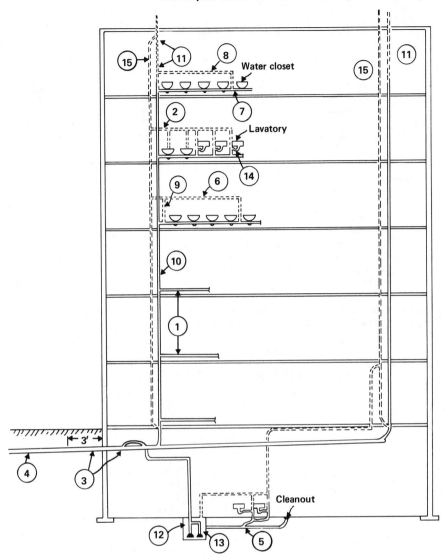

5.3—3   Elements of building drainage-and-vent system.

Key

1. *Branch Interval* — A distance along a soil or waste stack corresponding in general to a story height, but in no case less than 8 ft, within which the horizontal branches from one floor or story of a building are connected to the stack.

2. *Branch Vent* — A vent connecting one or more individual vents with a vent stack or stack vent.

3. *Building Drain* — That part of the lowest piping of a drain system that receives the discharge from soil, waste, and other drainage pipes inside the walls of the building and conveys it to the building sewer beginning 3 ft outside the building wall.

4. *Building Sewer* — That part of the drainage system that extends from the end of the building drain and conveys its discharge to a public sewer, private sewer, individual sewage-disposal system, or other point of disposal.

5. *Building Subdrain* — That portion of a drainage system that cannot drain by gravity into the building sewer.

6. *Circuit Vent* — A branch vent that serves two or more traps and extends from the downstream side of the highest fixture connection or a horizontal branch to the vent stack.

7. *Horizontal Branch Drain* — A drainpipe extending laterally from a soil or waste stack or building drain, with or without vertical sections or branches, which receives the discharge from one or more fixture drains and conducts it to the soil or waste stack or to the building drain.

5.3—3   Elements of building drainage-and-vent system (continued).

8. *Loop Vent* — A circuit vent that loops back to connect with a stack vent instead of a vent stack.

9. *Relief Vent* — A vent so planned as to permit additional circulation of air between drainage and vent systems.

10. *Stack* — A general term for any vertical line of soil, waste, vent, or inside conductor piping. This does not include vertical fixture and vent branches that do not extend through the roof or pass through less than two stories before being reconnected to a vent stack or stack vent.

11. *Stack Vent* — The extension of a soil or waste stack above the highest horizontal drain connected to the stack.

12. *Sump* — A tank or pit that receives sewage or liquid waste, located below the normal grade of the gravity system and which must be emptied by mechanical means.

13. *Sump Pump* — A mechanical device other than an ejector or bucket for removing sewage gases or liquid waste from a sump.

14. *Trap* — A fitting or device that provides a liquid seal to prevent the emission of sewer gases without materially affecting the flow of sewage or waste water through it.

15. *Vent Stack* — A vertical vent pipe installed to provide circulation of air to and from the drainage system.

Source: R.G. Bond et al., Eds., *Environmental Health and Safety in Health-Care Facilities,* Macmillan, New York, © 1973, 288. With permission.

5.3—4   MAXIMUM DISTANCE OF FIXTURE TRAP FROM VENT

Size of fixture drain, inches	Distance from trap to vent, feet
1¼	2.5
1½	3.6
2	5.0
3	6.0
4	10.0

Source: Report of Public Health Service Technical Committee on Plumbing Standards, PHS Pub. No. 1038, Public Health Service, U.S. Department of Health, Education, and Welfare, Washington, D.C., 1962, 73.

5.3—5   COMPARISON OF KENNEL-CLEANING AND CAGE-CLEANING WASTEWATERS[a]

Type of waste	Total animal weight, tons	Average composition of waste water per ton of animal			
		BOD,[b] pounds	Flow, gallons	TS,[b] pounds	VS,[b] pounds
Dog kennels	0.8	4.2	910	11.0	7.8
Animal cages	32.0	0.4	730	4.8	1.5

[a] Values depend on removal of gross feces before washing kennels and removing bedding with loose feces and absorbed urine several times between washings.

[b] BOD—biochemical oxygen demand; TS—total solids; VS—volatile solids.

Source: N.A. Jaworski and J.L.S. Hickey, *J. Water Pollut. Control Fed.,* 34, 40, 1962. With permission.

5.3-6   SUMMARY OF WASTES DISCHARGED FROM ELI LILLY LABORATORIES, GREENFIELD, INDIANA

			Treatment results		
Plant	Animals housed	Treatment provided	Initial BOD,[a] mg/l	Final BOD, mg/l	Reduction, %
238	240 cattle, 250 sheep	Bedding material removed to manure slab; waste water and wastes discharged to 20,000-gal underground tank where waste solids are digested; overflow to aerated lagoon; clarified effluent distributed over stone bed	225	75	67
237	10,000-15,000 chickens	Litter removed manually, hauled to manure slab; waste water to sewer, underground tanks, and stone filter bed	100	50-60	40-50
236	200 finishing pigs, 200 growing pigs, 20 sows and litters	Waste removal mechanically by power-driven barn cleaners; washings go to drain, vibrator screen (solids from screen hauled to manure slab), concrete holding tanks, aeration tank, clarifier; flow 5,000 gpd	510	75	85
226	4000-5000 mice and rats 400 dogs and cats, 200 other animals including rabbits, monkeys, etc.	Practically all waste goes to sewer, then to vibrating screen (solids to manure slab), concrete basins for aeration and clarification; sludge digested; lagoons with air sparger; flow 175,000 gpd	120	6	95
219	Horses, pigs, dogs, chickens	Bedding material removed; primary plant with surge tank clarifier, a primary clarifier, Imhoff tank used as digester, high rate trickling filter, final clarifier; flow 80,000 gpd	150	35	77

[a] BOD—biochemical oxygen demand.

Adapted from T.W. Bloodgood, *Treatment of Animal Wastes at the Greenfield Laboratories: 21st Industrial Wastes Conference*, Pub. No. 121, Purdue University Extension Service, Lafayette, Ind., 1966, 56, in R.G. Bond et al., Eds., *Environmental Health and Safety in Health-Care Facilities*, Macmillan, New York, © 1973, 294. With permission.

5.4 WATER SUPPLY DISTRIBUTION AND TREATMENT

5.4–1   HOSPITAL WATER CONSUMPTION

	Consumption, gallons per bed per day		
Time interval	Average[a]	Maximum	Minimum
Monthly readings	241	277	185
Daily readings	247	307	193
Hourly readings	305	578[b]	

[a] The number of beds for which the hospital was designed was used as the denominator.
[b] Peak hourly flow.

Source: Reprinted from P. E. Searcy and T. deS. Furman, *J. Am. Water Works Assoc.*, 53, 1111, 1961 by permission of the Association. Copyrighted 1961 by the American Water Works Association, Inc., 6666 West Quincy Avenue, Denver, CO 80235.

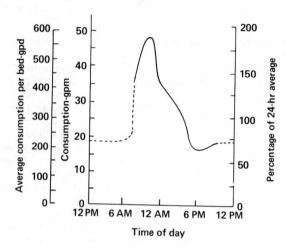

5.4–2   Hourly distribution of flow in hospitals.

Source: Reprinted from P. E. Searcy and T. deS. Furman, *J. Am. Water Works Assoc.*, 53, 1111, 1961 by permission of the Association. Copyrighted 1961 by the American Water Works Association, Inc., 6666 West Quincy Avenue, Denver, CO 80235.

5.4–3   PHYSICAL CHARACTERISTICS FOR DRINKING WATER

Substance	PHS drinking water standards[1]	Ideal water[2]
Turbidity, units	5	0.1
Color, units	15	3
Threshold odor number	3	No change on carbon contact

Source: R.G. Bond et al., Eds., *Environmental Health and Safety in Health-Care Facilities,* Macmillan, New York, © 1973, 300. With permission.

REFERENCES

1. Drinking Water Standards, PHS Pub. No. 956, Public Health Service, U.S. Department of Health, Education, and Welfare, Washington, D.C., 1962.
2. E.L. Bean, *J. Am. Water Works Assoc.*, 54, 1313, 1962.

5.4–4   MAXIMUM APPROVED CONCENTRATIONS OF CHEMICAL SUBSTANCES IN WATER

Concentrations in mg/l[a]

Substance	PHS drinking water standards[1]	Ideal water[2]
Alkylbenzenesulfonate (ABS)	0.5	0.20
Arsenic (As)	0.01	0.01
Carbon chloroform extract (CCE)	0.2	0.04
Chloride (Cl)	250.0	—
Copper (Cu)	1.0	0.2
Cyanide (CN)	0.01	0.01
Fluoride (F)	(See Table 5.4–6)	—
Iron (Fe)	0.3	0.05
Manganese (Mn)	0.05	0.01
Nitrate (NO_3)	45.0[b]	5[c]
Phenols	0.001	0.0005
Sulfate (SO_4)	250.0	—
Total dissolved solids	500.0	200
Zinc (Zn)	5.0	1.0

[a] These substances should not be present in quantities above those given when other more suitable supplies are or can be made available.

[b] When nitrates exceed these amounts, the public should be warned of the possible danger of the use of the water for infant feeding.

[c] As N.

Source: R.G. Bond et al., Eds., *Environmental Health and Safety in Health-Care Facilities,* Macmillan, New York, © 1973, 301. With permission.

REFERENCES

1. Drinking Water Standards, PHS Pub. No. 956, Public Health Service, U.S. Department of Health, Education, and Welfare, Washington, D.C., 1962.
2. E.L. Bean, *J. Am. Water Works Assoc.,* 54, 1313, 1962.

5.4–5   MAXIMUM ALLOWABLE CONCENTRATIONS OF CHEMICAL SUBSTANCES IN WATER

Concentrations in mg/I[a]

Substance	PHS drinking water standards[1]	Ideal water[2]
Arsenic (As)	0.05	0.01
Barium (Ba)	1.0	0.5
Cadmium (Cd)	0.01	0.01
Chromium (Cr^{6+})	0.05	0.01
Cyanide (CN)	0.2	0.01
Fluoride (F)	(See Table 5.4–6)	–
Lead (Pb)	0.05	0.03
Mercury (Hg)[b]	0.005	–
Selenium (Se)	0.01	0.01
Silver (Ag)	0.05	0.02

[a] Presence of the substances above the concentrations shown constitutes grounds for rejection of the supply.
[b] Added in 1970.

Source: R.G. Bond et al., Eds., *Environmental Health and Safety in Health-Care Facilities,* Macmillan, New York, © 1973, 301. With permission.

REFERENCES

1. Drinking Water Standards, PHS Pub. No. 956, Public Health Service, U.S. Department Of Health, Education, and Welfare, Washington, D.C., 1962.
2. E.L. Bean, *J. Am. Water Works Assoc.,* 54, 1313, 1962.

5.4–6 ALLOWABLE AND RECOMMENDED CONCENTRATIONS OF FLUORIDE IN DRINKING WATER

Annual average of maximum daily air temperatures[a]		Fluoride concentrations (mg/l); recommended control limits		
°F	°C	Lower	Optimum	Upper
50.0–53.7	10.0–12.1	0.9	1.2	1.7
53.8–58.3	12.2–14.6	0.8	1.1	1.5
58.4–63.8	14.7–17.7	0.8	1.0	1.3
63.9–70.6	17.8–21.5	0.7	0.9	1.2
70.7–79.2	21.6–26.3	0.7	0.8	1.0
79.3–90.5	26.4–32.3	0.6	0.7	0.8

[a] Based on temperature data obtained for at least 5 years.

Source: R.G. Bond et al., Eds., *Environmental Health and Safety in Health-Care Facilities,* Macmillan, New York, © 1973, 301. With permission.

REFERENCES

Drinking Water Standards, PHS Pub. No. 956, Public Health Service, U.S. Department of Health, Education, and Welfare, Washington, D.C., 1962.

5.4–7 ORGANIC CHEMICAL SUBSTANCES IN DRINKING WATER

Substance[a]	Maximum permissible limit, mg/l
Aldrin	0.017
Chlordane	0.003
DDT	0.042
Dieldrin	0.017
Endrin	0.001
Heptachlor	0.018
Heptachlor epoxide	0.018
Lindane	0.056
Methoxychlor	0.035
Organic phosphates and carbamates[b]	0.100
Toxaphene	0.005
Herbicides (e.g., 2,4-D; 2,4,5-T; 2,4,5-TP)	0.100

[a] Conventional treatment has little effect on these dissolved pesticides.
[b] Expressed as parathion equivalents in cholinesterase inhibition.

Source: *Manual for Evaluating Public Drinking Water Supplies,* Bureau of Water Hygiene, Public Health Service, U.S. Department of Health, Education, and Welfare, Cincinnati, 1969.

5.4–8   RADIOLOGIC SUBSTANCES IN DRINKING WATER

| | Permissible concentrations, pCi/l | |
Substance	PHS drinking water standards[1]	Ideal water[2]
Strontium-90	10	5
Radium 226	3	3
Gross β and γ activity[a]	1000[a]	100

[a] In the known absence of ^{90}Sr and alpha emitters.

Source: R.G. Bond et al., Eds., *Environmental Health and Safety in Health-Care Facilities,* Macmillan, New York, © 1973, 302. With permission.

REFERENCES

1. Drinking Water Standards, PHS Pub. No. 956, Public Health Service, U.S. Department of Health, Education, and Welfare, Washington, D.C., 1962.
2. E.L. Bean, *J. Am. Water Works Assoc.,* 54, 1313, 1962.

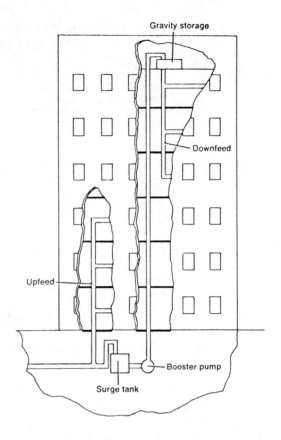

5.4–9   Combination upfeed-downfeed system for tall buildings.

Source: R.G. Bond et al., Eds., *Environmental Health and Safety in Health-Care Facilities,* Macmillan, New York, © 1973, 304. With permission.

5.4–10   HOT-WATER-USE REQUIREMENTS FOR HOSPITALS

| | Use | | |
	Clinical	Dietary	Laundry
Quantity (gallons per hour per bed)	6½	4	4½
Temperature (degrees Fahrenheit)	125	180	180

Source: General Standards of Construction and Equipment for Hospital and Medical Facilities, PHS Pub. No. 930-A-7, Public Health Service, U.S. Department of Health, Education, and Welfare, Washington, D.C., 1969, 22.

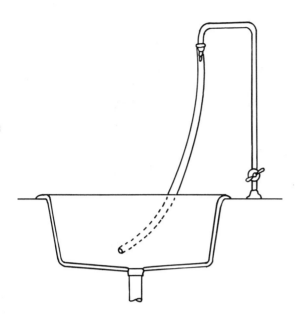

5.4–11 Cross-connection to laboratory sink.

Source: Water Supply and Plumbing Cross-connections, PHS Pub. No. 957, Public Health Service, U.S. Department of Health, Education, and Welfare, Washington, D.C., 1963.

5.4–12 Nonpressure-type vacuum breaker with fixed orifice valve.

5.4—13   CROSS-CONNECTIONS IN HEALTH-CARE FACILITIES

Possible cross-connections	Location[a]	Contaminant	Usual methods for protection[b]
Cooling-tower makeup	8	Algae, bird droppings, vermin, bacterial slimes, copper, chromates, compounds of mercury, quats	Air gap
Auxiliary supply for fire	8	Untreated supply such as river water or unprotected well	Air gap to elevated storage reservoir; air gap to surge tank; separate piping system properly labeled
Auxiliary supply for lawn sprinkler or submerged outlets	8	Untreated supply such as river water or unprotected well	Same as auxiliary supply for fire
Water for fire protection from drinking-water system	8	Chemicals for corrosion protection, slime prevention, or scale prevention, antifreeze for freeze protection, foam for fire fighting	Most codes allow check valves if no chemicals are used in system; with chemicals, use air gap to elevated storage or air gap to surge tank
Makeup for feed water to boiler	8	Boiler additives such as hydrazine, morpholine, and cyclohexylamine	Air gap to surge tank; some codes allow backflow preventers, other than the reduced pressure backflow preventer where it is specified that toxic boiler-water additives will not be used
Bedpan hopper (clinic sink)	9	Sewage	Nonpressure-type vacuum breaker (vacuum breaker to hose attachment should be installed on nonpressure side of shutoff valve near ceiling)

[a] Key to locations:

1. Animal quarters	8. Mechanical
2. Autopsy	9. Nursing units
3. Central service and supply	10. Pharmacy
4. Food service	11. Photographic laboratory
5. General	12. Surgery
6. Laboratory	13. X-ray
7. Laundry	14. Other

[b] The methods for protection given in the table do not include the "reduced-pressure backflow preventer" or the "barometric loop," devices used to protect the potable water supply to several fixtures or items of equipment.

5.4–13   CROSS-CONNECTIONS IN HEALTH-CARE FACILITIES (continued)

Possible cross-connections	Location[a]	Contaminant	Usual methods for protection[b]
Flush-valve toilets and urinals	5	Sewage	Nonpressure-type vacuum breaker
Autoclaves	1,2,3,6, 9,10,12 13	Contents of contaminated autoclave; sewage can also contaminate contents of autoclave if there is direct connection to sewer	Nonpressure-type vacuum breaker on water line to steam generators or condensers; indirect connection (air gap) to sewer
Specimen tanks	2,6	Microbial contaminants from tissues or chemical preservatives	Air gap or nonpressure-type vacuum breaker installed above spill level of tank
Instrument sterilizers	3,9,12	Microbial contaminants from instruments; sewage can contaminate instruments with direct connection to sewer	Air gap on water supply and to sewer
Pipette washers	3,6	Microbial or chemical contents of pipettes	Nonpressure-type vacuum breaker located above spill level of washer
Aspirators	1,2,3,6, 10,12	Chemical or biologic material being aspirated, and sewage	Nonpressure-type vacuum breaker (some codes do not allow water-operated aspirators)
Cage washers (wash water and water to chemical feeders)	6	Wash compounds or chemicals from feeders, contaminated water from wash tank; sewage can contaminate contents of washer if directly connected to sewer	Air gaps or nonpressure-type vacuum breakers (there may be a water-supply line to one or more wash tanks; provide indirect connection to waste either an air gap or air break)
Glassware washers (wash water and supply to chemical feeders)	1,3,6,10	Wash compounds or chemicals from feeders, contaminated water from wash tanks, sewage can contaminate contents of washer if directly connected to sewer	Same as for cage washers
Dishwasher (wash water and supply to chemical feeders)	4	Wash compounds or chemicals from feeders, contaminated water from wash tanks; sewage can contaminate contents of washer if directly connected to sewer	Same as for cage washers
Laundry wash machines (under rim or bottom inlets)	7	Contaminated wash water, wash compounds, and chemical wash additives; sewage can contaminate contents of washer if directly connected to sewer	Same as for cage washers
Photographic-processing tanks and machines (water supplies to vats and chemical feeders)	11,13	Chemicals such as acetic acid, potassium ferricyanide or aromatic organic chemicals	Nonpressure-type breakers or air gaps

5.4–13   CROSS-CONNECTIONS IN HEALTH-CARE FACILITIES (continued)

Possible cross-connections	Location[a]	Contaminant	Usual methods for protection[b]
Makeup water and chemical feed for swimming pools	8	Contaminated water from pools; chemicals such as acid, caustic chlorine; sewage from direct connection to sewer	Air gap (break tank) for makeup water; air gaps or nonpressure-type vacuum breakers for chemical feeders; indirect (air gap or air break) to sewer
Makeup to chilled-water systems for air-conditioning and hot-water systems for heating	8	Chemicals for prevention of corrosion such as chromates	Air gap and repump (some codes allow lesser protection such as a backflow preventer where toxic chemicals are not used for corrosion prevention)
Water for water-sealed pumps for chilled-water system and hot-water system	8	Chemicals for prevention of corrosion such as chromates	Air gap
Priming lines to hydraulic elevator pumps	8	Sewage when there is direct connection to sewer	Air gap on drain line from hydraulic cylinder to sewer
Priming lines to sump ejectors or sewage lift pumps	8	Sewage	Air gap
Cooling water to air compressor	8	Sewage	Air gap to sewer
Potato peeler	4	Food particles and contaminants from ptoatoes	Air gap or nonpressure vacuum breaker
Hydrotherapeutic equipment (arm bath, leg bath, pool, etc.)	9	Contaminated bath water	Nonpressure vacuum breakers or air gap
Garbage grinder (supply to bowl above and to chamber of units)	1,4	Aminal wastes, food wastes, sewage	Air gap on preflush and one or more vacuum breakers depending on number of water lines to unit
Bidets	9	Contaminated water in basin or sewage	Air gap or nonpressure vacuum breaker
Garbage-can washer	8	Garbage	Nonpressure vacuum breaker
Pressure and steam cookers	4	Contents of cooker may be contaminated with sewage if directly connected to sewer	Indirect (air-gap) connection to sewer
Water-operated insecticide and herbicide sprayers	8	Chemicals such as 2,4,D, 2,4,5,T, malathion, lindane, etc.	Use a substitute type of sprayer with mechanical or electrical means for providing pressure
X-ray-developing equipment	13	Toxic developing solutions	Nonpressure vacuum breaker or air gap to developing vats
Bathtubs	9	Bath water	Air gap
Coffee urns	4	Hot coffee and possibly sewage if there is a direct connection to sewer	Air gap on water supply; indirect connection to sewer

5.4—13   CROSS-CONNECTIONS IN HEALTH-CARE FACILITIES (continued)

Possible cross-connections	Location[a]	Contaminant	Usual methods for protection[b]
Dental cuspidors	12,14	Sputum, blood, etc.	Air gap or vacuum breaker
Drinking fountains (submerged inlet or precooler)	5	Wasted water or backed-up sewage	Air gap and discontinue use of single wall precooler (water wasted is used to precool incoming water)
Water connection to grease trap	1,4	Sewage	Nonpressure-type vacuum breaker on water lines and air gap to waste
Pressure-instrument washer-sterilizer	12	Washing compound and microbial contaminants from instruments; possible contamination of contents with sewage if directly connected to sewer	Nonpressure-type vacuum breaker; indirect (air-gap) waste connection to sewer
Automatic animal waterers	1	Water contaminated by animals	Air gap
Water connection to sewers for flushing of sewers	5	Sewage	Break connection and use mechanical cleaning methods
Hydraulically operated equipment that uses city water pressure directly (example: sink that moves up and down)	8,9	Sewage	Nonpressure vacuum breaker installed above level of sink (for line that discharges from hydraulic cylinder to waste)
Supply to brine tank for water softener	8	Salt solution	Air gap
Chemical-feed devices for regeneration of demineralizers	8	Caustic or acid	Nonpressure vacuum breaker upstream of aspirator
Ice machines	4,9	Potable ice contaminated by direct connection to sewer	Indirect waste to sewer
Flush-type floor drain	1,12	Sewage	Nonpressure vacuum breaker
Bedpan washer hose	9	Sewage	Nonpressure vacuum breaker located above level to which hose can be lifted
Bedpan washer (may include facilities for steaming)	9	Sewage	Nonpressure vacuum breaker
Water sterilizer or still	3,6,10 / 12	Acids or other cleaners; contamination of product with sewage	Disconnect water supply during cleaning with acids or caustics; air-gap connection to sewer
Supply to water-sealed pump for house vacuum system	8	Blood, sputum, chemicals, etc., that are inadvertently introduced into the vacuum system	Air gap on water line to water-sealed pump
Sitz bath	9	Contaminated bath water	Air gap
Water-cooled X-ray transformers and targets	13	Sewage	Air gap to waste

Source: R.G. Bond et al., Eds., *Environmental Health and Safety in Health-Care Facilities*, Macmillan, New York, © 1973, 312. With permission.

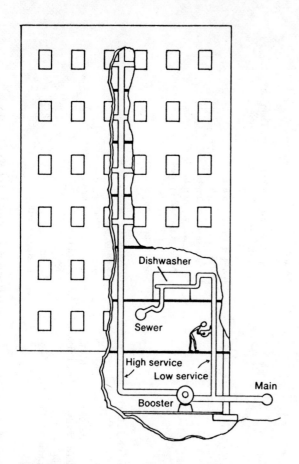

5.4–14   Booster pump causes negative pressure.

Source: Water Supply and Plumbing Cross-Connections, PHS Pub. No. 957, Public Health Service, U.S. Department of Health, Education, and Welfare, Washington, D.C., 1963.

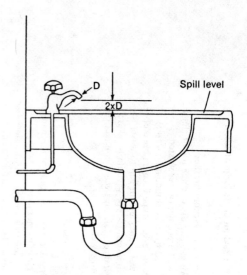

5.4–15   Air gap on lavatory.

Source: Water Supply and Plumbing Cross-Connections, PHS Pub. No. 957, Public Health Service, U.S. Department of Health, Education, and Welfare, Washington, D.C., 1963.

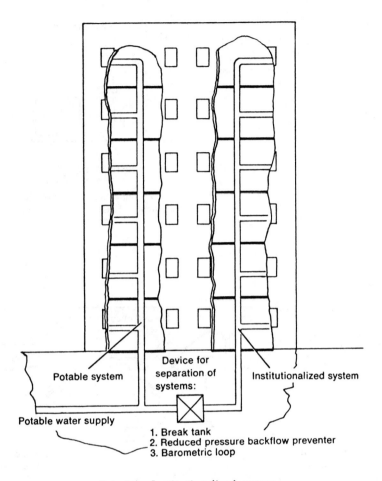

Potable system

Device for
separation of
systems:

Institutionalized system

Potable water supply

1. Break tank
2. Reduced pressure backflow preventer
3. Barometric loop

5.4–16   Institutionalized system.

Source: R.G. Bond et al., Eds., *Environmental Health and Safety in Health-Care Facilities,* Macmillan, New York, © 1973, 317. With permission.

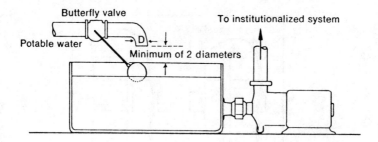

5.4—17   Break tank and booster pump.

Source: Water Supply and Plumbing Cross-Connections, PHS Pub. No. 957, Public Health Service, U.S. Department of Health, Education, and Welfare, Washington, D.C., 1963.

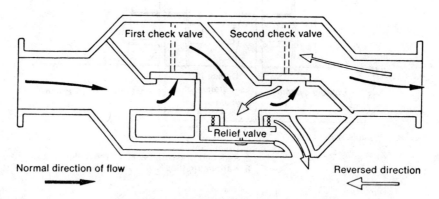

5.4—18   Reduced-pressure-type backflow preventer.

Source: Water Supply and Plumbing Cross-Connections, PHS Pub. No. 957, Public Health Service, U.S. Department of Health, Education, and Welfare, Washington, D.C., 1963.

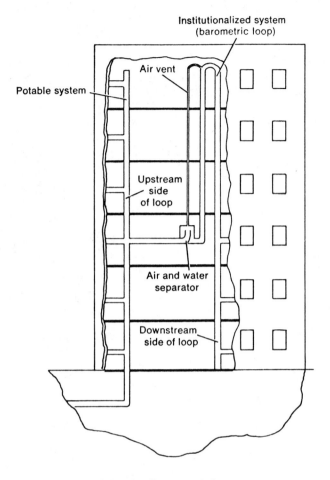

5.4–19   Barometric loop.

Source: R.G. Bond et al., Eds., *Environmental Health and Safety in Health-Care Facilities,* Macmillan, New York, © 1973, 319. With permission.

5.4–20   SIZE OF TYPICAL MEMBRANE PORES AND MICROBIAL CONTAMINANTS

Membrane pores and contaminants	Size	
	Micrometers	Angstroms
Membrane filters	0.22–0.45	2200–4500
Hemodialysis membranes	0.003–0.009	30–90
Reverse-osmosis membrane	0.02–0.05	200–500
Viruses	0.01–0.10	100–1000
Bacteria	0.50–1.00	5000–50,000
Bacterial pyrogens	0.05–1.00	500–10,000

Source: R.G. Bond et al., Eds., *Environmental Health and Safety in Health-Care Facilities,* Macmillan, New York, © 1973, 329. With permission.

5.4–21 SUMMARY OF WATER STANDARDS FROM PHARMACOPEIA

Classification of water and water use	Reaction	Heavy metals	Zinc	Foreign volatile matter	Total solids	Bacteriological purity	Pyrogens	pH	Chloride	Antimicrobial agents	Oxidizable substances	Ammonia	Calcium	Carbon dioxide	Sulfate
Water – clear, colorless, odorless liquid	(1)[a]	(2)	(3)	(4) No odor when heated near to boiling point	(5) <0.1%	(6) USPHS Standards for potable water	–	–	–	–	–	–	–	–	–
Water for injection – clear, colorless, odorless liquid prepared by distillation. Intended for use as solvent for the preparation of parenteral solutions.	–	(23)	–	–	(25) <10.0 ppm	(16) Sterility	(7) Pyrogen free	(17) 5.0–7.0	(18)	–	(24)	(20) <0.3 ppm	(21)	(22)	(19)
Bacteriostatic water for injection – clear, colorless liquid, odorless or having odor of antimicrobial substance. Consists of sterile water for injection containing one or more antimicrobial	–	(23)	–	–	(9) <40 ppm	(12) Sterility	(11) Pyrogen free	(8) 4.5–7.0	(13) <0.5 ppm	(10)	(14) Note: Water for formulation meets requirements for water for injection.	(20) Note: Water for formulation meets requirements for water for injection.	(21)	(22)	(19)

5.4–21 SUMMARY OF WATER STANDARDS FROM PHARMACOPEIA (continued)

Classification of water and water use	Re-action	Heavy metals	Zinc	Foreign volatile matter	Total solids	Bacteriological purity	Pyrogens	pH	Chloride	Anti-microbial agents	Oxidizable substances	Ammonia	Calcium	Carbon dioxide	Sulfate
Bacteriostatic water for injection – (continued) agents; prepared by distillation.															
Sterile water for injection – clear, colorless, odorless liquid; *water for injection* suitably sterilized and packaged; prepared by distillation.	—	(23) —	—	—	(15) <30 ml: <40 ppm 30–100 ml: <30 ppm >100 ml: <20 ppm	(16) Sterility	(7) Pyrogen free	(17) 5.0–7.0	(13) <0.5 ppm	—	(14)	(20) <0.3 ppm	(21)	(22)	(19)
Purified water – clear, colorless, odorless liquid; obtained by distillation or ion-exchange.	—	(23) —	—	—	(25) <10.0 ppm	(6) USPHS Drinking Water Standards	—	(17) 5.0–7.0	(18)	—	(24)	(20) <0.3 ppm	(21)	(22)	(19)

a Key:

1. Reaction (*Note:* For this test, use indicator test solutions specified for pH determinations.) – Add two drops of methyl red test solution[b] to 10 ml of *water* in a test tube; no pink or red color is produced. To another 10-ml portion of *water* in a test tube add two drops of phenolphthalein test solution: no pink or red color is produced.

2. Heavy metals – Adjust 40 ml of *water* with diluted acetic acid to a pH of 3.0 to 4.0 (using short-range pH indicator paper), add 10 ml of freshly prepared hydrogen sulfide test solution, and allow the liquid to stand for 10 min; the color of the liquid, when viewed downward over a white surface, is not darker than the color of a mixture of 40 ml of the same *water* with the same amount of diluted acetic acid as was added to the sample and 10 ml of *purified water*, matched color-comparison tubes being used for the comparison.

3. Zinc – To 50 ml contained in a test tube add three drops of glacial acetic acid

5.4–21   SUMMARY OF WATER STANDARDS FROM PHARMACOPEIA (continued)

and 0.5 ml of potassium ferrocyanide test solution; the solution shows no more turbidity than that produced by 50 ml of *purified water* in a similar test tube, treated in the same manner, and viewed downward over a dark surface.

4. Foreign volatile matter – When heated nearly to the boiling point and agitated, it evolves no odor.

5. Total solids – Evaporate 100 ml on a steam bath to dryness, and dry the residue at 105°C for 1 hr; not more than 100 mg of residue remains (0.1%).

6. Bacteriological purity – It meets the United States Public Health Service regulations for potable water with respect to bacteriological purity.

7. Pyrogen – When previously rendered isotonic by the addition of 900 mg of pyrogen-free sodium chloride for each 100 ml, it meets the requirements of the Pyrogen Test.

8. pH, between 4.5 and 7.0, determined potentiometrically in a solution prepared by the addition of 0.30 ml of saturated potassium chloride solution to 100 ml of *bacteriostatic water for injection.*

9. Total solids – Evaporate 30 ml on a steam bath to dryness, and dry the residue at 105°C for 1 hr or at such higher temperature and for such period of time as may be required to drive off any volatile added substance. Weigh the residue, and correct for any nonvolatile substances declared on the label; the weight does not exceed 1.3 mg (40 ppm).

10. Antimicrobial agent(s) – It meets the requirements under *Antimicrobial Agents – Effectiveness,* Page 845 of the original source of this table, and meets the labeled claim for content of the antimicrobial agent(s), as determined by the method set forth under *Antimicrobial Agents – Content,* Page 902 of the original source.

11. Pyrogen – When previously rendered isotonic by the addition of 900 mg of pyrogen-free sodium chloride for each 100 ml, it meets the requirements of the *Pyrogen Test,* Page 886 of the original source, the test dose being 5 ml/kg, injected very slowly.

12. Sterility – It meets the requirements under *Sterility Tests,* Page 851 of the original source.

13. Chloride – To 20 ml in a color-comparison tube add five drops of nitric acid and 1 ml of silver nitrate test solution, and gently mix; any turbidity formed within 10 min is not greater than that produced in a similarly treated control consisting of 20 ml of *special distilled water*[c] containing 10 mcg of Cl (0.5 ppm), viewed downward over a dark surface with light entering the tubes from the side.

14. Oxidizable substances – To 100 ml add 10 ml of diluted sulfuric acid and heat to boiling. Add 0.2 ml of 0.1 N potassium permanganate, and boil for 5 min; the pink color does not completely disappear.

15. Total solids – Proceed as directed in the test for *Total Solids* (see Item 25). The following limits apply for *sterile water for injection* in glass containers: up to and including 30 ml size, 40 ppm; from 30 ml up to and including 100-ml size, 30 ppm; and for larger sizes, 20 ppm.

16. Sterility – It meets the requirements under *Sterility Tests,* Page 851 of the original source.

17. pH between 5.0 and 7.0, determined potentiometrically in a solution prepared by the addition of 0.30 ml of saturated potassium chloride solution to 100 ml of *purified water.*

18. Chloride – To 100 ml add five drops of nitric acid and 1 ml of silver nitrate test solution; no opalescence is produced.

19. Sulfate – To 100 ml add 1 ml of barium chloride test solution; turbidity is produced.

20. Ammonia – To 100 ml add 2 ml of alkaline mercuric-potassium iodide test solution; any yellow color produced immediately is not darker than that of a control containing 30 mcg of added NH_3 in *special distilled water*[c] (0.3 ppm).

21. Calcium – To 100 ml add 2 ml of ammonium oxalate test solution; no turbidity is produced.

22. Carbon dioxide – To 25 ml add 25 ml of calcium hydroxide test solution; the mixture remains clear.

23. Heavy metals – Adjust 40 ml of *purified water* with diluted acetic acid to a pH of 3.0 to 4.0 (using short-range pH indicator paper), add 10 ml of freshly prepared hydrogen sulfide test solution, and allow the liquid to stand for 10 min; the color of the liquid, when viewed downward over a white surface, is not darker than the color of a mixture of 50 ml of the same *purified water* with the same amount of diluted acetic acid as was added to the sample, matched color-comparison tubes being used for the comparison.

24. Oxidizable substances – To 100 ml add 10 ml of diluted sulfuric acid and heat to boiling. Add 0.1 ml of 0.1 N potassium permanganate, and boil for 10 min; the pink color does not completely disappear.

25. Total solids – Evaporate 100 ml on a steam bath to dryness, and dry the residue at 105°C for 1 hr; not more than 1 mg of residue remains (10 ppm).

b Methyl red solution: Dissolve 24 mg of methyl red sodium in sufficient purified water to make 100 ml. If necessary, neutralize the solution with 0.02 N sodium hydroxide so that the titration of 100 ml of *special distilled water,* containing five drops of indicator, does not require more than 0.02 ml of 0.02 N sodium hydroxide to effect the color change of the indicator, which should occur at a pH of 5.6.

c Special distilled water: The water used in these tests has a specific conductivity, determined at 20°C, just prior to use, of not greater than 1 μmho. It meets the requirements of the test for heavy metals under purified water, and is free from copper. The water may be prepared by redistilling once-distilled water from a still of such proportions that the steam flows at low velocity when the still is operating at rated capacity. All parts of the still in contact with the water or steam are of borosilicate glass. Prior to distillation add, for each 1000 ml of water contained in the still, 1 drop of phosphoric acid that previously has been diluted with an equal volume of distilled water and boiled to the point of appearance of dense fumes. Reject the first 10 to 15% of the distillate, and retain the next 75%.

Source: *Pharmacopeia of the United States of America,* 18th rev., United States Pharmacopeial Convention, Washington, D.C., 1970.

5.4-22   WATER FOR PARENTERAL SOLUTIONS

Resistance (specific, ohms)	500,000
pH	5.7–6.0
Residue after evaporation (1 hr at 105°C)	1.0 ppm
Chlorides (Cl)	0.1 ppm
Ammonia (NH_3)	0.1 ppm
Heavy metals (as Pb)	0.01 ppm

Source: J.J. Perkins, *Principles and Methods of Sterilization in Health Sciences,* 2nd ed., 1969. Courtesy of Charles C Thomas, Publisher, Springfield, Ill.

5.4-23   MEDICAL CARE WATER USES AT THE UNIVERSITY OF MINNESOTA HOSPITALS

Type of water used	Water uses
Sterile distilled water	Filling of humidity reservoirs in isolettes,[a] dilution of infant formula,[a] ultrasonic nebulizers (tracheotomy),[a] ultrasonic nebulizers,[a] MAI respirators,[a] Pleura-Evac (chest suction device),[a] filling of infant heating pads[a]
Sterile distilled water for normal saline	Rinsing of catheters,[a] filling of trap on Emerson Suction Devices (closed drainage system)[a]
Normal saline (30-cc bottles)	Lavage,[a] irrigation of plugged IV's,[a] nebulizers on intermittent positive pressure breathing (IPPB) machines[a]
Normal saline	Wound irrigation, bladder irrigation, nasal tube irrigation,[a] gastric tube irrigation,[a] irrigation of GI bleeder, burn patient irrigation
Bacteriostatic water or normal saline (30-cc vials)	Dilution of injectable preparations
Nonsterile distilled water or sterile distilled water	Oxygen bubble jets attached to oxygen system,[a] oxygen bubble jets attached to tracheotomy[a]
Nonsterile distilled	Ultrasonic nebulizers (non-tracheotomy),[a] filling of instrument sterilizer on nursing station,[a] lubrication of endotracheal tubes prior to sterilization[a]
Tap water (municipal)	Filling of room humidifiers,[a] filling of water bath for heart-lung machine[a]
Tap water wash followed by tap water rinse in washer-claves	Washing of surgical instruments
a. Tap water stagnant soak b. Tap water running soak c. Distilled water-detergent wash d. Distilled water rinse prior to sterilization	Reprocessing needles and syringes[a]

[a] No recognized standard for water quality.

5.4–23   MEDICAL CARE WATER USES AT THE UNIVERSITY OF MINNESOTA HOSPITALS (continued)

Type of water used	Water uses
Wash cycle and initial rinse with tap water followed by rinse with distilled water. When machine is used for equipment that does not come into contact with bloodstream, then final rinse is with tap water.	Equipment washer – reusable syringes, surgical instruments, etc. which may come into contact with bloodstream[a]
Pyrogen-free sterile distilled water	Preparation of injectables
Distilled water from pharmacy system[b]	Preparation of irrigating solution, i.e., irrigating solution for flushing body cavities after transplant surgery; nonsterile preparations, i.e., solution used in mist tents for cystic fibrosis patients; final rinse in bottle washer for (a) bottles for non-sterile germicides and anti-septics, (b) bottles (500- and 1000-ml) reused for irrigating solutions, and (c) serum vials for water for injection; rinsing of 100-gal mixing tanks used by pharmacy;[a] filling of bottles of nonsterile distilled water distributed to nursing stations
Purchased water for injection	Eye solution formulation
Water from a reverse osmosis unit	Water for kidney dialysis[a]

[b] A specially designed style for production of pyrogen-free water.

5.4-24   SUMMARY OF SPECIAL PURPOSE WATER STANDARDS

Number	Classification of water	Specific conductance or resistance	Silicate (or SiO_2)	Total heavy metals	Organics	Arsenic	Total solids	Bacteriologic	Pyrogens	Oxygen	pH	Iron	Copper	Ammonia	Carbon dioxide	Hardness	Sodium	Ref.
1	Reagent	0.5 meg-ohm·cm at 25°C (min)	0.01 ppm	0.01 ppm (as Pb)	Pass consumption of permanganate test (500 ml, 1 ml H_2SO_4, 0.03 ml 0.1 N $KMnO_4$; 1 hr at room temp)	—	—	—	—	—	—	—						1
2	Reagent water, Type I,[a] ASTM;[b] prepared by distillation of feed water with maximum conductivity of 20 μmho/cm at 25°C followed by mixed bed deionization and 0.2-μm membrane filter	0.06 μmho/cm at 25°C (max), or 16.66 meg-ohm·cm at 25°C (min). *Note:* use in line probe for test.	—	—	Pass consumption of permanganate test (500-ml sample, 1.0 ml H_2SO_4, 0.20 ml 0.01 N $KMnO_4$; 1 hr at room temp)	—	0.1 ppm	—	—	—	6.8–7.2 at 25°C. *Note:* use in line probe for test.	—						2

[a] The dissolved or particulate organic contamination normally ranges from 1 to 5 μg/l. Concentration of nonionized dissolved gases may exceed 10 mg/l.

[b] ASTM: American Society for Testing and Materials.

5.4–24   SUMMARY OF SPECIAL PURPOSE WATER STANDARDS (continued)

Number	Classification of water	Specific conductance or resistance	Silicate (or SiO₂)	Total heavy metals	Organics	Arsenic	Total solids	Bacteriologic	Pyrogens	Oxygen	pH	Iron	Copper	Ammonia	Carbon dioxide	Hardness	Sodium	Ref.
3	Reagent water, Type II, ASTM; prepared by double distillation or by still with special baffles	1.0 μmho/cm at 25°C (max), or 1.0 meg-ohm·cm at 25°C (min). Note: use in-line probe for test.	—	—	Pass consumption of permangate test (500-ml sample, 1.0 ml H₂SO₄, 0.20 ml 0.01 N KMnO₄; 1 hr at room temp)	—	0.1 ppm	Autoclave to insure sterility	Free of pyrogens subject to test	—	6.6–7.2 at 25°C Note: use in-line probe for test.	—	—	—	—	—	—	2
4	Reagent water, Type III, ASTM; prepared by distillation, ion-exchange, or reverse osmosis followed by polishing with 0.45-μm membrane filter	1.0 μmho/cm at 25°C (max), or 1.0 meg-ohm·cm at 25°C (min).	—	—	Pass consumption of permanganate test (500-ml sample, 1.0 ml H₂SO₄, 0.20 ml 0.01 N KMnO₄; 10 min at room temp)	—	1.0 mg/l	—	—	—	6.5–7.5	—	—	—	—	—	—	2
5	Reagent water, Type IV, ASTM; prepared by distil-	5.0 μmho/cm at 25°C (max). or 0.20	—	—	Pass consumption of permanganate test	—	2.0 mg/l	—	—	—	5.0–8.0	—	—	—	—	—	—	2

5.4–24 SUMMARY OF SPECIAL PURPOSE WATER STANDARDS (continued)

Number	Classification of water	Specific conductance or resistance	Silicate (or SiO₂)	Total heavy metals	Organics	Arsenic	Total solids	Bacteriologic	Pyrogens	Oxygen	pH	Iron	Copper	Ammonia	Carbon dioxide	Hardness	Sodium	Ref.
5	lation, ion-exchange, reverse osmosis, or electro-dialysis	megohm·cm at 25°C (min)			(500-ml sample, 1.0 ml H_2SO_4, 0.2 ml of 0.01 N $KMnO_4$; 10 min at room temp.)													
6	Reagent water, Type I, CAP[c]	0.1 μmho/cm at 25°C (max), or 10 megohms·cm at 25°C (min).	0.01 mg/l	0.01 mg/l	Pass consumption of permanganate test (500-ml sample, 1 ml H_2SO_4, 0.2 ml 0.1 N $KMnO_4$; 1 hr at room temp)	—	—	Suggest periodic testing using thioglycolate broth	—	—	6.0–7.0	—	—	0.1 mg/l	3 mg/l	Limit of detectability of test	0.1 mg/l	3
7	Reagent water, Type II, CAP	2 μmho/cm at 25°C (max) or 0.5 megohm·cm at 25°C (min)	0.01 mg/l	0.01 mg/l	Pass consumption of permanganate test (500-ml sample, 1 ml	—	—	Suggest periodic testing using thioglycolate broth	—	—	6.0–7.0	—	—	0.1 mg/l	3 mg/l	Limit of detectability of test	0.1 mg/l	3

c CAP: College of American Pathologists.

5.4—24 SUMMARY OF SPECIAL PURPOSE WATER STANDARDS

Number	Classification of water	Specific conductance or resistance	Silicate (or SiO$_2$)	Total heavy metals	Organics	Arsenic	Total solids	Bacteriologic	Pyrogens	Oxygen	pH	Iron	Copper	Ammonia	Carbon dioxide	Hardness	Sodium	Ref.
8	Reagent water, Type III, CAP	5 μmho/cm at 25°C (max) or 0.2 megohms·cm at 25°C (min)	0.01 mg/l	0.01 mg/l.	H$_2$SO$_4$, 0.2 ml 0.1 N KMnO$_4$; 1 hr at room temp) Pass consumption of permanganate test (500-ml sample, 1 ml H$_2$SO$_4$, 0.2 ml 0.1 N KMnO$_4$; 1 hr at room temp)	—	—	Suggest periodic testing using thioglycolate broth	—	—	6.0–7.0	—	—	0.1 mg/l	3 mg/l	Limit of detectability of test	0.1 mg/l	3
9	Water for hemodialysis (additions and exceptions to USP "Purified Water")	1 megohm·cm at 25°C (min)	—	0.1 ppm	—	0.1 ppm	—	—	—	—	—	—	—	—	—	—	—	4
10	Biopure water	—	—	—	—	—	1.0 ppm	Sterility	Pyrogen free	—	—	—	—	—	—	—	—	5
11	Water for semiconductor manufacture	15 megohm·cm at 25°C (min)	—	—	1.0 ppm	—	30 ppb	8 org/ml	—	0.2 ppm	—	—	—	—	—	—	—	5

5.4–24   SUMMARY OF SPECAL PURPOSE WATER STANDARDS (continued)

Number	Classification of water	Specific conductance or resistance	Silicate (or SiO_2)	Total heavy metals	Organics	Arsenic	Total solids	Bacteriologic	Pyrogens	Oxygen	pH	Iron	Copper	Ammonia	Carbon dioxide	Hardness	Sodium	Ref.
12	Feed water for supercritical boilers	—	20 ppb	—	—	—	500 ppb (total dissolved)	—	—	—	—	—	—	—	—	—	—	5
13	Water for once-through boilers (Supercritical)	10 meg-ohm·cm at 25°C	10 ppb	—	—	—	50 ppb (total dissolved)	—	—	5 ppb	≅9.6	5 ppb	5 ppb	—	—	—	—	6
14	Water for once-through boilers (subcritical)	—	20 ppb	—	—	—	500 ppb (total dissolved)	—	—	5 ppb	≅9.2	10 ppb	10 ppb	—	—	—	—	6
15	Water for semiconductor manufactured	15–18 meg-ohm·cm at 25°C	—	—	1.0 ppm	—	1.0–1.5 ppm	8 org/ml	—	200 ppb	—	—	—	—	—	—	—	6
16	Electron tube manufacture	1 meg-ohm·cm at 25°C	—	—	—	—	—	Very low	—	—	—	5 ppb	5 ppb	—	—	—	—	6

d Suspended solids – should be less than 30% plugging of membrane filter in 15 min or less than 100 particles/ml.

REFERENCES

1. L. Kenyon, *Reagent Chemicals, ACS Specifications*, 4th ed., American Chemical Society, Washington, D.C., 1968, 631.
2. *Annual Book of ASTM Standards, Part 23: Water, Atmospheric Analysis*, American Society for Testing and Materials, Philadelphia, 1972, Standard D1193-72, p. 203.
3. A.R. Stier and L.K. Miller, *Pathologist*, 26, 41, 1972.
4. Tentative Standards for Hemodialysis of the American Association for the Advancement of Medical Instrumentation, September 1970.
5. *The Barnstead Basic Book on Water*, Barnstead Co. (division of Sybron Corp.), 225 Rivermoor St., Boston, MA 02132, 1971.
6. C. Calmon and G.P. Simon, Ultrapure Water – A Survey of Industrial Requirements and Specifications, Liberty Bell Corrosion Course, NACE, Drexel Institute of Technology, Philadelphia, September 16, 1965.

5.4–25   WATER CLASSIFICATION BASED ON CONDUCTIVITY AND TREATMENT METHOD

Classification of water	Method of production	Specific resistance (ohm-cm at 25°C)
Low purity Types III and IV	Distillation, ion exchange, or reverse osmosis	200,000– 1,000,000
High purity Type II	Distillation	500,000– 2,000,000
Ultrapure Type I	Distillation with mixed-bed polishing, followed by 0.2-μm filtration	10,000,000– 16,700,000
Absolute purity	Conductivity stills; water digested with oxidizing agents to break down organics	18,300,000[a]

[a]   Difficult to maintain because of reabsorption of CO_2 from the atmosphere.

Adapted from ASTM Specifications (Committee D-19), College of American Pathologists Specifications for Reagent Grade Water, and the *Barnstead Basic Book on Water,* Barnstead Co. (Division of Sybron Corp.), 225 Rivermoor St., Boston, MA 02132. With permission.

5.4–26   WATER CLASSIFICATION BASED ON CONDUCTIVITY

Grade	Specific resistance (ohm·cm at 25°C)	Electrolytes (ppm)
Low purity	<300,000	1.4
Average purity	300,000–1,000,000	0.4–1.4
High purity	100,000–10,000,000	0.02–0.4
Ultrapure	10,000,000–17,000,000	0.002–0.02
Theoretically pure	18,000,000	0.00

Adapted from J.J. Perkins, *Principles and Methods of Sterilization in Health Sciences,* 2nd ed., 1969. Courtesy of Charles C Thomas, Publisher, Springfield, Ill.

5.4–27   STANDARDS FOR WATER USED IN BIOMEDICAL AND MEDICAL CARE

1. Resistivity – dissolved ionics – will be divided into ten classes as follows (all at 25°C):
   - <20,000 ohm
   - 20,000–50,000 ohm
   - 50,000–100,000 ohm
   - 100,000–200,000 ohm
   - 200,000–500,000 ohm
   - 500,000 ohm–1 megohm
   - 1–5 megohm
   - 5–10 megohm
   - 10–15 megohm
   - 15–18 megohm
2. Pyrogens
   - a. Pyrogenic
   - b. Nonpyrogenic
3. Bacteriological
   - a. Sterile
   - b. Less than 50 col/ml
   - c. Less than 500 col/ml
4. Particulate matter (micrometers)
   - a. 0–0.05 absolute
   - b. 0.05–0.1 absolute
   - c. 0.1–0.25 absolute
   - d. 0.25–0.5 absolute
   - e. 0.5–1 absolute
   - f. 1–5 nominal
   - g. 5–10 nominal
   - h. 10–25 nominal
5. Organics

Source: N. I. McClelland, personal communication (National Sanitation Foundation, Ann Arbor, Mich., March 1972).

5.4–28   CATEGORIES OF BACTERIOLOGICAL QUALITY ON THE BASIS OF TOTAL COUNT

Category of quality	Total numbers of organisms per milliliter	Examples of possible uses
1a	Sterile	Inhalation therapy
1b	Sterile (pyrogen free)	Injection
2	1–10	Dialysate preparation
3	11–100	Most laboratory uses
4	101–1000	Most laboratory glassware final rinsing
5	>1000	Some laboratory glassware washing

Source: R.L. DeRoos et al., Study of Critical Quality Requirements for High-Purity Water in Biomedical Research and Medical Facilities at the National Institutes of Health, Division of Environmental Health, School of Public Health, University of Minnesota, Minneapolis, September 1974.

5.4—29   EXAMPLES OF BACTERIOLOGICAL QUALITY PREDICTED ON BASIS OF TREATMENT AND SYSTEM

Category	Quality, org/ml	Water system
1a	Sterile	Autoclaved, bottled water
1b	Sterile and pyrogen free	Distillation (special still) followed by autoclaving
2	1—10	From local still with protected storage
3	11—100	Building distilled water system with heavy use
4	101—1000	Building deionized water system with heavy use
5	<100	Building deionized system, poorly maintained

Source: R.L. DeRoos et al., Study of Critical Quality Requirements for High-Purity Water in Biomedical Research and Medical Facilities at the National Institutes of Health, Division of Environmental Health, School of Public Health, University of Minnesota, Minneapolis, September 1974.

5.4–30   COMPOSITE WATER STANDARD FOR LABORATORIES

Grade of water	Specific resistance, megohm·cm, 25°C	Organics	Silicates (or SiO$_2$),[a] mg/l	Total heavy metals,[a] mg/l as Pb	Total solids,[a] mg/l	pH,[a] at 25°C	Ammonia,[a] mg/l	Carbon dioxide,[a] mg/l	Sodium,[a] mg/l
Type I	16.66 (ASTM[b])	Pass test 1 hr (ASTM)	0.01 (CAP[c])	0.01 (CAP)	0.1 (ASTM)	6.8–7.2 (ASTM)	0.1 (CAP)	3.0 (CAP)	0.1 (CAP)
Type II[d]	1.0 (ASTM)	Pass test 1 hr (ASTM)	0.01 (CAP)	0.01 (CAP)	0.1 (ASTM)	6.6–7.2 (ASTM)	0.1 (CAP)	3.0 (CAP)	0.1 (CAP)
Type III	1.0 (ASTM)	Pass test 1 hr (ASTM)	0.01 (CAP)	0.01 (CAP)	1.0 (ASTM)	6.5–7.5 (ASTM)	0.1 (CAP)	3.0 (CAP)	0.1 (CAP)
Type IV	0.2 (ASTM)	Pass test 10 minutes	–	–	2.0 (ASTM)	5.0–8.0 (ASTM)	–	–	–

Notes: Method of preparation as suggested for ASTM types of water in Table 5.4–26 or equivalent as determined by test of water as it is used (following storage and distribution). Method of preparation was given for ASTM standards.

a   Values given are maximum allowable.
b   ASTM: American Society for Testing and Materials.
c   CAP: College of American Pathologists.
d   Free of pyrogens subject to test.

Source: R.L. DeRoos et al., Study of Critical Quality Requirements for High-Purity Water in Biomedical Research and Medical Facilities at the National Institutes of Health, Division of Environmental Health, School of Public Health, University of Minnesota, Minneapolis, September 1974.

5.4—31   WATER USES

Type I   Where methods require maximum accuracy and precision for enzymology, flame photometry, blood gas and pH determinations, electrolytes and inorganic ions, and reference buffer solutions. Atomic absorption most often requires Type I water.

Type II   For most analytical procedures, including many of those listed above for Type I water depending on reagents used, methodology, and sensitivity. Use for most serological, hematological, and microbiological procedures. ASTM specifies (according to their recommended procedure for production) that this water be used where organic free water is needed.

Type III   For most general laboratory procedures including most qualitative analyses, urinalyses, parasitology, and histological procedures. This would include rinsing of glassware.

Type IV   For synthetic test solutions and make-up of large volumes of solutions requiring moderate purity water.

REFERENCES

Annual Book of ASTM Standards, Part 23, Water, Atmospheric Analysis, American Society for Testing and Materials, Philadelphia, 1972, Standard D1193-72, p. 203. A. R. Stier and L. K. Miller, *Pathologist,* 26, 41, 1972.

5.4-32 SUMMARY OF BACTERIOLOGICAL DATA FROM VARIOUS SYSTEMS

System	Number of samples	Number and percent of samples above measurable range[a]	Number and percent of samples with >100 org/ml	Average number of org/ml	95% confidence interval for average number of org/ml
Central distilled	108	20 (19%)	48 (44%)	225	123–327
Central deionized	36	14 (39%)	34 (94%)	1370	381–1860
Laboratory distilled	31	5 (16%)	9 (29%)	188	–11–387
Laboratory deionized	126	28 (22%)	34 (27%)	635	424–845

[a] Range endpoints used for values above or below measurable range, e.g., 1 for values reported as <1 and 3000 for values reported as >3000.

Source: R.L. DeRoos et al., Study of Critical Quality Requirements for High-Purity Water in Biomedical Research and Medical Facilities at the National Institutes of Health, Division of Environmental Health, School of Public Health, University of Minnesota, Minneapolis, September 1974.

5.4-33 SUMMARY OF BACTERIOLOGICAL DATA FROM LEASED LABORATORY DEIONIZERS BY LEASOR

Leasor	Number of samples	Number and percent of samples above measurable range[a]	Number and percent of samples with >100 org/ml	Average number of org/ml	95% confidence interval for average number of org/ml
A	55	0 (0%)	0 (0%)	7.79	6.36–9.22
B	55	28 (51%)	34 (62%)	1443	1168–1718

[a] Range endpoints used for values above or below measurable range, e.g., 1 for values reported as <1 and 3000 for values reported as >3000.

Source: R.L. DeRoos et al., Study of Critical Quality Requirements for High-Purity Water in Biomedical Research and Medical Facilities at the National Institutes of Health, Division of Environmental Health, School of Public Health, University of Minnesota, Minneapolis, September 1974.

5.4–34   TOTAL BACTERIAL COUNTS (ORGANISM/ml) FOR VARIOUS HIGH PURITY WATER SYSTEMS AT NIH

October 4, 1972–December 18, 1972

System	Bldg.	Location	10/4/72	10/16/72	10/31/72	11/7/72	11/13/72	11/20/72	11/27/72	12/5/72	12/11/72	12/18/72	Mean	Standard deviation
Central distilled	4	1st fl. tap	43	56	1620	6	31	160	14	13	9	2	195.40	502.73
	10	Rm. B2N255	45	26	61	38	18	1	0	5	11	1	20.60	21.33
	30	1st fl. tap	18	<1	41	<1	0	1	0	0	3	2	6.70	13.27
	30	Hall tap Rm 308	<1	<1	3	<1	0	1	0	0	0	<1	.80	.91
Central deionized	36	1st fl. tap	56	27	300	131	5060	77	106	244	12	797.5	681.05	1555.96
	41	Tap on tank	41	522	392	44	84	43	54	72	154	10.6	141.66	173.18
	29	Rm. B-1	69	960	556	168	492	44	118	985	375	113	388.0	355.91
	30	Rm. B-29	145	2340	2820	3480	3620	580	1882	1310	1160	2175	1951.20	1131.11
Lab distilled	9	Basement	<1	<1	<1	<1	0	0	0	0	0	0	.40	.26
	10	Rm. 2B41	5	25	39	10	82	–	–	–	–	–	32.20	15.63
Lab deionized	30	Rm. 308	2	4	1	<1	2	0	1	1	3	<1	1.50	1.33
	37	Rm. 3C09	–	–	–	–	<1	3	9	7	3	<1	3.66	3.58
Mean			38.5	360.1	530.3	68.0	782.5	75.8	181.9	219.8	144.0	258.7		
Standard deviation			42.8	724.4	896.8	109.3	1638.6	165.0	536.8	441.7	338.0	642.0		

Source: R.L. DeRoos et al., Study of Critical Quality Requirements for High-Purity Water in Biomedical Research and Medical Facilities at the National Institutes of Health, Division of Environmental Health, School of Public Health, University of Minnesota, Minneapolis, September 1974.

5.4–35   TOTAL BACTERIAL COUNTS (ORGANISMS/ml) FOR VARIOUS HIGH-PURITY WATER SYSTEMS AT NIH

August 6–21, 1973

System	Bldg.	Location	8/6/73	8/6/73	8/7/73	8/7/73	8/9/73	8/10/73	8/14/73	8/21/73	Mean	Standard deviation
Central distilled	5	3rd fl. hall tap	>300	>300	>300	28	44[a]	31	625	13.5	205.18	216.36
	5	1st fl. hall tap	>300	>300	>300	83	110[a]	190	460	31.5	221.81	143.46
	10	Rm. 13-2	>300	>300	>300	>300	130[a]	465	125	185	263.12	112.82
	10	13th fl, W. hall tap	>300	42	>300	14	95[a]	87	65	725	203.50	238.18
	10	13th fl, E. hall tap	>300	>300	>300	>300	380[a]	555	210	350	336.87	100.82
	10	2nd fl, E. hall tap	>300	>300	>300	>300	1300[a]	780	400	345	503.12	361.02
Central deionized	10	Media Dept.	>300	>300	>300	>300	>3000	>3000	>3000	3800	1750.00	1608.02
	29	Rm. B-1	>300	>300	>300	>300	5000	>3000	>3000	>3000	1900.00	1919.08
Lab distilled	9	Basement tap	>300	>300	140	>300	>300	>3000	750	39	641.12	975.07
	13	Rm. 3W-86	42	5.8	11.25	12	46	205	66.5	146	66.81	71.94
Mean			274.2	244.8	255.1	193.7	1040.5	1131.3	950.2	229.4		
Standard deviation			81.6	116.7	99.4	138.6	1671.3	1309.0	1325.2	240.4		

[a]Change from filtering 1 ml to 0.1 ml.

Source: R.L. DeRoos et al., Study of Critical Quality Requirements for High-Purity Water in Biomedical Research and Medical Facilities at the National Institutes of Health, Division of Environmental Health, School of Public Health, University of Minnesota, Minneapolis, September 1974.

5.4–36   TOTAL BACTERIAL COUNTS (ORGANISMS/ml) FOR VARIOUS LEASED LABORATORY DEIONIZERS AT NIH

July 18, 1973–August 22, 1973

Room	7/18/73	7/23/73	7/24/73	7/25/73	7/25/73	7/27/73	8/1/73	8/8/73	8/17/73	8/22/73	Mean	Standard deviation
303	8.50	12.95	7.50	10.40	6.50	23.00	5.25	4.60	22.00	3.00	10.01	6.75
304	7.15	16.00	5.80	9.25	4.30	9.30	1.35	8.95	4.50	21.15	11.38	10.30
309	1.50	1.05	2.60	1.40	6.30	0.95	2.15	1.70	3.60	8.50	2.89	2.40
313	1.80	11.08	2.60	3.95	4.30	5.95	0.60	1.30	1.40	9.10	4.09	3.55
328	18.75	8.80	1.20	15.65	11.00	8.80	6.15	12.25	2.70	16.50[b]	10.54	5.65
329	12.60	73.00	0	32.00	14.00	75.50	78.00	91.50	30.00	21.00	44.60	32.08
229	>300	80.50	335	690.00	>3000	1840	53.00	>3000[a]	>3000	>3000	1586.22	1304.55
234	170	>3000	2150	>3000	>3000	>3000	>3000	>3000	>3000	>3000	2742.72	853.28
239	>300	>3000	>3000	>3000	>3000	>3000	>3000	>3000	>3000	>3000	2754.54	814.08
124	10.70	15.00	162	24.40	60.00	8.40	24.40	97.50	>300	>300	91.66	113.39
Mean	83.10	621.90	651.70	678.70	910.60	797.20	617.10	921.80	936.40	937.90		
Standard deviation	125.00	1253.70	1242.40	1241.50	1441.90	1292.60	1256.20	1434.50	1426.80	1425.70		

[a] Organic filter changed since last reading.
[b] Complete unit changed since last reading.

Source: R.L. DeRoos et al., Study of Critical Quality Requirements for High-Purity Water in Biomedical Research and Medical Facilities at the National Institutes of Health, Division of Environmental Health, School of Public Health, University of Minnesota, Minneapolis, September 1974.

5.4–37   COSTS PER GALLON FOR WATER TREATED BY VARIOUS METHODS

Method of production and delivery	Cost per gallon, $
Bottled distilled water	0.40
1 GPH electrically heated still	0.06
1 GPH steam heated still	0.035
100 GPH steam heated still	0.02
Distilled water central system	0.015
Bottled deionized water	0.25
Rented deionized system	0.06
Large purchased deionizer system	0.002
Reverse osmosis system	0.0005
Reverse osmosis followed by deionization	0.0007

Source: H.H. Hadley, Usable Water, paper presented at the Scientific Products Seminar and Products Show, Minneapolis, April 15, 1973.

5.5   EMERGENCY AND DISASTER PLANNING

5.5–1   REFERENCES FOR EMERGENCY AND DISASTER PLANNING

Title	Publisher, publication number, date	Address of publisher[a]
Checklist for a Hospital Civil Disturbance Preparedness Progress	American Hospital Association, 1968	2
Checklist for Developing a Packaged Disaster Hospital Readiness Plan	U.S. Public Health Service, Division of Emergency Health Services, Pub. No. 1071–F–16, 1968	5
Checklist for Hospital Disaster Planning	American Hospital Association, 1964	2
Disaster Management, A Guide for Hospital Administrators	American Hospital Association, 1971	2
Emergency Health Services, Preparedness Checklist for Hospitals	U.S. Public Health Service, Division of Emergency Health Services, Pub. No. 1071–G–3, 1969	5
Emergency Health Services, Preparedness Planning	U.S. Public Health Service, Division of Emergency Health Services, Pub. No. 1071–A–3, 1965	5
Emergency Medical Supplies: Hospital Reserve Disaster Inventory (HRDI)	U.S. Public Health Service, Division of Emergency Health Services, 1969	5
Establishing the Packaged Disaster Hospital	U.S. Public Health Service, Division of Emergency Health Services, Pub. No. 1071–F–1, 1966	5
Guide to Developing an Industrial Disaster Medical Service	American Medical Association: reprinted from *Arch. Environ, Health,* 15, 119, 1967	3
Hospital Planning for National Disaster	U.S. Public Health Service, Division of Emergency Health Services, Pub. No. 1071–G–1, 1968	5
Manual for Protection of Public Water Supplies from Chemical Agents	U.S. Public Health Service, Division of Emergency Health Services, Pub. No. 1071–J–1, 1966	5
Operation of Generators in the Packaged Disaster Hospital	U.S. Public Health Service, Division of Emergency Health Services, Pub. No. 1071–F–5, 1968	5
Orientation Manual on Disaster Preparedness for Pharmacists	U.S. Public Health Service, Division of Emergency Health Services, Pub. No. 1071–D–7, 1965	5
Preparing the Hospital Plant for Emergencies	U.S. Public Health Service, Division of Emergency Health Services, Pub. No. 1071–G–2, 1967	5
Principles of Disaster Planning for Hospitals	American Hospital Association, 1967	2
Public Water Supply Facilities: Emergency Preparedness Checklist	U.S. Public Health Service, Division of Emergency Health Services, 1963	5
Radiologic Aspects of Disaster Planning	American College of Radiology, Committee on Radiologic Aspects of Disaster Planning, 1962	1
Readings in Disaster Planning for Hospitals	American Hospital Association, 1966	2
Safe Drinking Water in Emergencies	U.S. Public Health Service. Environmental Control Administration, Pub. No. 387, 1964	4

a Names and adresses of publishers

1. American College of Radiology, 29 North Wacker Drive, Chicago, IL 60606
2. American Hospital Association, 840 North Lake Shore Drive, Chicago, IL 60611
3. American Medical Association, 535 North Dearborn Street, Chicago, IL 60610
4. U.S. Department of Health, Education, and Welfare, Information Office, Environmental Protection Agency, 12720 Twinbrook Parkway, Rockville, MD 20852
5. U.S. Department of Health, Education, and Welfare, Information Office, Division of Emergency Health Services, Health Services and Mental Health Administration, Public Health Service, 5600 Fishers Lane, Rockville, MD 20852

Adapted from U.S. Department of Health, Education, and Welfare, *Emergency Health Services: Selected Bibliography,* PHS Pub. No. 1071-A-1, Public Health Service, U.S. Department of Health, Education, and Welfare, Washington, D.C., 1970, in R.G. Bond et al., Eds., *Environmental Health and Safety in Health-Care Facilities,* Macmillan, New York, © 1973, 341. With permission.

5.5–2   DISASTER SITUATIONS AND RESPONSES

Disaster situations	Hospital response
1. *Internal disasters* Disasters within the hospital (fire, explosion, etc.)	*Evacuation* of patients from threatened or affected areas *Expansion* of treatment areas to care for casualties
2. *External disasters, minor* Community disasters involving relatively small numbers of casualties (storm, fire, flood, hurricane, explosion, train wreck, etc.)	
3. *External disasters, major* Community disasters involving large numbers of casualties (storm, fire, flood, hurricane, explosion, train wreck, epidemic, etc.)	*Evacuation* of reception and treatment areas to care for inpatient and outpatient casualties *Evacuation* of some inpatients to free beds for casualties
4. *Disaster threats* Disaster threatening either the hospital or the whole community (fire in buildings near or adjacent to the hospital; impending storm, hurricane, tornado, flood, etc.; warning of enemy attack)	Precautionary *evacuation,* either partial or total "Alert" notification to staff and outside cooperating agencies Preparation of reserve equipment and supplies
5. *Disasters in other communities* Disasters in communities nearby	*Expansion* to receive casualties and/or inpatients transferred from the stricken community Send personnel and supplies, upon request, to provide medical support in the affected community

Source: American Hospital Association, *Principles of Disaster Planning for Hospitals,* AHA, Chicago, 1967. With permission.

5.5–3   CHECKLIST FOR HOSPITAL DISASTER PLANNING

Basic Assumptions

		Yes	No
1.	Does the hospital have a disaster plan?	___	___
2.	Has a disaster committee been appointed for the hospital?	___	___
3.	Has the plan been approved by the hospital governing board or appropriate authority?	___	___
4.	Have appropriate portions of the hospital disaster plan been officially incorporated into the local civil defense survival plan?	___	___

General Provisions

1.	Does the plan take into account all the disaster situations possible in the community in which the hospital is located?	___	___
2.	Is there a separate and specific internal fire and explosion plan?	___	___
3.	Is there an awareness by hospital personnel of the specific phases of the areawide plan of diaster operation applicable to the community in which the hospital is located?	___	___
4.	Does the plan meet the requirements imposed by geographical location?	___	___
5.	Does the plan include provision for handling of increased patient loads through a variety of routines designed to meet different situations?	___	___
6.	Does the plan provide for extensive minimal hospital functions over a prolonged period?	___	___

Organization for Disaster Operations

1.	Does the disaster organization include lines of authority and responsibility, and provide for success?	___	___
2.	Are all key personnel included in the plan?	___	___
3.	Is each operating department covered by a specific disaster operations plan according to each phase of the following basic provisions: (1) internal accidents or diaster, (2) local large-scale disaster, (3) total disaster including enemy action, (4) evacuation and survival?	___	___
4.	Is a functioning medical chief of disaster operations designated and readily available? Does he have designated deputies and alternates equally available?	___	___

5.5—3   CHECKLIST FOR HOSPITAL DISASTER PLANNING (continued)

Organization for Disaster Operations (continued)

	Yes	No
5. Is the plan for each department individually published, in the hands of each responsible department head, and known to all personnel of each specific department?	___	___
6. Is the assignment of key responsibilities fixed within each department?	___	___
7. Is the organization for evening, night, weekend, and holiday routines established so that the disaster plan may be activated as promptly as during day shifts of the normal work week?	___	___
8. Have evening and night personnel been included in the formulation of plans for disaster operations?	___	___
9. Have the following services been provided for and responsible officers and personnel appointed:		
a. Administrative services?	___	___
b. Records?	___	___
c. Information and communications?	___	___
d. Social service?	___	___
e. Food service?	___	___
f. Engineering and maintenance service?	___	___
g. Transportation?	___	___
h. Supply?	___	___
i. Decontamination?	___	___
j. Safeguarding personal effects?	___	___
10. Have the following areas been designated in or around the hospital:		
a. First aid station?	___	___
b. Receiving and sorting (triage) areas?	___	___
c. Shock or holding ward?	___	___
d. Psychotic ward?	___	___
e. Treatment and nursing areas?	___	___
f. Emergency blood bank?	___	___
g. Pharmacy?	___	___
h. Temporary morgue?	___	___
i. Decontamination area?	___	___

Headquarters for Disaster Operations

	Yes	No
1. Is a designated area established for headquarters operations?	___	___
2. Is adequate space provided for the following disaster officials:		
a. Medical chief?	___	___
b. Administrative services chief?	___	___
c. Supply, stores, and procurement officer?	___	___
d. Communications officer?	___	___
e. Nursing officer?	___	___
f. Personnel reserve officer?	___	___
g. Public relations and press officer?	___	___
h. Transportation officer?	___	___
i. Food and housing officer?	___	___
j. Police and security control officer?	___	___
k. Laundry, housekeeping, and sanitation officer?	___	___
l. Buildings and grounds officer?	___	___

Alerting

	Yes	No
1. Does the plan provide for prompt activation of the hospital to emergency status upon the occurrence of a disaster?	___	___
2. Is either the administrator or his representative on duty or available immediately by telephone at all times?	___	___
3. Is there an organized procedure for reporting internal disasters?	___	___
4. Is there a procedure for alerting the hospital switchboard?	___	___
5. Is it tied in with civil defense communications?	___	___
6. Does the hospital have a radio system to substitute for usual telephone service?	___	___
7. Does the switchboard maintain a roster of those to be immediately alerted in emergency circumstances?	___	___

5.5–3   CHECKLIST FOR HOSPITAL DISASTER PLANNING (continued)

Alerting (continued)

Yes   No

8.   Does each department have a fan-out system for contacting responsible individuals or alternatives in response to an emergency?

Activation of Plan

1.   Does the plan provide that the first administrative person notified should consult with designated individuals within the disaster organization?
2.   Does the plan authorize and direct a designated individual or his alternate to make the necessary value judgment promptly?
3.   Is authority to activate any phase of the disaster program assigned to an individual who is backed up by several responsible deputies or alternatives?

Receipt and Control of Casualties

1.   Is the receiving and sorting area accessible and in close proximity to the areas of the hospital in which definitive care will be given?
2.   Is the area equipped with portable auxiliary power for illumination and other electrical equipment, or can power be supplied from hospital circuits?
3.   Does the area allow for retention, segregation, and processing of incoming casualties?
4.   Are radiological monitors and radiation detection instruments assigned to the area?
5.   Is the sorting staff organized and directed by an experienced physician?
6.   Are sufficient equipment, supplies, and apparatus available in an organized manner to permit prompt and efficient patient movement?
7.   Will the entire area be policed adequately?

Decontamination Procedures

1.   Is there a decontamination area properly labeled, staffed, and controlled in or near the receiving and sorting area?
2.   Will contaminated casualties be separated in the staging area and specifically processed?
3.   Will contaminated items such as clothing, bedding, stretchers, etc. be isolated and properly tagged?
4.   Will contaminated casualties be monitored after decontamination prior to admission into the hospital?

Traffic Flow and Control

1.   Will the elevators be manned and controlled?
2.   Is there a traffic control chart showing patient movement to and from special treatment areas, i.e., operating room, delivery room, urology, radiology, and from these to bed areas or to the discharge areas?
3.   Will all entrances and exits be controlled?
4.   Will receiving docks, driveways, and entrances to storerooms be kept open for traffic?

Discharge Procedures

1.   Is a bed census made periodically to determine the number of patients subject to:
     a.   Immediate discharge?
     b.   Transfer to convalescent facilities?
     c.   Continued care within the hospital?
2.   Is a review program established in the plan for early and prompt review of all treated casualties indicating level of care required, where they could best be housed, and when they should be discharged to ambulatory care?
3.   Is there an organized discharge routine sufficiently streamlined to handle large numbers of patients upon short notice?
4.   Have provisions been made for evacuation of the patients by utilizing private cars, station wagons, and all available transport belonging to the hospital staff?
5.   Are procedures established for the orderly disposition of casualties to their homes or other resources following their definitive care?

5.5–3   CHECKLIST FOR HOSPITAL DISASTER PLANNING (continued)

Consolidation of Patient Areas

Yes   No

1. Is there a predetermined schedule signifying which wards, room arrangements, classrooms, etc. will be used for housing emergency casualties?
2. Are these areas schematically defined in the plan showing bed, mattress, or cot location; avenues of travel; source of supplies and utilities?
3. Are specific personnel assignments established and organized by classification for the opeation of these areas?
4. Is there a schedule showing clinical segregation of patients by area?
5. Does each area have a preestablished complement of bedside items, a store of drugs, dressings, instrument trays, etc.?
6. Are reserve supplies of linens, bedding, and emergency clothing included in the plan?

Reallocation of Patients to Expanded Bed Areas

1. Is there a method for delegating authority and decision-making responsibility for such transfers?
2. Is there a schedule delineating professional staff assignment to continuation care areas?
3. Is there a sequence for patient transfers along preestablished routes?
4. Is there a "time sequence" built into the plan designating approximate moving times, assigned personnel, and priority of patients to specific locations?

Expansion of Patient Areas

1. Are there emergency stores of equipment and supplies located adjacent to each ancillary department?
2. Is there a plan by which personnel will be assigned to activate expanded ancillary units?
3. Does the plan establish a system of priority and does it differentiate ancillary clinical treatment facilities as minor and major, ambulatory and nonambulatory?

Evacuation and Survival Planning

1. Is there a survival ward? If so, how adequate are physical safety, emergency supplies, substitute utilities, accessibility of location?
2. Is there a predetermined schedule for discharge of patients and evacuation of nonambulatory patients and those requiring continuing care?
3. Has a plan been developed for the evacuation or relocation of the hospital if the hospital is endangered?

Emergency and Reserve Supply Availability

1. Are bulk, easily prepared food items locally available?
2. Are emergency pharmacy stocks for treatment of burns, as determined by treatment policy, in reserve storage?
3. Are intravenous solutions, essential drugs, plasma expanders, and medical gases available to meet the requirements of a potential 100% increase in patient load, or is there a secondary source of supply?
4. Are there sufficient quantities of suitable chemical disinfectants on hand if needed for instrument sterilization?
5. Is there a predetermined procedure for unlocking and issuing critical storeroom items?
6. Is there a separate set of emergency keys for the stores department, pharmacy, and laundry?
7. Is there a Civil Defense Emergency Hospital prepositioned in the area served by the hospital?

Maintenance of Physical Plant

1. Has the hosptial sufficient standby power to operate elevators, communication system, laboratory and X-ray equipment, and other necessary electrical appliances?
2. Is there reserve fuel to operate the boiler plant for at least 14 days?
3. Is there a system of substitute fuel for the power plant?

5.5–3   CHECKLIST FOR HOSPITAL DISASTER PLANNING (continued)

Maintenance of Physical Plant (continued)

		Yes	No
4.	Has provision been made for additional laundry services?	___	___
5.	Has a plan for water conservation been included in the overall disaster operations plan?	___	___
6.	Are there facilities for storage of water?	___	___
7.	Is there an independent hospital source of water?	___	___
8.	Is a chlorination procedure included in the plan for use in event of contamination of water or conversion to a substitute source of supply?	___	___
9.	Are additional maintenance supplies and personnel available to support expanded hospital activities?	___	___

Personnel Protection and Plant Safety

		Yes	No
1.	Are there designated fallout shelter areas, properly marked and stocked, in the hospital or adjacent thereto, for the protection of hospital personnel and patients?	___	___
2.	Are evacuation routes diagrammed and properly designated?	___	___
3.	In case of fire within the hospital during emergency disaster operations, is a fire-control system included in the plan?	___	___
4.	Are personnel instructed in techniques of self-preservation during an attack?	___	___

Internal Communication Systems

		Yes	No
1.	Is there an organized messenger system to substitute for telephone or other electrical systems in the event of a complete power failure?	___	___
2.	Is there a messenger pool and will it be properly supervised?	___	___
3.	Will messenger personnel be provided with schematic area layout maps showing key areas for disaster operations?	___	___
4.	Are "walkie-talkie" sets available for internal communications?	___	___

Vertical Transportation

		Yes	No
1.	In the absence of power will there be an alternative to elevators, i.e., canvas stretchers, blanket brigades, etc.?	___	___
2.	Are building maintenance supplies available, such as extension ladders, rope ladders, or block and tackle, which may be used to evacuate patients, personnel, and supplies from above ground level areas?	___	___

Food Service

		Yes	No
1.	Are facilities for mass field-station-type feeding of personnel and ambulatory casualties included in the plan?	___	___
2.	Are there preestablished disaster menus and food dispensing routines?	___	___
3.	Is the auxiliary kitchen assigned sufficient staff to prepare and distribute large quantities of food?	___	___
4.	Is the kitchen staff familiar with accepted chemical means of sanitizing or sterilizing eating and drinking utensils should the means of heating water be disrupted?	___	___
5.	Does the food service staff know how and where to requisition food during a major disaster?	___	___
6.	Are sanitation facilities properly organized for handling food, utensils, and food storage areas?	___	___
7.	Are waste and garbage facilities planned for?	___	___

Sewage and Waste Disposal Control

		Yes	No
1.	In event of breakdown of garbage and sewage removal systems, are sufficient substitute procedures included in the plan?	___	___
2.	Is a substitute incinerator or a method of fill and cover for waste disposal provided?	___	___
3.	Are insecticides and other normally available rodent and insect destroyers in reserve supply?	___	___

5.5–3   CHECKLIST FOR HOSPITAL DISASTER PLANNING (continued)

Records and Record Keeping

Yes   No

1. Is there an emergency record and tagging system for casualties?   _____   _____
2. Are the medical records and the admissions departments organized to handle an increased number of patients?   _____   _____
3. Is there a system for retention and safekeeping of valuable personal items removed from casualties?   _____   _____
4. Does the admitting area provide status reports at stated intervals to both the headquarters and the public relations desk?   _____   _____

Emergency Morgue Facilities

1. Are building areas such as warehouses, lecture halls, empty stores, or freight depots designated for holding temporarily the increased number of cadavers?   _____   _____
2. Is the disposition of cadavers and their legal clearance correlated with duties of properly constituted authorities?   _____   _____

Public Relations and Press Releases

1. Is the individual responsible for the release of information centrally located in disaster headquarters?   _____   _____
2. Is the individual responsible supplied with adequate prompt information and provided with specifically assigned messengers?   _____   _____
3. Is there a method whereby information requested with respect to a particular casualty can be readily assembled and the medically responsible individual contacted?   _____   _____

Personnel Identification, Orientation, and Disaster Training

1. Has each employee been supplied with an identification card that will be recognized by local civil defense authorities?   _____   _____
2. Does the plan include provision for training of volunteers to augment the present hospital staff?   _____   _____
3. Are all new employees given instruction in disaster operations and a copy of the plan under which they are to function?   _____   _____
4. Is general interest maintained by frequent distribution of literature and showing of appropriate movies or slides?   _____   _____

Volunteers

1. Is there a plan for the participation of volunteers in disaster situations?   _____   _____
2. Is there an organized program to train volunteers for their service roles in disaster situations?   _____   _____
3. Is there a center in the hospital to which volunteers will be referred in the event of a disaster?   _____   _____
4. Has a plan been made for the assignment or gracious rejection of volunteers not previously associated with the hospital?   _____   _____

Liaison with Civil Defense Agencies and Other Organizations

1. Does the plan include details relating to procurement of supplies, services, and assistance from adjoining regions?   _____   _____
2. Is the method for contact and communication with civil defense agencies, etc. included in the plan, i.e., are name, title, code number, location, etc. kept on file?   _____   _____
3. Does the plan provide for expansion of the hospital by using nearby buildings such as churches, schools, halls, etc. for use as auxiliary hospitals, or wards, or decontamination stations?   _____   _____
4. Does the plan call for coordinating the hospital activities with the following divisions of civil defense through the local civil defense director and medical service chief:
    a.   Emergency welfare division, for additional food and nonmedical personnel to assist in the dietary service?   _____   _____
    b.   Transportation division, for any additional transportation needed?   _____   _____
    c.   Communications division, for any additional communication facilities?   _____   _____
    d.   Utility division, for additional utilities (water, gas, and electric)?   _____   _____

5.5—3   CHECKLIST FOR HOSPITAL DISASTER PLANNING (continued)

Liaison with Civil Defense Agencies and Other Organizations (continued)

		Yes	No
e.	Radiological safety division, for radiation detection?	_____	_____
f.	Law enforcement division, for protection and maintenance of order?	_____	_____
g.	Traffic control division, to direct traffic and route ambulances?	_____	_____
h.	Engineering division, for removal of heavy debris and clearance of access routes to hospital?	_____	_____
i.	Supply division, for procurement of supplies beyond those available in the area?	_____	_____
j.	Mortuary division, for disposition of remains?	_____	_____
5.	Does the plan provide for coordination with other hospitals and health facilities in the event of a disaster?	_____	_____

Frequency of Review and Evaluation of Existing Disaster Plan

		Yes	No
1.	Is the plan reviewed by a permanently assigned planning committee at least annually?	_____	_____
2.	Are test exercises or drills conducted periodically? If so, how frequently?	_____	_____
3.	Are critiques held after each test exercise?	_____	_____

Source: American Hospital Association, Chicago, 1964. With permission.

Section 6

Nursing Homes

_____

6. NURSING HOMES

6-1 PERCENT OF BEDS OCCUPIED IN HOMES FOR THE AGED AND CHRONICALLY ILL BY PRIMARY TYPE OF SERVICE AND TYPE OF OWNERSHIP OF HOME

United States, April–June 1963 and June–August 1969

Type of service	April–June 1963				June–August 1969			
	All types of ownership	Proprietary	Nonprofit	Government	All types of ownership	Proprietary	Nonprofit	Government
All types of service	88.2	89.0	90.3	84.2	90.9	91.2	91.2	88.6
Nursing care	90.6	90.6	90.8	90.1	91.0	91.5	90.3	88.5
Personal care with nursing	88.0	88.3	91.8	79.9	91.6	90.7	93.3	88.2
Personal care	82.6	82.9	86.4	72.2	87.4	87.0	85.3	92.3

Source: *Vital Health Stat.,* 12, 1, 1974.

6-2 NUMBER OF ADMISSIONS TO HOMES FOR THE AGED AND CHRONICALLY ILL AND NUMBER OF ADMISSIONS PER BED BY PRIMARY TYPE OF SERVICE AND TYPE OF OWNERSHIP OF HOME

United States, 1962 and 1968

Type of service and type of ownership	Number of admissions		Number of admissions per bed	
	1962[a]	1968	1962[a]	1968
All homes	402,896	968,750	0.8	1.1
Type of service				
Nursing care	264,955	876,645	0.8	1.2
Personal care with nursing	93,529	69,373	0.5	0.5
Personal care	44,412	22,732	0.8	0.5
Type of ownership				
Proprietary	281,545	788,142	0.8	1.3
Nonprofit	57,761	140,374	0.4	0.7
Government	63,590	40,234	0.7	0.5

[a] These figures were selected from a published report on Resident Places Survey-1, conducted in April, May, and June of 1963, *Vital and Health Statistics,* Series 12, No. 4. The 1962 figures were adjusted to exclude hospitals in order that that the figures would be compatible with 1968 figures. Bed sizes were not compared because data by bed size for 1962 were not available.

Source: *Vital Health Stat.,* 12, 1, 1974.

6–3   NUMBER AND PERCENT DISTRIBUTION OF HOMES FOR THE AGED AND CHRONICALLY ILL BY PRIMARY TYPE OF SERVICE, TYPE OF OWNERSHIP, AND BED SIZE OF HOME

United States, April–June 1963 and June–August 1969

Type of service, type of ownership, and bed size	April–June 1963		June–August 1969	
	Number	Percent	Number	Percent
All homes	16,370	100.0	18,391	100.0
Type of service				
Nursing care	7,834	47.9	11,576	62.9
Personal care with nursing	4,968	30.3	3,768	20.5
Personal care	3,568	21.8	3,047	16.6
Type of ownership				
Proprietary	13,428	82.0	14,161	77.0
Nonprofit	2,012	12.3	2,847	15.5
Government	930	5.7	1,383	7.5
Bed size				
Less than 30 beds	10,502	64.2	8,100	44.0
30–49 beds	3,059	18.7	3,574	19.4
50–99 beds	1,921	11.7	4,573	24.9
100 beds or more	888	5.4	2,144	11.7

Source: *Vital Health Stat.*, 12, 1, 1974.

6–4   NUMBER OF DISCHARGES, PERCENT DISCHARGED ALIVE, AND PERCENT DISCHARGED DEAD FROM HOMES FOR THE AGED AND CHRONICALLY ILL BY PRIMARY TYPE OF SERVICE AND TYPE OF OWNERSHIP OF HOME

United States, 1962 and 1968

Type of service and type of ownership	Number of discharges		Percent discharged alive		Percent discharged dead	
	1962[a]	1968	1962[a]	1968	1962[a]	1968
All homes	378,326	900,521	60.9	66.9	39.1	33.1
Type of service						
Nursing care	253,156	803,365	59.6	66.1	40.4	33.9
Personal care with nursing	86,106	68,887	56.9	64.7	43.1	35.3
Personal care	39,064	28,269	77.9	92.7	22.1	7.3
Type of ownership						
Proprietary	263,602	716,589	59.0	66.7	41.0	33.3
Nonprofit	53,515	145,362	58.0	69.0	42.0	30.1
Government	61,209	38,568	71.2	58.3	28.8	41.7

[a] These figures were selected from a published report on Resident Places Survey-1, conducted in April, May, and June of 1963, *Vital and Health Statistics*, Series 12, No. 4. The 1962 figures were adjusted to exclude hospitals in order that the figures would be compatible with 1968 figures. Bed sizes were not compared because data by bed size for 1962 were not available.

Source: *Vital Health Stat.*, 12, 1, 1974.

6–5   PROPORTIONS (PERCENT) OF ADMISSIONS AND DISCHARGES BY BED SIZE OF HOME AND SELECTED FORMER PLACE OF RESIDENCE AND PLACE PATIENT DISCHARGED TO

United States, 1967 for Admissions and 1968 for Discharges

Bed size	Admissions				Discharges			
	General hospital	Other hospital	Patient's home	Another nursing home	General hospital	Other hospital	Patient's home	Another nursing home
All homes	53.9	8.5	25.9	9.4	25.1	3.5	39.6	11.0
1–24 beds	25.2	15.0	38.2	16.2	21.4	13.8	32.0	16.5
25–49 beds	44.3	10.0	31.0	12.3	27.5	4.1	36.6	14.6
50–99 beds	56.9	8.0	24.7	8.8	23.7	2.5	37.7	10.1
>100 beds	59.8	7.1	22.8	7.7	26.2	3.3	44.6	9.3

Source: *Vital Health Stat.*, 12, 1, 1974.

6–6   PERCENT OF HOMES FOR THE AGED PROVIDING ROUTINE SERVICES

United States, June–August 1969

Routine service	Percent
Supervision of medications	66.9
Medications administered in accordance with physician's orders	98.5
Rub and massage	80.8
Help with dressing	90.0
Help with correspondence or shopping	90.4
Help with walking or getting about	86.0
Help with eating	81.9

Source: *Vital Health Stat.*, 12, 1, 1974.

6–7   PERCENT OF HOMES FOR THE AGED PROVIDING CONTRACTUAL SERVICES BY SELECTED CONTRACT SERVICES

United States, June–August 1969

Contractual services	Percent of homes providing
Physician's (M.D. or D.O.)	45.0
Dental	18.9
Pharmaceutical	31.1
Physical therapy	22.4
Occupational therapy	14.4
Recreational therapy	23.8
Speech therapy	5.6
Social work	24.9
Dietary	33.9
Food	31.8
Housekeeping	34.6
Other	27.0

Source: *Vital Health Stat.*, 12, 1, 1974.

6–8   PERCENT DISTRIBUTION OF EMPLOYEES IN HOMES FOR THE AGED AND CHRONICALLY ILL BY STATUS OF EMPLOYEES, ACCORDING TO GEOGRAPHIC REGION AND BED SIZE OF HOME

United States, 1969[a]

Employee status	Geographic region					Bed size			
	All homes	Northeast	North central	South	West	Less than 30 beds	30–99 beds	100–199 beds	200 beds or more
	Percent distribution								
All statuses	100.0	100.0	100.0	100.0	100.0	100.0	100.0	100.0	100.0
Professional	24.0	29.1	20.7	22.8	24.5	31.6	23.8	22.0	20.6
Nonprofessional	76.0	70.9	79.3	77.2	75.5	68.4	76.2	78.0	79.4

[a] Excluding Alaska and Hawaii.

Source: *Vital Health Stat.*, 12, 1, 1974.

6—9 NUMBER OF HOMES FOR THE AGED AND CHRONICALLY ILL AND PERCENT PROVIDING
ROUTINE PERSONAL SERVICES BY TYPE OF HOME, BED SIZE, AND TYPE OF OWNERSHIP

United States, June—August 1969[a]

Type of home, bed size, and type of ownership	All homes	Supervision of medications	Medication administered in accordance with physician's orders	Rub and massage	Help with dressing	Help with correspondence or shopping	Help with walking or getting about	Help with eating
					Percent			
Total	18,391	66.9	98.5	80.8	90.0	90.4	86.0	81.9
Type of home								
Nursing care	11,576	60.0	99.5	93.1	96.6	91.5	95.6	95.4
Personal care with nursing	3,768	72.2	98.8	79.3	90.1	87.1	86.4	82.0
Personal care	3,047	86.8	94.3	35.6	65.0	90.4	49.2	30.5
Bed size								
Less than 30 beds	8,100	71.9	97.1	68.8	82.8	88.6	72.9	66.0
30—49 beds	3,574	63.4	99.6	86.1	93.5	90.7	95.4	91.9
50—99 beds	4,573	62.0	99.8	93.0	97.3	93.2	97.3	95.9
100—199 beds	1,835	64.9	99.3	91.1	96.8	90.6	96.4	95.9
200 beds or more	309	63.4	99.7	91.3	94.2	93.5	94.5	93.2
Type of ownership								
Proprietary	14,161	64.0	98.1	79.9	90.2	90.5	85.6	81.4
Nonprofit	2,847	77.4	99.8	87.6	94.6	90.5	93.6	88.7
Government	1,383	74.8	99.7	75.3	79.0	90.2	74.6	73.6

[a] Excluding Alaska and Hawaii.

Routine personal services

Source: *Vital Health Stat.,* 12, 1, 1974.

6–10   PERCENT OF BEDS OCCUPIED IN HOMES
FOR THE AGED AND CHRONICALLY ILL
BY TYPE OF SERVICE AND OWNERSHIP
AND BED SIZE OF HOME

United States, June–August 1969[a]

Type of service and type of ownership	All bed sizes	Less than 30 beds	30–99 beds	100–299 beds	300 beds or more
			Percent		
All types of services	90.9	90.2	91.7	90.3	88.3
Proprietary	91.2	90.3	92.0	90.0	90.4
Nonprofit	91.2	89.0	90.6	92.4	89.4
Government	88.6	91.0	92.1	86.8	86.8
Nursing care	91.0	92.9	91.7	89.8	87.6
Proprietary	91.5	93.6	92.0	89.9	89.2
Nonprofit	90.3	89.6	90.1	91.2	86.5
Government	88.5	88.2	93.2	86.7	87.6
Personal care with nursing	91.6	88.7	92.3	93.4	89.4
Proprietary	90.7	87.9	93.0	93.6	97.9
Nonprofit	93.3	93.8	91.6	94.3	93.0
Government	88.2	91.3	91.1	86.6	84.6
Personal care	87.4	86.8	87.7	89.3	96.6
Proprietary	87.0	86.7	86.2	92.0	99.1
Nonprofit	85.3	80.3	97.1	82.3	93.1
Government	92.3	94.2	86.9	99.0	–

[a] Excluding Alaska and Hawaii.

Source: *Vital Health Stat.*, 12, 1, 1974.

6-11 NUMBER OF ADMISSIONS TO AND DISCHARGES FROM HOMES FOR THE AGED AND CHRONICALLY ILL, NUMBER OF ADMISSIONS PER BED, PERCENT DISCHARGED ALIVE, AND PERCENT DISCHARGED DEAD BY TYPE OF SERVICE, BED SIZE, AND TYPE OF OWNERSHIP OF HOME

United States, 1968[a]

Type of service, bed size, and type of ownership	Number of admissions	Number of admissions per bed	Number of discharges	Percent discharged alive	Percent discharged dead
Total	968,750	1.1	900,522	66.9	33.1
Type of service					
Nursing care	876,645	1.2	803,365	66.1	33.9
Personal care with nursing	69,373	0.5	68,887	64.7	35.3
Personal care	22,732	0.5	28,269	92.7	7.3
Bed size					
Less than 30 beds	81,648	1.2	90,073	64.3	35.7
30–49 beds	143,993	1.0	125,705	67.0	33.0
50–99 beds	387,070	1.2	354,848	67.7	32.3
100–199 beds	280,844	1.2	265,462	65.5	34.5
200 beds or more	75,195	0.8	64,434	71.3	28.7
Type of ownership					
Proprietary	788,142	1.3	716,589	66.7	33.3
Nonprofit	140,374	0.7	145,362	69.9	30.1
Government	40,234	0.5	38,568	58.3	41.7

[a] Excluding Alaska and Hawaii.

Source: *Vital Health Stat.*, 12, 1, 1974.

6–12   NUMBER AND PERCENT DISTRIBUTION OF ADMISSIONS TO HOMES FOR THE AGED AND CHRONICALLY ILL BY PATIENT'S FORMER PLACE OF RESIDENCE ACCORDING TO TYPE OF SERVICE, BED SIZE, AND TYPE OF OWNERSHIP OF HOME

United States, 1967[a]

Type of service, bed size, and type of ownership	All admissions		Patient's former place of residence					
	Number	Percent	Mental hospital	General hospital	Other hospital	Patient's home	Another nursing home	Other places
			Percent distribution					
Total	801,013	100.0	3.5	53.9	5.0	25.9	9.4	2.4
Type of service								
Nursing care	681,797	100.0	2.3	59.2	5.2	22.4	9.1	1.8
Personel care with nursing	85,090	100.0	7.7	26.6	4.0	46.7	10.1	4.9
Personal care	34,126	100.0	15.9	16.1	4.6	41.9	13.8	7.7
Bed size								
1–24 beds	59,659	100.0	10.0	25.2	5.0	38.2	16.2	5.4
25–49 beds	127,714	100.0	4.3	44.3	5.7	31.0	12.3	2.3
50–99 beds	257,693	100.0	2.4	56.9	5.6	24.7	8.8	1.4
100 beds or more	355,947	100.0	2.8	59.8	4.3	22.8	7.7	2.5
Type of ownership								
Government	47,915	100.0	10.3	39.0	4.0	31.1	8.8	6.8
Nonprofit	109,415	100.0	2.2	41.1	3.4	40.8	7.7	4.5
Proprietary	643,683	100.0	3.1	57.1	5.3	22.9	9.7	1.7

[a] Excluding Alaska and Hawaii.

Source: *Vital Health Stat.*, 12, 1, 1974.

6—13 NUMBER AND PERCENT DISTRIBUTION OF LIVE DISCHARGES FROM HOMES FOR THE AGED AND CHRONICALLY ILL BY PLACE PATIENT DISCHARGED TO ACCORDING TO TYPE OF SERVICE, BED SIZE, AND TYPE OF OWNERSHIP OF HOME

United States, 1968[a]

Type of service, bed size, and type of ownership	All live discharges		Place patient discharged to						
	Number	Percent	General or short-stay hospital	Long-term specialty hospital	Mental hospital	Another nursing home	Personal care or domiciliary home	Patient's home or family	Other places
			Percent						
Total	602,192	100.0	25.1	1.8	1.7	11.0	3.8	39.6	17.0
Type of service									
Nursing care	531,385	100.0	25.5	1.3	1.3	10.5	3.9	40.5	17.1
Personal care with nursing	44,601	100.0	27.2	2.7	4.4	15.0	3.2	28.2	19.4
Personal care	26,204	100.0	13.9	10.4	7.0	14.7	1.7	40.7	11.6
Bed size									
Less than 25 beds	42,466	100.0	21.4	7.2	6.6	16.5	2.0	32.0	14.2
25—49 beds	99,737	100.0	27.5	1.9	2.2	14.6	5.5	36.6	11.7
50—99 beds	240,210	100.0	23.7	1.2	1.3	10.1	3.6	37.7	22.4
100 beds or more	219,783	100.0	26.2	1.2	1.1	9.3	3.4	44.6	14.1
Less than 30 beds	57,950	100.0	23.8	5.8	5.6	16.7	3.7	30.9	13.4
30—49 beds	84,249	100.0	26.9	1.9	2.0	14.2	5.0	38.3	11.7
50—99 beds	240,210	100.0	23.7	1.2	1.3	10.1	3.6	37.7	22.4
100—199 beds	173,857	100.0	26.5	1.1	1.0	9.9	2.8	44.8	13.8
200 beds or more	45,925	100.0	25.1	1.6	1.8	6.9	5.6	43.6	15.4
Type of ownership									
Proprietary	478,123	100.0	25.5	1.8	1.7	11.3	3.4	38.9	17.5
Nonprofit	101,582	100.0	22.3	1.6	1.3	9.7	4.4	45.7	15.0
Government	22,484	100.0	28.0	2.0	5.4	10.7	8.9	28.1	16.9

[a] Excluding Alaska and Hawaii.

Source: *Vital Health Stat.*, 12, 1, 1974.

6–14   NUMBER OF HOMES FOR THE AGED AND CHRONICALLY ILL AND PERCENT PROVIDING CONTRACTUAL SERVICES ON A REGULAR BASIS THROUGH CONTRACT OR OTHER FEE ARRANGEMENTS BY TYPE OF HOME, BED SIZE, AND TYPE OF OWNERSHIP

United States, June—August 1969[a]

Type of home, bed size, and type of ownership	Number of homes	Contractual service													
		Physician's service (M.D. or D.O.)	Dental service	Pharmaceutical service	Physical therapy	Occupational therapy	Recreational therapy	Speech therapy	Social work	Dietary service (dietitian)	Food service (meal preparation)	House-keeping	Other
		Percent											
Total	18,391	45.0	18.9	31.1	22.4	14.4	23.8	5.6	24.9	33.9	31.8	34.6	27.0
Type of home													
Nursing care	11,576	51.6	23.1	38.4	32.0	19.3	30.0	8.5	29.5	48.3	33.8	36.5	18.7
Personal care with nursing	3,768	40.8	16.4	22.2	8.9	7.6	16.9	0.9	21.1	15.3	34.2	35.0	32.8
Personal care	3,047	25.0	6.0	14.4	2.4	3.9	6.9	0.1	12.3	2.0	21.3	26.5	51.6
Bed size													
Less than 30 beds	8,100	22.6	7.6	16.3	4.6	5.1	9.8		14.3	11.4	24.3	27.9	40.5
30–49 beds	3,574	50.5	21.5	32.6	20.4	13.2	25.9	5.9	24.6	35.9	36.4	39.6	21.7
50–99 beds	4,573	51.8	27.8	46.0	40.7	23.0	36.5	8.0	35.3	57.9	37.6	39.3	15.9
100–199 beds	1,835	57.8	35.9	51.8	53.3	30.7	43.8	18.7	44.0	64.6	39.2	40.8	8.4
200 bed or more	309	67.6	50.1	58.3	55.7	45.6	40.1		42.1	60.2	45.3	43.0	12.0
Type of ownership													
Proprietary	14,161	42.5	16.6	30.8	22.3	13.6	23.2	5.8	25.0	34.0	29.9	33.2	28.1
Nonprofit	2,847	57.2	28.1	34.6	28.0	21.1	28.5	5.0	24.3	42.4	39.8	39.4	19.1
Government	1,383	45.2	22.8	27.2	11.5	7.7	15.8		25.7	14.5	34.6	38.7	32.2

[a] Excluding Alaska and Hawaii.

Source: *Vital Health Stat.,* 12, 1, 1974.

6–15 NUMBER AND PERCENT DISTRIBUTION OF EMPLOYEES IN HOMES FOR THE AGED AND CHRONICALLY ILL BY STATUS OF EMPLOYEES ACCORDING TO SEX OF EMPLOYEES AND BED SIZE OF HOME

United States, 1969[a]

Employee status	All employees	Sex[b] Male	Sex[b] Female	Less than 30 beds	30–99 beds	100–199 beds	200 beds or more
				Number			
All statuses	564,783	68,134	494,570	71,487	292,721	147,905	52,669
Professional	135,658	16,149	119,105	22,557	69,764	32,488	10,849
Nurses[c]	94,507	1,088	93,186	11,700	50,311	24,635	7,862
Other professional[d]	41,151	15,060	25,919	10,857	19,453	7,854	2,987
Nonprofessional	429,124	51,986	375,465	48,930	222,957	115,417	41,821
Nurse's aides, etc.[e]	243,927	12,744	230,372	26,517	130,099	67,050	20,261
Clerical	16,786	1,578	15,173	782	8,005	5,366	2,633
Food service	85,919	9,818	75,774	10,771	45,378	21,780	7,991
Housekeeping	71,747	24,398	47,018	9,340	34,943	18,598	8,867
Other nonprofessional	10,745	3,447	7,128	1,520	4,532	2,624	2,070
				Percent distribution			
All statuses	100.0	100.0	100.0	100.0	100.0	100.0	100.0
Professional							
Nurses[c]	69.7	6.7	78.2	51.9	72.1	75.8	72.5
Other professional[d]	30.3	93.3	21.8	48.1	27.9	24.2	27.5
Nonprofessional							
Nurse's aides, etc.[e]	56.8	24.5	61.4	54.2	58.3	58.1	48.4
Clerical	3.9	3.0	4.0	1.6	3.6	4.6	6.3
Food service	20.0	18.9	20.2	22.0	20.4	18.9	19.1
Housekeeping	16.7	46.9	12.5	19.1	15.7	16.1	21.2
Other nonprofessional	2.5	6.6	1.9	3.1	2.0	2.3	4.9

[a] Excluding Alaska and Hawaii.

[b] Figures for "all employees" include 2046 employees (0.38%) for whom sex was unknown; these employees not included in the subtotals for males or females.

[c] Includes registered nurses, licensed practical nurses or vocational nurses, and practical nurses.

[d] Includes administrators, physicians (M.D. or D.O.), dentists, registered occupational therapists or other occupational therapist assistants, qualified physical therapists or physical therapist assistants, recreation therapists, dietitians or nutritionists, registered medical record librarians or other medical record librarians and technicians, social workers, speech therapists, and other professional persons.

[e] Includes nurse's aides, orderlies, and student nurses.

Source: *Vital Health Stat.*, 12, 1, 1974.

6–16   NUMBER AND PERCENT DISTRIBUTION OF EMPLOYEES IN HOMES FOR THE AGED AND CHRONICALLY ILL BY STATUS OF EMPLOYEES ACCORDING TO GEOGRAPHIC REGION AND TYPE OF OWNERSHIP OF HOME

United States, 1969[a]

Employee status	All employees	Geographic region				Type of ownership		
		Northeast	North central	South	West	Government	Nonprofit	Proprietary
Number								
All statuses	564,783	148,446	187,062	146,184	83,091	48,409	135,137	381,237
Professional	135,658	43,262	38,686	33,368	20,342	10,881	30,064	94,714
Nurses[b]	94,507	32,447	25,523	23,526	13,011	7,424	20,927	66,157
Other professional[c]	41,151	10,815	13,163	9,842	7,332	3,457	9,137	28,558
Nonprofessional	429,124	105,183	148,376	112,816	62,748	37,528	105,074	286,522
Nurse's aides, etc.[d]	243,927	56,051	84,164	68,162	35,549	19,301	47,990	176,636
Clerical	16,786	4,592	4,859	4,603	2,732	1,625	5,410	9,752
Food service	85,919	21,752	30,948	20,990	12,230	7,078	24,981	53,860
Housekeeping	71,747	19,327	25,103	16,988	10,329	7,758	22,920	41,069
Other nonprofessional	10,745	3,462	3,303	2,072	1,908	1,766	3,774	5,206
Percent distribution								
All statuses	100.0	100.0	100.0	100.0	100.0	100.0	100.0	100.0
Professional								
Nurses[b]	69.7	75.0	66.0	70.5	64.0	68.2	69.6	69.8
Other professional[c]	30.3	25.0	34.0	29.5	36.0	31.8	30.4	30.2
Nonprofessional								
Nurse's aides, etc.[d]	56.8	53.3	56.7	60.4	56.6	51.4	45.7	61.6
Clerical	3.9	4.4	3.3	4.1	4.4	4.3	5.1	3.4
Food service	20.0	20.7	20.9	18.6	19.5	18.9	23.8	18.8
Housekeeping	16.7	18.4	16.9	15.1	16.5	20.7	21.8	14.3
Other nonprofessional	2.5	3.3	2.2	1.8	3.0	4.7	3.6	1.8

a Excluding Alaska and Hawaii.

b Includes registered nurses, licensed practical nurses or vocational nurses, and practical nurses.

c Includes administrators, physicians (M.D. or D.O.), dentists, registered occupational therapists or other occupational therapist assistants, qualified physical therapists or physical therapist assistants, recreation therapists, dietitians or nutritionists, registered medical record librarians or other medical record librarians and technicians, social workers, speech therapists, and other professional persons.

d Includes nurse's aides, orderlies, and student nurses.

Source: *Vital Health Stat.*, 12, 1, 1974.

6–17   DISTRIBUTION OF HOMES IN THE RESIDENT PLACES SURVEY-3 UNIVERSE AND DISPOSITION OF SAMPLE HOMES ACCORDING TO PRIMARY STRATA (TYPE OF SERVICE AND BED SIZE OF HOME)

United States[a]

Type of service and bed size	Number of homes in sample				
	Universe[b] (sampling frame)	All homes	Out of scope or out of business	In scope and in business	
				Nonresponding homes	Responding homes
All types of services	21,301	2088	153	81	1854
Nursing care	10,480	1289	48	66	1175
Less than 15 beds	858	21	4	2	15
15–24 beds	1,756	88	13	3	72
25–49 beds	3,448	260	16	10	234
50–99 beds	3,166	477	4	24	449
100–199 beds	1,062	316	9	24	283
200–299 beds	126	64	1	2	61
300 beds or more	64	63	1	1	61
Personal care with nursing	3,608	402	35	7	360
Less than 15 beds	941	24	6	–	18
15–24 beds	767	37	9	–	28
25–49 beds	828	62	7	1	54
50–99 beds	612	92	3	3	86
100–199 beds	332	100	6	2	92
200–299 beds	82	41	1	–	40
300 beds or more	46	46	3	1	42
Personal care	4,725	183	42	3	138
Less than 15 beds	2,937	60	16	–	44
15–24 beds	988	40	11	–	29
25–49 beds	561	35	5	–	30
50–99 beds	183	24	3	1	20
100–199 beds	48	17	5	2	10
200–299 beds	6.	5	2	–	3
300 beds or more	2	2	–	–	2
"Births"[c]	2,488	214	28	6	181
Unknown bed size[d]	473	–	–	–	–
Less than 15 beds	304	6	2	–	4
15–24 beds	255	11	3	–	8
25–49 beds	492	31	3	1	27
50–99 beds	681	83	4	3	76
100–199 beds	241	58	7	1	50
200–299 beds	30	13	3	–	10
300 beds or more	12	12	6	–	6

[a] Excluding Alaska and Hawaii.

[b] The universe for the RPS-3 sample consisted of the nursing and personal care homes included in the Master Facility Inventory and the Agency Reporting System.

[c] Births consist of those homes which were assumed to be in scope of RPS-3 but for which current data were not available.

[d] Births of unknown bed size were inadvertently excluded from frame.

Source: *Vital Health Stat.*, 12, 1, 1974.

6–18   APPROXIMATE STANDARD ERRORS, EXPRESSED IN PERCENTAGE POINTS, OF PERCENTAGES OF HOMES SHOWN IN THIS REPORT

Base of estimated percent (number of homes)	Estimated percent						
	1 or 99	5 or 95	10 or 90	20 or 80	30 or 70	40 or 60	50
	Standard error						
200	2.4	5.3	7.2	9.6	11.0	11.8	12.1
500	1.5	3.3	4.6	6.1	7.0	7.5	7.6
1,000	1.1	2.3	3.2	4.3	4.9	5.3	5.4
2,000	0.8	1.7	2.3	3.0	3.5	3.7	3.8
3,000	0.6	1.4	1.9	2.5	2.9	3.0	3.1
4,000	0.5	1.2	1.6	2.2	2.5	2.6	2.7
5,000	0.5	1.1	1.4	1.9	2.2	2.4	2.4
6,000	0.4	1.0	1.3	1.8	2.0	2.2	2.2
7,000	0.4	0.9	1.2	1.6	1.9	2.0	2.0
8,000	0.4	0.8	1.1	1.5	1.7	1.9	1.9
9,000	0.4	0.8	1.1	1.4	1.6	1.8	1.8
10,000	0.3	0.7	1.0	1.4	1.6	1.7	1.7
20,000	0.2	0.5	0.7	1.0	1.1	1.2	1.2

Source: *Vital Health Stat.*, 12, 1, 1974.

6–19   APPROXIMATE STANDARD ERRORS, EXPRESSED IN PERCENTAGE POINTS, OF PERCENTAGES OF BEDS, ADMISSIONS, DISCHARGES, DAYS OF CARE, AND EMPLOYEES SHOWN IN THIS REPORT

Base of estimated percent	Estimated percent						
	1 or 99	5 or 95	10 or 90	20 or 80	30 or 70	40 or 60	50
	Standard error						
1,000	3.3	7.3	10.0	13.4	15.3	16.4	16.7
2,000	2.4	5.2	7.1	9.5	10.8	11.6	11.8
5,000	1.5	3.3	4.5	6.0	6.9	7.3	7.5
10,000	1.1	2.3	3.2	4.2	4.8	5.2	5.3
20,000	0.7	1.6	2.2	3.0	3.4	3.7	3.7
30,000	0.6	1.3	1.8	2.4	2.8	3.0	3.1
40,000	0.5	1.2	1.6	2.1	2.4	2.6	2.6
50,000	0.5	1.0	1.4	1.9	2.2	2.3	2.4
80,000	0.4	0.8	1.1	1.5	1.7	1.8	1.9
100,000	0.3	0.7	1.0	1.3	1.5	1.6	1.7
200,000	0.2	0.5	0.7	0.9	1.1	1.2	1.2
500,000	0.1	0.3	0.4	0.6	0.7	0.7	0.7
900,000	0.1	0.2	0.3	0.4	0.5	0.5	0.6

Source: *Vital Health Stat.*, 12, 1, 1974.

Index

INDEX